79 Structure and Bonding

Editors:
M. J. Clarke, Chestnut Hill
J. B. Goodenough, Oxford · J. A. Ibers, Evanston
C. K. Jørgensen, Genève · D. M. P. Mingos, Oxford
J. B. Neilands, Berkeley · G. A. Palmer, Houston
D. Reinen, Marburg · P. J. Sadler, London
R. Weiss, Strasbourg · R. J. P. Williams, Oxford

Complexes, Clusters and Crystal Chemistry

With contributions by
R. Brec   N. E. Brese   J. Darriet
R. G. Denning   M. Drillon   M. Evain
J. E. McGrady   D. M. P. Mingos
M. O'Keeffe   A. L. Rohl   P. Zanello

With 268 Figures and 82 Tables

Springer-Verlag
Berlin Heidelberg GmbH

ISBN 978-3-662-14979-9   ISBN 978-3-540-46694-9 (eBook)
DOI 10.1007/978-3-540-46694-9

This work is subject to copyright. All rights are reserved, whether the whole or part of the material is concerned, specifically the rights of translation, reprinting, re-use of illustrations, recitation, broadcasting, reproduction on microfilms or in other ways, and storage in data banks. Duplication of this publication or parts thereof is only permitted under the provisions of the German Copyright Law of September 9, 1965, in its version of June 24, 1985, and a copyright fee must always be paid.

© Springer-Verlag Berlin Heidelberg 1992
Originally published by Springer-Verlag Berlin Heidelberg New York in 1992
Softcover reprint of the hardcover 1st edition 1992

The use of general descriptive names, trade marks, etc. in this publication, even if the former are not especially identified, is not to be taken as a sign that such names, as understood by the Trade Marks and Merchandise Marks Act, may accordingly be used freely by anyone.

Typesetting: Macmillan India Ltd, Bangalore-25; India

51/3020-5 4 3 2 1 0 – Printed on acid-free paper

Editorial Board

Professor *Michael J. Clarke*, Boston College, Department of Chemistry,
Chestnut Hill, Massachusetts 02167, U.S.A.

Professor *John B. Goodenough*, Inorganic Chemistry Laboratory,
University of Oxford, South Parks Road, Oxford OX1 3QR, Great Britain

Professor *James A. Ibers*, Department of Chemistry, Northwestern University,
Evanston, Illinois 60201, U.S.A.

Professor *Christian K. Jørgensen*, Dépt. de Chimie Minérale de l'Université,
30 quai Ernest Ansermet, CH-1211 Genève 4

Professor *David Michael P. Mingos*, Imperial College of Science, Technology and
Medicine, Department of Chemistry, South Kensington, London SW7 2AY,
Great Britain

Professor *Joe B. Neilands*, Biochemistry Department, University of California,
Berkeley, California 94720, U.S.A.

Professor *Graham A. Palmer*, Rice University, Department of Biochemistry,
Wiess School of Natural Sciences, P.O. Box 1892, Houston, Texas 77251, U.S.A.

Professor *Dirk Reinen*, Fachbereich Chemie der Philipps-Universität Marburg,
Hans-Meerwein-Straße, D-3550 Marburg

Professor *Peter J. Sadler*, Birkbeck College, Department of Chemistry,
University of London, London WC1E 7HX, Great Britain

Professor *Raymond Weiss*, Institut Le Bel, Laboratoire de Cristallochimie et de
Chimie Structurale, 4, rue Blaise Pascal, F-67070 Strasbourg Cedex

Professor *Robert Joseph P. Williams*, Wadham College, Inorganic Chemistry
Laboratory, Oxford OX1 3QR, Great Britain

Attention all "Structure and Bonding" readers:

A file with the complete volume indexes Vols. 1 through 79 in delimited ASCII format is available for downloading at no charge from the Springer EARN mailbox. Delimited ASCII format can be imported into most databanks.

The file has been compressed using the popular shareware program "PKZIP" (Trademark of PK ware Inc., PKZIP is available from most BBS and shareware distributors).

This file is distributed without any expressed or implied warranty.

To receive this file send an e-mail message to:
SVSERV@DHDSPRI6.BITNET.

The message must be: "GET/STBO/SB__V1.ZIP".

SPSERV is an automatic data distribution system. It responds to your message. The following commands are available:

HELP	returns a detailed instruction set for the use of SVSERV,
DIR (*name*)	returns a list of files available in the directory "name",
INDEX (*name*)	same as "DIR",
CD ⟨*name*⟩	changes to directory "name",
SEND ⟨*filename*⟩	invokes a message with the file "filename",
GET ⟨*filename*⟩	same as "SEND".

Table of Contents

Moments of Inertia in Cluster and Coordination Compounds
   D. M. P. Mingos, J. E. McGrady, A. L. Rohl......... 1

Progress in Polymetallic Exchange-Coupled Systems, some Examples in Inorganic Chemistry
   M. Drillon, J. Darriet.......................... 55

Stereochemical Aspects Associated with the Redox Behaviour of Heterometal Carbonyl Clusters
   P. Zanello................................... 101

Electronic Structure and Bonding in Actinyl Ions
   R. G. Denning 215

A New Approach to Structural Description of Complex Polyhedra Containing Polychalcogenide Anions
   M. Evain, R. Brec.............................. 277

Crystal Chemistry of Inorganic Nitrides
   N. E. Brese, M. O'Keeffe........................ 307

Author Index Volumes 1–79 379

Moments of Inertia in Cluster and Coordination Compounds

David M.P. Mingos, John E. McGrady and Andrew L. Rohl

Imperial College of Science, Technology and Medicine, Department of Chemistry, South Kensington, London SW7 2AY, Great Britain

The calculated moments of inertia of molecules may be used to define the shapes of coordination compounds and cluster molecules. This shape analysis may be related either to electronic factors responsible for distorting the geometry of the molecule or used to define the location of the molecule on the rearrangement coordinate linking alternative polytopal forms.

1 Introduction	2
2 Moments of Inertia	4
2.1 Definition	4
2.2 Cubic and Icosahedral Geometries	6
2.3 Bipyramids	6
2.4 Prisms and Antiprisms	7
2.5 The Effect of Size and Connectivity	8
3 Alkali Metal Clusters	10
3.1 Introduction	10
3.2 3-Dimensional Clusters	16
3.3 Planar Clusters	20
4 Structural Correlation Analysis of Coordination Compounds and their Polytopal Rearrangement Pathways	23
4.1 Introduction	23
4.2 5-Coordinate Complexes	23
4.3 6-Coordinate Complexes	32
4.4 7-Coordinate Complexes	38
4.5 8-Coordinate Complexes	44
5 References	53

1 Introduction

The precise definition of molecular geometry, and its relationship to electronic properties has been a central concern of valence theory since X-ray crystallographic and electron diffraction techniques began to reveal the diversity of molecular structures in both the solid state and the gas phase. Simple theoretical approaches to bonding problems generally relate electronic structure to idealised polyhedral shapes. For example the Polyhedral Skeletal Electron Pair Theory [1] treats clusters in terms of idealised polyhedra with triangular faces (deltahedra), which maximise electron delocalisation within the cluster. In a similar way, Valence Shell Electron Pair Repulsion Theory [2] discusses the structures of molecular compounds in terms of the same deltahedra, which minimise electron pair repulsions. When the number of valence electron pairs is equal to the number of ligands then all vertices of the polyhedron are occupied by ligands. As the number of electron pairs is increased, the polyhedral shape is retained, but lone pairs successively replace ligands at the vertices. Structures which do not correspond exactly to any one polyhedral form are loosely described as distortions from one or other of these idealised forms. This imprecise terminology is confusing, and obviously does not allow a quantitative measure of the degree of distortion of structure relative to idealised polyhedra. Many real molecules deviate greatly from idealised forms, and it is desirable to be able to quantitatively relate the distorted form to the idealised geometries [3]. Examples of the importance of a quantitative approach to the problem can be drawn from the fields of cluster and coordination chemistry, upon which this review is based.

Bürgi and Dunitz [4, 5] recognised that solid state structures of coordination compounds frequently deviate from the idealised geometries, and that the structures lie along a reaction coordinate describing the lowest energy pathway on the energy hypersurface. Their principle of 'Structural Correlation' is based on the assumption that the crystal framework freezes the molecule at some point along the reaction coordinate away from the energy minimum for an isolated molecule. There are considerable problems involved in representing the structures of these distorted coordination compounds, since the molecule has 3N-6 internal degrees of freedom, where N is the coordination number. This problem becomes more acute as the coordination number increases.

Muetterties and Guggenberger [6] developed a method first proposed by Porai-Koshits and Aslanov [7] which uses the dihedral angles between the normals to the faces of the polyhedron as shape defining parameters. In polyhedra with a large number of vertices, there is a large set of dihedral angles (equal to the number of edges of the polyhedron), not all of which are independent. Muetterties and Guggenberger identified subsets of dihedral angles which they described as shape determining. These angles are not unique, and were chosen because:

i) They differentiate limiting structures.

ii) They correlate between the 2 idealised geometries.

iii) They correspond to features readily identifiable in the structure.

If the polyhedron is close to the limiting idealised structure then the shape determining edges are easy to establish, but more often it is necessary to calculate all dihedral angles, and identify the smallest set of angles which will define the shape.

An illustrative example is provided by coordination polyhedra with 5 vertices, where the limiting polyhedra are the trigonal bipyramid (point group D_{3h}) and the square pyramid (point group C_{4v}).

In defining the limiting polyhedra the following assumptions were introduced by Muetterties and Guggenberger:

i) The ratio of bond lengths (axial and equatorial) was set to unity.

ii) The axial-central atom-equatorial bond angle in the square pyramid was set to 102° (observed values range from 100° to 104°).

The resultant edge lengths and dihedral angles were then calculated by simple trigonometry.

The arrows indicate the parameters which correlate with each other in the 2 structures. The shape determining dihedral angles as defined by Muetterties and Guggenberger are underlined, and the associated edges are shown as double lines in Fig. 1.

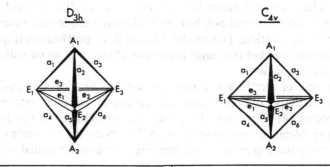

D_{3h}	C_{4v}
Re/Ra=1	Re/Ra=1
$a_1=a_3=a_4=a_6$=1.414R ---------	----------> $a_1=a_3=a_4=a_6$=1.386R
$a_2=a_5$=1.414R $e_1=e_2$=1.732R --------	----------> $e_1=e_2=a_2=a_5$=1.554R
e_3=1.732R ---------	----------> e_3=1.956R
$\delta a_1=\delta a_3=\delta a_4=\delta a_6$=101.5° --------	----------> $\delta a_1=\delta a_3=\delta a_4=\delta a_6$=119.8°
$\underline{\delta e_1}=\underline{\delta e_2}$=53.1° $\delta a_2=\delta a_5$=101.5° --------	----------> $\underline{\delta e_1}=\underline{\delta e_2}=\delta a_2=\delta a_5$=75.7°
$\underline{\delta e_3}$=53.1°	----------> $\underline{\delta e_3}$=0.0°

Fig. 1. Dihedral angles of 5 vertex polyhedra

In conclusion, the dihedral angle technique can be useful, particularly in smaller molecules, to define reaction coordinates between idealised forms. The method has the following drawbacks:

i) The method is ambiguous in some cases
ii) It uses only selected portions of the available data, for example one particular dihedral angle, which correlates between 2 idealised forms, to plot reaction coordinates representing the rearrangement from one idealised geometry to another.
iii) The method does not lend itself to quantifying the degree of distortion in those cases where the structure does not lie on the reaction coordinate connecting the specified idealised structures.
iv) Cluster compounds do not generally have the atoms lying on a single sphere and therefore some of the basic assumptions have to be modified.

In another, less widely used methodology, Bürgi [4] proposed the use of angles subtended at the centre of the molecule by the atoms as shape defining parameters. This suffers from similar problems as the dihedral method, in that a lot of information is not used, and it is not generally useful for defining distortions in clusters.

Dollase [8] approached the problem by relating the distorted structure to an optimal idealised polyhedron. The optimal idealised polyhedron was found by rotation, translation and scaling of an idealised form to obtain the least squares best fit. The average distance between the atoms on the idealised polyhedron and those on the distorted polyhedron is taken as a single parameter to describe the degree of distortion. This model is useful if the polyhedron is quite close to one idealised form, but the single parameter obtained gives no indication of the direction of distortion.

Moments of inertia are commonly used in physics and engineering for defining the shapes of objects, and in chemistry for the interpretation of microwave spectra [9]. Gavezzotti [10] has used moments of inertia to rationalise packing problems in organic systems. In this review the moments of inertia are used to study some geometric and bonding problems associated with inorganic molecules.

2 Moments of Inertia

2.1 Introduction

The moment of inertia of a rigid body about any axis passing through its centre of mass is defined as

$$I = \sum_i m_i r_i^2$$

where m is the mass of a particle and r is the perpendicular distance from the plane. The locus of points a distance $I^{-1/2}$ radially from the centre of mass in the direction of the axis of rotation defines the surface of an ellipsoid, and the 3 mutually perpendicular axes of this ellipsoid coincide with the 3 principal axes of inertia of the molecule. In highly symmetric molecules, the principal axes can readily be identified, as they coincide with the symmetry axes of the molecule. In less symmetric molecules, the principal axes of inertia cannot be established by symmetry, so the calculation of principal moments of inertia is more complex.

The principal moments of inertia can be related to the best plane in the molecule, which can be found by minimising the sum of the squares of the distances to the plane,

$$S = \sum_i \omega_i [x_i l + y_i m + z_i n - d]^2$$

with respect to l, m, n under the condition that

$$l^2 + m^2 + n^2 = 1$$

where ω is a weighting factor, x, y, z are coordinates of atoms, l, m, n are the direction cosines of the unit normal, and d is the origin to plane distance. If ω, the weighting factor, is taken to be the mass of the atom, then the problem is equivalent to finding the principal moments of inertia. The best procedure [11] is to carry out a principal axis transformation of the symmetric matrix

$$\begin{bmatrix} \Sigma mX^2 & \Sigma mXY & \Sigma mXZ \\ \Sigma mXY & \Sigma mY^2 & \Sigma mYZ \\ \Sigma mXZ & \Sigma mYZ & \Sigma mZ^2 \end{bmatrix}$$

X, Y and Z are the atomic coordinates after the origin has been moved to the centroid. The eigenvalues, $\lambda(i)$, correspond to the weighted sums of squares of the distances from the best, worst, and intermediate planes.

$$\lambda(1) = \sum_i m_i X_i'^2 \quad \lambda(2) = \sum_i m_i Y_i'^2 \quad \lambda(3) = \sum_i m_i Z_i'^2$$

The primed co-ordinates are those in the inertial reference plane. In the calculations described in this review the atoms are assumed to have unit mass, so that the resultant eigenvalues are given by

$$Ix = \Sigma X_i'^2 = R_{1x}^2, \; Iy = \Sigma Y_i'^2 = R_{1y}^2, \text{ and } Iz = \Sigma Z_i'^2 = R_{1z}^2$$

where R_{1x}, R_{1y} and R_{1z} are the radii of gyration of the molecule. The results therefore reflect the geometric distribution of atoms, and are independent of their identity. In subsequent discussions, the calculated moments of inertia are identical to R_i^2 unless otherwise stated.

The following section will focus on some idealised polyhedra, and relate the moments of inertia to geometric and symmetry features more familiar to chemists. When the atoms lie on the surface of a sphere, and the molecule belongs to a high symmetry point group, there are precise mathematical relationships between the positions of the atoms on the surface, and its principal moments of inertia. The z axis is chosen to lie along the unique symmetry axis.

2.2 Cubic and Icosahedral Geometries

In these high symmetry point groups, X, Y and Z belong to the same irreducible representation so the moments of inertia are equal

$$Ix = Iy = Iz$$

In addition

$$Ix = \sum_i x_i^2, Iy = \sum_i y_i^2, Iz = \sum_i z_i^2$$

Therefore

$$Ix + Iy + Iz = \sum_i (x_i^2 + y_i^2 + z_i^2) = na^2$$

where a is the radius of the sphere which passes through all the atoms, and n is the number of atoms in the molecule.

It follows that

$$Ix = Iy = Iz = \frac{na^2}{3}$$

Table 1 summarises the calculated moments of inertia for some clusters with cubic symmetry and edge lengths of 1.6 Å. This bond length corresponds to a typical B-B distance in a boron hydride cluster. The dimensions a = b = c for the corresponding sphere are also summarised in the Table. It is noteworthy that the moments of inertia do not distinguish these high symmetry polyhedra from a sphere. Higher order moments are required to make this distinction.

2.3 Bipyramids

Bipyramidal molecules belong to D_{nh} point groups, where Z and (X, Y) no longer belong to the same irreducible representation. This means that the moment ellipsoid is no longer spherical and so $Iz \neq Ix = Iy$. The atoms lie on a surface defined by an ellipsoid with semi major and semi minor axes of dimension $a = b \neq c$. The corresponding relationships between the dimensions

Table 1. Moments of inertia for cubic and icosahedral polyhedra

Polyhedron	Ix(= Iy = Iz)	a(= b = c)(Å)
Tetrahedron	1.280	0.979
Octahedron	2.560	1.131
Icosahedron	9.262	1.522

of the ellipsoid and the moments of inertia are:

$$Ix = Iy = \frac{(n-2)a^2}{2}$$

$$Iz = 2c^2$$

The calculated moments of inertia and radii of these molecules, with edge lengths of 1.6 Å are given in Table 2.

For a trigonal bipyramid a < c, so the ellipsoid is prolate, and the moments of inertia reflect this shape, as $Ix = Iy < Iz$. In contrast, a pentagonal bipyramid is oblate, and $Ix = Iy > Iz$.

It is apparent from the equations above that the moments of inertia do not depend only on the dimensions of the ellipsoid which passes through all the atoms. For example, for the trigonal bipyramid $Ix = Iy = Iz$ when

$$\frac{3}{2}a^2 = 2c^2 \text{ i.e. } \frac{a}{c} = \frac{2}{\sqrt{3}} \simeq 1.155$$

which corresponds to an oblate ellipsoid. The moments of inertia therefore represent not only the surface which is defined by the atoms, but also the distribution of atoms on that surface. Specifically, molecules with moments of inertia $Iz > Ix = Iy$ are described as prolate because on average their atoms are distributed closer to the poles, and those with $Iz < Ix = Iy$ are described as oblate because their atoms are distributed on average closer to the equator.

2.4 Prisms and Antiprisms

Prisms and antiprisms have a minimum of D_{nh} or D_{nd} point group symmetry respectively, and in some cases, such as the octahedron (trigonal antiprism) and cube (square prism), the symmetry is higher. For these polyhedra, the corresponding moments of inertia are:

$$Iz = n\cos^2\theta c^2$$

$$Ix = Iy = \frac{n\sin^2\theta b^2}{2}$$

θ is the angle that the atoms make with the polar (Z) axis.

Table 2. Moments of inertia for bipyramidal polyhedra

Polyhedron	Ix(= Iy)	Iz	a(= b)(Å)	c(Å)
Trigonal bipyramid	1.280	3.413	0.924	1.306
Pentagonal bipyramid	4.631	1.415	1.361	0.841

Table 3. Moments of inertia for prisms and antiprisms

Polyhedron	Ix(= Iy)	Iz	a(= b)(Å)	c(Å)	θ
Octahedron	2.560	2.560	1.131	1.131	54.74°
Trigonal prism	2.560	3.840	1.222	1.222	49.11°
Squre antiprism	5.120	3.620	1.316	1.316	59.26°
Cube	5.120	5.120	5.120	1.386	54.74°

The calculated moments of inertia, radii and θ angle for some of these polyhedra are given in Table 3.

It is notable that in these examples, although the atoms are symmetry equivalent, the moments of inertia are not necessarily equal. Information regarding the distortion of the cluster from sphericity is contained solely in the term involving θ, which describes the distribution of atoms on this spherical surface. The eclipsed and staggered nature of the distribution of atoms about the equator in prisms and antiprisms leads to different θ values for the 2 classes of polyhedra. For 6 vertex polyhedra, all moments of inertia are equal when θ = 54.74°, which is the case for an octahedron (a trigonal antiprism). In the corresponding trigonal prism θ = 49.11° leading to Iz > Ix = Iy. The moments of inertia therefore reflect information about the distribution of atoms on this surface, rather than the shape of the surface. The distribution of atoms in a trigonal prism is less equal than that in an octahedron, and the moments of inertia reflect this asymmetry (Ix = Iy < Iz for a trigonal prism, Ix = Iy = Iz for an octahedron).

Expressions for more complex species can be defined from the above expressions, because each atom makes an independent contribution to the moments of inertia. For example a bicapped antiprism has Ix and Iy identical to a square antiprism, as the addition of atoms on the z axis cannot influence Ix or Iy, so Iz = Iz (square antiprism) + $2c^2$, where c is the distance of the capping atoms from the origin of the polyhedron. The term $2c^2$ is identical to the contribution made by the axial atoms in a trigonal bipyramid.

Table 4 summarises the calculated moments of inertia for a range of clusters with commonly observed geometries. The moments of inertia provide a convenient and effective way of defining the oblate and prolate nature of the clusters.

2.5 Effect of Size and Connectivity

It is evident from the above discussions that the magnitude of the moments of inertia are influenced not only by the shape of the polyhedron, but by the number of atoms which are present on the surface. For polyhedra with cubic and icosahedral symmetry Ix + Iy + Iz = na^2, where a is the radius of the sphere and n is the number of atoms. The radius, a, is a function of nuclearity as well as n, since the sphere expands as the nuclearity inceases. This precludes

Table 4. Summary of moments of inertia for some common geometries

Nuclearity	Geometry	Shape	Ix	Iy	Iz
3	Triangle	Oblate	1.280	1.280	0.000
4	Tetrahedron	Spherical	1.280	1.280	1.280
4	Square	Oblate	2.560	2.560	0.000
5	Trigonal Bipyramid	Prolate	1.280	1.280	3.413
6	Octahedron	Spherical	2.000	2.000	2.000
6	Trigonal Prism	Prolate	2.560	2.560	3.840
7	Pentagonal Bipyramid	Oblate	4.631	4.631	1.415
8	Cube	Spherical	5.120	5.120	5.120
8	Square Antiprism	Oblate	5.120	5.120	3.620
8	Dodecahedron	Prolate	3.410	3.410	6.730
9	Tricapped Trigonal Prism	Oblate	6.370	6.370	3.840
10	Bicapped Square Antiprism	Prolate	5.120	5.120	10.13
10	Pentagonal Prism	Oblate	9.262	9.262	8.944
12	Icosahedron	Spherical	9.262	9.262	9.262

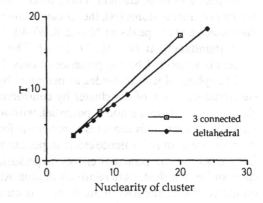

Fig. 2. Relationship between size and moments of inertia for 3-connected and deltahedral clusters

a simple relationship between the moments of inertia and the nuclearity. The role of size, and the connectivity can be simply illustrated by defining a 'size parameter', T

$$T = \sqrt{Ix} + \sqrt{Iy} + \sqrt{Iz}$$

Figure 2 illustrates a plot of T against nuclearity for both deltahedral and 3 connected polyhedra with edge lengths of 1.6Å.

The tetrahedron is both deltahedral and 3 connected so represents a common point for both classes. The slopes of the lines for the two classes differ because for a given nuclearity, the 3 connected polyhedra have larger surface areas than the corresponding deltahedral polyhedra, which are more highly connected.

The moments of inertia of a polyhedral molecule therefore contain information concerning its dimensions, shape, and the distribution of atoms on its surface. In the remainder of this review the applications of moments of inertia to a range of problems in cluster and coordination chemistry are examined. Since many valence theory arguments depend on defining the relationship between molecular shape and the molecule's electronic structure, the moments of inertia provide a convenient medium for such discussions.

3 Alkali Metal Clusters

3.1 Introduction

The study of bare metal clusters is central to the understanding of the links between solid state chemistry and that of discrete molecular species. Alkali metal clusters have been studied in molecular beams [12, 13], and the theoretical models proposed have attempted to interpret the abundances observed in the mass spectra of these clusters. These spectra show large abundances for specific numbers of metal atoms (N), the so-called 'magic numbers'. Neutral alkali metal mass spectra show peaks at $N = 2, 8, 20, 40, 58$, whereas cationic species show large abundances at $N = 19, 21, 35, 41$. The theoretical study of alkali metal clusters is simplified by the presence of only 1 valence electron per atom.

The spherical jellium model, as proposed by Knight et al [14], describes the electronic structure of the cluster by considering the energies of the N valence electrons moving in a smooth potential, without specifically taking account the atomic positions. This model is appropriate for alkali metal clusters because of the presence of only 1 valence electron per atom. This means that the bonding is fairly weak, so the potential energy hypersurface is fairly flat, and the atomic cores within the cluster are relatively mobile. Also the effective charge remaining on the core after the loss of an electron is quite small, so no localised areas of high potential occur.

A complementary approach is the Tensor Surface Harmonic Theory [19], based on the linear combination of atomic orbitals (LCAO) model, which explicitly incorporates the atomic positions. A set of atomic cores on the surface of a sphere are considered, and a basis set of s atomic orbitals used. If only these s orbitals are used, then the results are identical to the spherical jellium model. The three most stable orbitals are respectively 1, 3 and 5 fold degenerate, leading to closed shells at 2, 8 and 18 electrons.

The potential used by Knight was the so-called Woods-Saxon potential, although the conclusions are relatively insensitive to the precise form of the potential. The Schrödinger equation for this system is separable into radial and angular parts, and the wavefunctions are given by:

$$\Psi = F_{nl}(r)Y_{lm}(\theta, \phi)$$

where $Y_{lm}(\theta, \phi)$ are spherical harmonics, and $F_{nl}(r)$ depends on the form of the potential. The results of the calculation give the order of energy levels as

$$1s < 1p < 1d < 2s < 1f < 2p < 1g$$

This suggests that closed shell configurations will occur for 2, 8, 18, 20, 34, 40 and 58 electrons.

This model tends to predict more magic numbers than are actually seen in the mass spectrum. The reason for this may have geometric origins since only certain nuclearities can adopt highly spherical structures which coincide with a closed shell of electrons. When the cluster cannot adopt a highly spherical structure, a splitting of the jellium shells occurs, leading to some instability. A further and more crucial limitation is that the spherical jellium model provides no direct information on the structure, even for clusters with closed shells.

Extensions to the spherical jellium model have been made to incorporate deviations from sphericity. Clemenger [15] replaced the Woods-Saxon potential with a perturbed harmonic oscillator model, which enables the spherical potential well to undergo prolate and oblate distortions. The expansion of a potential field in terms of spherical harmonics has been used in crystal field theory, and these ideas have been extended to the nuclear configuration in a cluster in the structural jellium model [16].

The nuclear-electron potential is given by

$$V = - \sum_{i=1}^{N} \frac{Z}{r_{ie}}$$

where r_{ie} is the distance between electron and nucleus, and Z is the nuclear charge.

The term $1/r_{ie}$ can be expanded in terms of spherical harmonics [17], and the resultant expression separated into a spherical and non-spherical part.

$$V = V_0(r) + V(r, \theta, \phi)$$

The non-spherical part is treated as a perturbation of the system. The spherical part is similar in form to the Woods-Saxon potential, but now contains specific information about atomic positions.

In essence the model inter-relates the angular momentum of the electronic state of the cluster to specific cluster geometries. The overall stability of the cluster is determined by the number of nearest neighbour bonding interactions and the magnitudes of the splittings of the jellium shells and particularly the highest occupied shell. A cluster with a complete jellium shell, i.e. with 2, 8, 18, 20, ..., valence electrons has no net angular momentum and therefore this spherical electron distribution is best matched by a cluster geometry which is also spherical. This combination leads to a minimal splitting of those degenerate jellium shells. If the highest occupied jellium shell is incompletely filled then the cluster has net angular momentum. The maximum energy gap between the components of this shell is achieved when there is a match between the shape of the cluster and the lack of sphericity of the electron distribution. For example, a cluster with a pair of electrons with opposite spins in a P shell has a prolate electron distribution because the Pz component is occupied. The maximum stabilisation energy for this configuration occurs when this is matched with a prolate cluster geometry.

Clusters with a total of 3 to 7 valence electrons have an incomplete shell. Stabilisation energies occur from the splitting of orbitals which result from geometric distortions. The only perturbation terms involved in the non-spherical V(r, θ, φ) part are the V_2^m [16]. When the principal axis is of order 3 or more, only the V_2^0 term is needed since the V_2^m must belong to the totally symmetric representation of the point group considered. The V_2^0 perturbation corresponds to an oblate/prolate distortion of the cluster potential from spherical symmetry. We now show how this term generates 'oblate' and 'prolate' perturbations, even for clusters with a single shell of atomic cores arranged on a spherical surface.

The sign of the V_2^0 term is determined by the sign of the coefficient of V_2^0 in the expansion of the potential.

$$\text{Sgn}[\Delta E(lm)] = \text{Sgn}[(\Sigma Y_2^0(\theta i, \phi i))\langle lm|LM|lm\rangle]$$

The splitting of orbitals is reproduced in Fig. 3 for the case where $\Sigma Y_2^0(\theta_i, \phi_i)$ (abbreviated as Σ) is positive, which corresponds to a prolate distortion, and the case where Σ is negative, corresponding to an oblate distortion. In this way the geometry of the cluster, even with all the atomic cores assumed to be a distance r_0 from the centre of the cluster, may be classified as oblate or prolate according to the sign of Σ.

The sign of Σ may be easily seen to be positive for a trigonal bipyramid, which is therefore classified as prolate, zero for a cube or an octahedron which are therefore spherical, and negative for a pentagonal bipyramid which is thus oblate. More examples are listed in Table 5.

For a molecule with atoms of unit mass on a unit sphere the moments of inertia may be expressed in terms of the following spherical harmonic expansions:

$$Iz - Ix = \frac{\sqrt{5\pi}}{2} \sum_{i=1}^{N} Y_{20}(\theta_i, \phi_i)$$

$$Ix + Iy + Iz = \sqrt{4\pi} \sum_{i=1}^{N} Y_{00}(\theta_i, \phi_i)$$

Therefore, the V_2^0 term used in the perturbation theory expression and the moments of inertia are directly connected and the quantum mechanical and geometric considerations may be used interchangeably. The examples in Table 5 show that the sign of Σ parallels the differences in the calculated moments of inertia $Iz - Ix$. These moments of inertia have recently been used to classify the shapes of Na_n clusters [18].

The Jahn-Teller theorem indicates that when a degenerate shell is not completely filled, a distortion away from spherical symmetry occurs, leading to a loss of degeneracy. The splitting pattern produced by the Jahn-Teller type distortion is identical to that predicted by the crystal field approach used in the structural jellium model.

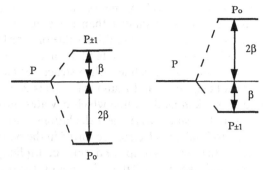

Fig. 3. The splitting of cluster p orbitals under prolate and oblate distortions

Table 5. Comparison of Σ and Ix-Iy for some common geometries

Nuclearity	Geometry	Shape	Σ	Iz − Ix
3	Triangle	Oblate	− 3	− 1.280
4	Tetrahedron	Spherical	0	0
4	Square	Oblate	− 4	− 2.560
5	Trigonal Bipyramid	Prolate	1	2.133
6	Octahedron	Spherical	0	0
6	Trigonal Prism	Prolate	1.714	1.280
7	Pentagonal Bipyramid	Oblate	− 1	− 3.216
8	Cube	Spherical	0	0
8	Square Antiprism	Oblate	− 1.730	− 1.500
8	Dodecahedron	Prolate	1.176	3.320
9	Tricapped Trigonal Prism	Oblate	− 1.286	− 2.530
10	Bicapped Square Antiprism	Prolate	2.270	5.010
10	Pentagonal Prism	Oblate	− 2.296	− 0.318
12	Icosahedron	Spherical	0	0

These simple theoretical studies on distorted clusters predict the gross nature of the distortion that may occur for a particular electron count, but do not quantify the extent of distortion.

Accurate ab initio calculations on lithium and sodium clusters have been performed by Fantucci et al [20, 21, 22, 23]. These studies have yielded detailed information about the equilibrium geometries, stabilities, and ionisation potentials of both neutral and cationic clusters. In this section the moments of inertia are used as a chemical tool to interpret the gross shape of the cluster. The deviation from sphericality is defined by the parameter L, derived from the moments of inertia by

$$L = \frac{I \text{ unique}}{I \text{ average}} = \frac{3Iz}{Ix + Iy + Iz}$$

For the case of spherical top molecules L = 1. In non-spherical structures I unique may be larger or smaller than I average, so L is greater than 1 for prolate structures, and tend to a limiting value of 3 for linear structures. L is less than 1 for oblate structures, and tend to a limiting value of 0 for planar structures. L is strictly only defined when structures have axial symmetry, since only then does Ix equal Iy, and I unique (= Iz) is defined. If this is not the case then I unique is taken as the value which deviates most from the mean.

Tables 6 and 7 summarise the calculations carried out on alkali metal clusters with 2 to 8 valence electrons reported in the papers of Fantucci et al. The tables contain the point group (and spin multiplicity), moments of inertia, L parameter, and, where available, the energy separation (ΔE) between that structure and the lowest energy structure for that stoichiometry. The structures include those of neutral, cationic and anionic clusters of lithium and sodium.

The potential energy hypersurface of these clusters is quite flat, and exhibits several local minima. The second best geometries lie less than 11 kcal above the

Table 6. Symmetries and calculated moments of inertia of lithium clusters

		Pt Group (Spin)	Ix	Iy	Iz	L	ΔE(Kcal mol^{-1})
Most Stable	Li$_3$	C$_{2v}$(2)	7.880	2.940	0.000	0.000	
Neutral	Li$_4$	D$_{2h}$(1)	16.359	3.618	0.000	0.000	
Clusters	Li$_5$	C$_{2v}$(2)	23.963	8.568	0.000	0.000	
	Li$_6$	C$_{5v}$(1)	20.021	20.021	0.515	0.038	
	Li$_6$	D$_{3h}$(1)	24.420	24.420	0.000	0.000	0.25
	Li$_7$	D$_{5h}$(2)	17.950	17.950	4.805	0.354	
	Li$_8$	T$_d$(1)	18.157	18.157	18.157	1.000	
Second Most	Li$_4$	C$_{2v}$(3)	8.904	7.450	1.254	0.214	10.51
Stable Neutral	Li$_4$	D$_{4h}$(3)	9.610	9.610	0.000	0.000	10.55
Clusters	Li$_5$	C$_{2v}$(2)	5.746	3.347	14.797	1.858	3.96
	Li$_5$	D$_{3h}$(4)	5.770	5.770	11.139	1.473	8.71
	Li$_6$	C$_{2v}$(1)	18.337	12.752	3.591	0.311	2.12
	Li$_6$	C$_{2v}$(3)	2.509	8.911	19.837	1.904	5.95
	Li$_7$	C$_{3v}$(2)	19.279	19.279	9.238	0.580	3.78
Most Stable	Li$_3^+$	D$_{3h}$(1)	5.042	5.042	0.000	0.000	
Cationic	Li$_4^+$	C$_{2v}$(2)	25.93	4.621	0.000	0.000	
Clusters	Li$_4^+$	D$_{2h}$(2)	17.880	4.263	0.000	0.000	0.26
	Li$_5^+$	D$_{3h}$(1)	4.200	4.200	17.287	2.019	
	Li$_5^+$	D$_{2h}$(1)	34.81	9.000	0.000	0.000	0.58
	Li$_6^+$	D$_{2h}$(2)	11.289	3.809	21.344	1.757	
	Li$_7^+$	D$_{5h}$(1)	21.340	21.340	3.538	0.230	
	Li$_8^+$	C$_5$(2)	23.272	18.809	9.524	0.554	
	Li$_9^+$	D$_{4d}$(1)	22.620	22.620	18.000	0.854	
Second Most	Li$_4^+$	D$_{3h}$(4)	17.061	17.061	0.000	0.000	21.64
Stable Cationic	Li$_7^+$	C$_{3v}$(1)	19.152	19.152	9.134	0.578	5.30
Clusters	Li$_9^+$	C$_{2v}$(1)	22.983	17.301	33.622	1.365	6.91
Anionic	Li$_6^-$	D$_{4h}$(2)	12.881	12.881	5.544	0.531	
Clusters	Li$_7^-$	C$_{3v}$(1)	11.537	11.537	20.917	1.426	
	Li$_7^-$	D$_{5h}$(1)	16.670	16.670	8.490	0.609	

Table 7. Symmetries and calculated moments of inertia of sodium clusters

		Point Group (Spin)	Ix	Iy	Iz	L	ΔE(Kcal mol^{-1})
Most Stable Neutral Clusters	Na$_3$	C$_{2v}$(2)	11.472	3.972	0.000	0.000	
	Na$_4$	D$_{2h}$(1)	22.700	5.281	0.000	0.000	
	Na$_5$	C$_{2v}$(2)	33.893	12.981	0.000	0.000	
	Na$_6$	C$_{5v}$(1)	28.665	28.665	0.515	0.027	
	Na$_6$	D$_{3h}$(1)	17.113	17.113	0.000	0.000	
	Na$_7$	D$_{5h}$(2)	25.850	25.850	8.201	0.411	
	Na$_8$	T$_d$(1)	27.332	27.332	27.332	1.000	
Most Stable Cationic Clusters	Na$_3^+$	D$_{3h}$(1)	6.771	6.771	0.000	0.000	
	Na$_4^+$	D$_{2h}$(2)	23.846	5.951	0.000	0.000	
	Na$_4^+$	C$_{2v}$(2)	35.759	6.195	0.000	0.000	0.58
	Na$_5^+$	D$_{2h}$(1)	47.693	11.963	0.000	0.000	
	Na$_6^+$	C$_s$(2)	15.536	5.131	32.311	1.830	
	Na$_7^+$	D$_{5h}$(1)	29.542	29.542	5.478	0.255	
	Na$_8^+$	C$_{2v}$(2)	36.373	31.620	14.310	0.522	
	Na$_9^+$	C$_{2v}$(1)	29.902	27.807	42.004	1.264	
Second Most Stable Cationic Clusters	Na$_5^+$	D$_{3h}$(1)	6.090	6.090	23.093	1.964	4.76
	Na$_6^+$	D$_{2h}$(2)	11.696	9.374	34.527	1.863	4.81
	Na$_9^+$	D$_{4d}$(1)	33.620	33.620	36.444	1.054	0.31

optimised geometry, and therefore represent energetically accessible states, possibly involved in skeletal rearrangement processes. Some molecules exhibit a second structure within 0.6 kcal of the optimum geometry. In this case the 2 structures are included together in the category of 'Most Stable Geometries'. This is the case for Na$_6$, Li$_4^+$, Na$_4^+$, Li$_5^+$ and Na$_9^+$.

Figure 4 illustrates a plot of T(= $\sqrt{Ix} + \sqrt{Iy} + \sqrt{Iz}$) against nuclearity for both lithium and sodium clusters. Figure 4 is similar to Fig. 2, and indicates that the parameter T increases approximately linearly with nuclearity. The effect of increasing the average bond length is illustrated by the difference between the sodium and lithium plots. The value of T for sodium is greater than that for lithium, and the gradient is greater, indicating that where the bond lengths are larger, the cluster size increases more rapidly with increasing nuclearity. Such plots may be useful for predicting the moments of inertia of cluster molecules not currently accessible to ab initio calculations.

It is immediately obvious from Tables 6 and 7 that the clusters fall into two clear categories: planar networks and 3-dimensional structures. Small clusters tend to favour planar geometries, being composed of deformed sections of the (111) lattice plane of the face centred cubic lattice. This has been rationalised by Fantucci et al. [20] in terms of the electronic structures of the clusters. More compact geometries in general favour cluster stability, as they maximise the number of nearest neighbours. However, the molecular orbitals in high symmetry close packed geometries retain degenerate orbital sets. In those clusters

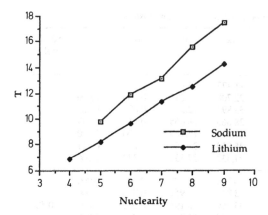

Fig. 4. Relationship between nuclearity and moments of inertia for sodium and lithium clusters

which do not have sufficient valence electrons to completely fill these degenerate orbitals a distortion occurs which removes the degeneracy. Small clusters therefore often adopt planar close packed geometries, where the valence orbital degeneracy is reduced.

3.2 3-Dimensional Structures

In this section the parameter L is used to quantify the distortions from sphericality, and define the relationship between cluster shape and the number of valence electrons. Figure 3 illustrated the splitting of the 1f shell under oblate and prolate distortions. The resultant orbital stabilisation energies in terms of β are shown in Table 8 for clusters with 4 to 8 valence electrons. Both high and low spin cases are considered. The most stable geometries predicted by this simple model, are summarised in Table 9.

The clusters described in Tables 6 and 7 are mostly low spin, but some triplet and quadruplet (high spin) cases occur, and in the cases of 4 and 6 valence electrons, the predicted distortion is different for low and high spin geometries. In Table 6 the most stable Li_4 cluster is a singlet (low spin) state, and has a prolate geometry (L > 1). However the second best geometries for Li_4 are both triplets, and have oblate geometries. A particularly striking example of the qualitative success of the moments of inertia methodology is the case of Li_6. The 2 second best geometries are close in energy and both belong to the same point group, one being a singlet state, and the other a triplet state. The moments of inertia indicate that the singlet state is oblate, as predicted in Table 9, while the triplet state is prolate, also in accordance with the predictions of Table 9. The parameter L therefore distinguishes two very similar structures as prolate and oblate, in agreement with the simple model.

Table 8. Calculated stabilisation energies for distorted clusters

No. of Electrons	Prolate		Oblate	
	High Spin	Low Spin	High Spin	Low Spin
4	β	4β	2β	2β
5	0	3β	0	3β
6	2β	2β	β	4β
7	β	β	2β	2β
8	0	0	0	0

Table 9. Predicted Geometries for clusters on the basis of Table 8

No. of Electrons	High Spin	Low Spin
4	Oblate	Prolate
5	No Preference	No Preference
6	Prolate	Oblate
7	Oblate	Oblate
8	Spherical	Spherical

Figure 5 shows a plot of average value of L, and the scatter of data points, against the number of valence electrons for all 3-dimensional low spin clusters.

Figure 5 indicates that 4 and 5 electron species adopt prolate geometries while 6 and 7 electron species adopt oblate geometries, and 8 electron species adopt approximately spherical gometries The driving force for distortion away from spherical geometries is greatest for 4 and 6 electron low spin clusters, where the orbital stabilisation is 4β, and this is reflected in Fig. 5, where the L values for 4 and 6 electron molecules deviate most from unity. The relationship between the degree of distortion and the orbital stabilisation energy is illustrated in Fig. 6. Negative values for the stabilisation energy indicate that an oblate geometry is adopted.

The degree of distortion increases approximately linearly with orbital stabilisation energy, for both prolate and oblate distortions, indicating that electronic factors determine the magnitude of the distortion. However, in all cases the orbital splitting effects must be balanced against close packing considerations. For example, the cube which has $I_x = I_y = I_z$ is not the preferred structure for 8 atom clusters because it does not maximise the nearest neighbour bonding interactions. The preferred structure is based on a tetracapped tetrahedron which satisfies both criteria since it is more closely packed and has tetrahedral symmetry. For closed shell clusters with 7 or 9 atoms and 8 valence electrons a spherical structure is not possible and therefore the nearest neighbour interactions predominate. For example, the preferred structure for M_9^+ is a centred square antiprism and M_7^- is a pentagonal bipyramid. Therefore the average value of L is approximately 1 for such clusters, but a spread of data points to either side of 1 occurs (Fig. 5).

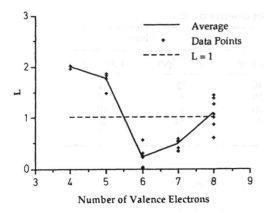

Fig. 5. Distortion parameter, L, for non-planar clusters with 4 to 8 electrons

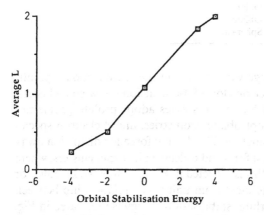

Fig. 6. Relationship between the average distortion parameter, L, and orbital stabilisation energy

Figure 7 illustrates the difference in moments of inertia for the most stable and second most stable structures. In all cases except for 5 electron clusters, where the average values of L are very similar, the optimised geometry is less spherical than the second best geometry. This again emphasises the dominant role of electronic factors in determining cluster geometry.

As described in earlier discussions, the L parameter quantifies distortions in the axial direction, and is only strictly defined when the cluster is symmetric about the unique axis (for the cluster to be symmetric about the principal axis, the principal axis of rotation must be at least a 3-fold axis). The 5 electron species illustrate the effect of a distortion perpendicular to the principal axis. The electronic structure of these clusters is $1s^2 \, 1p_0^2 \, 1p_\pm^1$, and consequently is orbitally degenerate. The Jahn-Teller theorem predicts that a distortion will occur to lift the degeneracy and lower the energy of the system. In order to lift the degener-

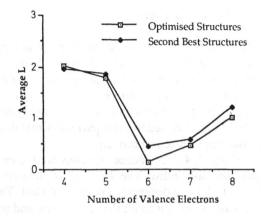

Fig. 7. Difference in average value of L for optimised and second most stable structures for clusters with 4 to 8 valence electrons

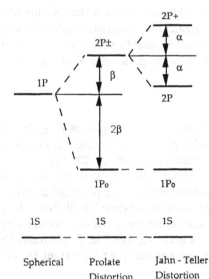

Fig. 8. Splitting of the $p \pm$ subshell by Jahn-Teller distortions

acy of the $p_{\pm 1}$ subshell, the distortion must be perpendicular to the principal (Z) axis, causing the order of the principal axis of rotation to be reduced to 2, and the degeneracy of Ix and Iy to be lifted. The effect of this distortion on the valence orbitals is illustrated schematically in Fig. 8.

The resultant orbital stabilisation energy for 5 electron clusters is $3\beta + \alpha$ instead of 3β for the axially symmetric clusters. Tables 6 and 7 show that the optimised 5 electron clusters have principal axes of order only 2, and consequently Ix ≠ Iy. 7 electron species show no such distortions. In this case the singly occupied orbital is $1p_0$, so Jahn-Teller distortions do not operate.

3.3 Planar Structures

The descriptions oblate and prolate are not applicable in this case, as all the clusters are oblate in the sense that Iz = 0. In the limit of a circle (Ix = Iy) can be considered as a basis for simple quantum mechanical calculations in the same sense as the spherical limit (Ix = Iy = Iz) in the case of 3-dimensional structures. The lowest energy and second lowest energy orbitals obtained from solution of Schrödinger's equation for a particle constrained to move on a circle are 1 and 2 fold degenerate respectively.

When 3, 4 or 5 valence electrons are present, the doubly degenerate shell is only partially filled. The Jahn-Teller theorem predicts that a distortion will occur to lift the degeneracy of this subshell. This distortion must be about the principal axis if it is to lift the degeneracy, and consequently the degeneracy of Ix and Iy will also be lost. The effect of a distortion of this type on the valence orbitals is shown in Fig. 9.

Table 10 shows the orbital stabilisation energies of clusters with 2 to 6 electrons in terms of the splitting parameter α. The driving force for distortion is greatest for 4 electron molecules with low spin configurations, where the $\pi \pm$ orbital is doubly occupied, and is zero for 2 and 6 electron species. For these clusters the parameter K is defined.

$$K = \frac{Ix}{Iy}$$

Figure 10 shows a plot of K against number of valence electrons for the most stable planar geometries. The results are taken from the ab initio calculations of Fantucci et al [20, 21, 22, 23].

Figure 10 illustrates that for all the 2 and 6 electron clusters there is no distortion about the principal axis (Ix = Iy) and consequently no splitting of the doubly degenerate subshell. For 3, 4 and 5 electron species the value of K rises to a maximum for 4 valence electrons, as predicted in Table 10.

Figure 10, like Fig. 5, illustrates that in these planar clusters the electronic effects are the most important factor in determining the degree of distortion

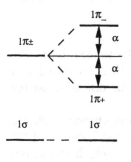

Fig. 9. Splitting of $\pi \pm$ orbitals by distortion from circular geometry

Table 10. Calculated orbital stabilisation energies (OSE) for planar clusters

No. of Electrons	OSE
2	0
3	α
4	2α
5	α
6	0

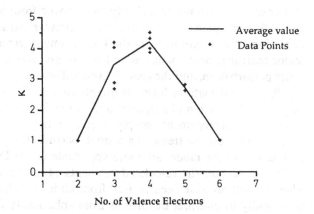

Fig. 10. Distortion parameter, K, for planar clusters with 4 to 8 valence electrons

from the most symmetric structure. The effect of geometric constraints is shown by the occurrence of data points in pairs in the Fig. 10. One of the pair corresponds to a sodium cluster, the other to the analogous lithium cluster. For 4 electron molecules, there are 4 data points, one pair corresponding to neutral clusters of nuclearity 4, while the other pair correspond to cationic structures of nuclearity 5. The additional atom in the 5 atom clusters imposes geometric constraints which cause the value of K to differ from those of other 4 atom clusters.

The only clusters not included in Fig. 10 are the 4 atom clusters with C_{2v} symmetry. For both sodium and lithium these form one of a pair of second best structures, the other structure having D_{2h} symmetry. The energy of these structures is very similar (the maximum separation between the two is 0.6 kcal mol^{-1}). These structures are illustrated in Fig. 11.

The reason for the deviation of the C_{2v} symmetry structures from the predictions of the simple model, while the D_{2h} structures conform to it, lies in the basic assumptions of the model. The basic structure of the clusters was assumed to be circular, and deviations from this are treated as small perturbations of the system, resulting in the orbital energies shown in Fig. 9. The structure of the

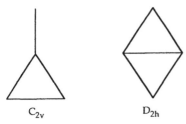

Fig. 11. Structures of 4 atom clusters with C_{2v} and D_{2h} symmetry

cluster with C_{2v} symmetry is clearly far removed from one of circular symmetry (for a 4 atom planar cluster the 'most circular' structure is a square). The distortion of a square to form the C_{2v} symmetry structure therefore involves major rearrangements of atoms and bonds, and so cannot be treated as a first order perturbation, and the energy levels will bear little resemblance to those in Fig. 9, and so deviations from the quantitative predictions of the model are not surprising. Distortion of a square to form the D_{2h} symmetry structure involves much less rearrangement, simply requiring the shortening of one diagonal distance. This can be treated as a small perturbation, and so the quantitative conclusions of the model are more applicable to the D_{2h} structural type.

In this section the calculated moments of inertia of alkali metal clusters, whose structures have been derived from ab initio calculations have been used successfully to interpret distortions from sphericity for 3-dimensional structures. The analysis has proved successful in relating these distortions to simple models of cluster bonding, and provides a useful tool for the prediction of stable cluster structures for as yet uncharacterised species. Distortions from circular to elliptical geometries for planar structures can be evaluated using a similar methodology.

The success of the moments of inertia as a useful structural parameter is at first surprising, since the mathematical expressions for the moments of inertia of idealised clusters are complex, do not relate to any recognisable feature, and vary from structure to structure. In this analysis, the moments of inertia are used as a parameter to describe only the gross deviation from sphericity. As was illustrated in Sect. 2.2, the shape of the ellipsoid enveloping the atoms is not sufficient to describe clusters as oblate or prolate, information about atomic positions is also required. The splitting of the orbitals, which is defined by the spherical harmonic perturbation, is dependent on the same factors which determine the moments of inertia, the cluster shape and distribution of atoms about the surface. The moments of inertia therefore provide a generalised shape parameter, without referring to specific features of the individual structures.

The structures of gold cluster compounds bear many similarities to those described above and are amenable to a similar structural/electronic analysis. Kanters and Steggerda have discussed how moments of inertia may be used to distinguish spherical and toroidal clusters [24, 25].

4 Structural Correlation Analyses of Coordination Compounds and Their Polytopal Rearrangements

4.1 Introduction

The principle of structural correlation proposed by Bürgi and Dunitz [4, 5] relates the observed crystal structures for coordination compounds to reaction or rearrangement pathways. Bürgi proposed that if a relationship between two independent structural parameters for a series of related structures is found, then it can be assumed that an inherent property of the fragment is responsible for this behaviour. More recently Bürgi and his coworkers have utilised statistical cluster and factor analyses to provide a more detailed analysis of structural correlations [26]. The inherent property of the fragment is the energy hypersurface. The concept of a hypersurface for a chemical reaction was proposed by Eyring and Polanyi [27]. The hypersurface describes the variaton of energy of the fragment as a function of internal parameters such as bond lengths and bond angles. This multi-dimensional surface can be projected onto a plane described by the variation of two appropriate parameters, and the valleys in the resultant surface define minimum energy pathways. The environment of the fragment in each crystal acts as a perturbation, and distorts the fragment away from the minimum energy geometry, along the minimum energy pathway. Each observed structure of the fragment represents the equilibrium between intramolecular effects favouring the minimum energy conformation and the intermolecular effects of the environment. Each observed structure may therefore be regarded as marking a point along the lowest energy path for either a ligand dissociation or a polytopal rearrangement.

This section deals with the application of moments of inertia for defining the geometries of coordination compounds and the structural correlations used to map the minimum energy paths of polytopal rearrangement processes in coordination compounds. The fluxionality of coordination compounds has long been recognised, and the pathways for rearrangement have been the subject of much debate.

4.2 5-Coordinate Complexes

Berry [28] proposed that the rapid intramolecular rearrangement which leads to the exchange of axial and equatorial ligands in trigonal bipyramidal complexes proceeds by the mechanism shown in Fig. 12. The transition state is a square pyramid, and both quantum mechanical [29] and empirical force field methods [30] have shown that this is an energetically feasible transition state, with an energy approximately 6 kcal/mol above the trigonal bipyramidal ground state for PF_5. Alternative mechanisms have been proposed [31], most notably the 'Turnstile Rotation' proposed by Ugi et al. [32] and illustrated in

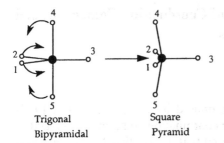

Trigonal Bipyramidal

Square Pyramid

Fig. 12. Berry pseudorotation rearrangement pathway for 5-coordinate molecules

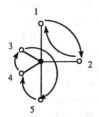

Fig. 13. The Turnstile rearrangement pathway for 5-coordinate molecules

Fig. 13. This may be described as an internal rotation of one apical and one equatorial ligand rotating as a pair about a local C_2 axis, while the other three rotate in the opposite direction about a local C_3 axis.

Atoms 1 and 2 rotate as a pair in one direction about a local C_2 axis, while atoms 3, 4 and 5 rotate about an approximate C_3 axis. The C_3 axis is not a perfect 3-fold axis, because in the trigonal bipyramidal ground state, one of the atoms is axial, and the other 2 are equatorial, and are symmetry distinct. The reaction pathway can be modelled by a rotation of 30° about the 3 fold axis of the group of 3 atoms relative to the pair of atoms, with a simultaneous distortion of the bond angles to attain a transition state where all the atoms in the group of 3 are identical.

The two mechanisms have very different transition states, and so should be distinguished by structural correlations if suitable molecules can be found to map the reaction coordinate.

Meakin [33] has criticised the Berry mechanism, arguing that it is only strictly applicable where C_{2v} symmetry is maintained throughout the rearrangement. This limits the mechanism to MX_5 species, where the symmetry changes from D_{3h} (trigonal bipyramid) to C_{4v} (square pyramid) along a pathway that retains C_{2v} symmetry, and MX_4Y species, where the structure varies from C_{2v} (Trigonal bipyramid) to C_{4v} (Square pyramid), along a pathway which retains C_{2v} symmetry if the Y atom acts as the pivotal ligand. Where the molecule contains different ligands the retention of strict C_{2v} symmetry is impossible, but when the electronic properties of the ligand are very similar the situation will approximate to that for the more symmetrical polyhedron. Meakin suggested that any one idealised mechanism is an over-simplification, and for certain structures mechanisms other than the Berry mechanism will have a lower activation energy, and so will occur preferentially. Previous structural correlations have concentrated on the Berry mechanism, and have used the dihedral angles proposed by Muetterties and Guggenberger as the structural parameter. In their initial work Muetterties and Guggenberger correlated variations in the smallest dihedral angle with the axial-basal angle, θ, for MX_5 species. Bürgi correlated the average dihedral angles $(\delta_{13} + \delta_{23})/2$ and $(\delta_{34} + \delta_{35})/2$ (See Chapter 1). Holmes [34] and Auf der Heyde and Nassimbini [35] extended the correlation to include all the dihedral angles. The sum, over all edges, of the changes in dihedral angles accompanying the trigonal bipyramid to square pyramid transition (Berry Pseudorotation) is given by

$$\Sigma_{ij}[\delta_{ij}(TBP) - \delta_{ij}(SP)] = 217.9°$$

If a compound C lies on the Berry coordinate, then

$$\Sigma_{ij}[\delta_{ij}(C) - \delta_{ij}(TBP)] + \Sigma_{ij}[\delta_{ij}(C) - \delta_{ij}(SP)] = 217.9°.$$

A plot of $\Sigma_{ij}[\delta_{ij}(C) - \delta_{ij}(TBP)]$ versus $217.9° - \Sigma_{ij}[\delta_{ij}(C) - \delta_{ij}(SP)]$ is linear. Consideration of a series of compounds with different ligands attached to the same central atom showed a spread of data about the theoretical Berry reaction coordinate. The method of Auf der Heyde and Nassimbini is more rigorous, as it uses all of the available information, whereas the previous approaches used only selected parameters. All three methods suffer from the fact that the relationship of structural data to only one reaction coordinate can be considered.

Alternative idealised co-ordination geometries are generally associated with different patterns of moment of inertia. For example, the 8 coordination geometries based on a cube, square antiprism and a dodecahedron are spherical, oblate and prolate respectively. The moments of inertia of a molecule may be calculated as the structure passes along any chosen reaction coordinate connecting the idealised structures, and the three parameters, Ix, Iy and Iz can be used to characterise the structure in 3-dimensional space. The moments of inertia of crystallographically characterised molecules can then be compared with theoretical plots for idealised reaction coordinates. In the study of polytopal rearrangements of this type, it is assumed that all the information necessary to describe the rearrangement is contained in the variation of bond angles, and the bond lengths are therefore assumed to maintain a constant value with all atoms lying on a sphere of unit radius. Muetterties and Guggenberger also used this assumption [6]. For real MX_5 species, the ratio of axial to equatorial bond lengths varies from 0.96 to 1.07.

In Sect. 1 it was demonstrated that for a molecule with atoms of unit mass

$$Ix + Iy + Iz = \sum_i [x_i^2 + y_i^2 + z_i^2]$$

If the metal atom lies at the centre of the sphere of unit radius, then

$$x_i^2 + y_i^2 + z_i^2 = r_i^2 = 1$$

and so $Ix + Iy + Iz = N$, where N is the coordination number. The centre of the sphere is defined as the average of all the coordinates, so when the ligands are symmetrically disposed about the central atom (for example in a trigonal bipyramid), the atom lies at the centre of the sphere. When the ligands are less symmetrically disposed about the central atom, significant deviations from Ix + Iy + Iz = N occur. Figure 14 shows the structures of the molecules PF_5 and CF_4, where the central atom lies at the centre of the sphere, and IF_5 and SF_4, which are more hemispherical and where the central atom does not coincide with the centre of mass.

Table 11 shows the calculated moments of inertia of these molecules, and the value of Ix + Iy + Iz. The hemispherical molecules SF_4 and IF_5 have Ix + Iy + Iz significantly less than N, reflecting the presence of the lone pair. Therefore the moments of inertia may be used to define the stereochemical effects of lone pairs in main group molecules, although this is not explored in detail in this review.

In most molecules described in this secton, Ix + Iy + Iz approximates very closely to N, and so when the moments of inertia are displayed in 3-dimensions, the points are approximately co-planar, lying near the plane Ix + Iy + Iz = N. To simplify the 3-dimensional plots, the projection of the points onto the Ix + Iy + Iz = N plane is illustrated. The projection onto this plane is calculated by performing a principal axis transformation.

$$\begin{bmatrix} \frac{1}{\sqrt{3}} & \frac{1}{\sqrt{3}} & \frac{1}{\sqrt{3}} \\ \frac{1}{\sqrt{2}} & \frac{-1}{\sqrt{2}} & 0 \\ \frac{1}{\sqrt{6}} & \frac{1}{\sqrt{6}} & \frac{-2}{\sqrt{6}} \end{bmatrix} \begin{bmatrix} Ix \\ Iy \\ Iz \end{bmatrix} = \begin{bmatrix} Iz' \\ Ix' \\ Iy' \end{bmatrix}$$

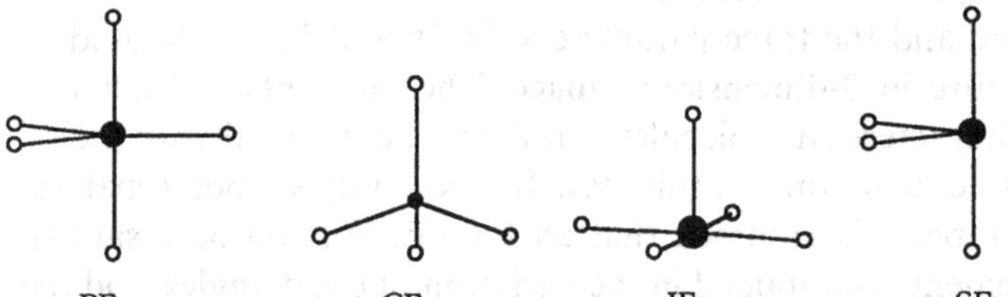

Fig. 14. Idealised structures of some simple coordination compounds

Table 11. Calculated moments of inertia for spherical and hemi-spherical molecules

Species	Ix	Iy	Iz	Ix + Iy + Iz
CF_4	1.333	1.333	1.333	4
SF_4	0.166	1.5	2	3.666
PF_5	1.5	1.5	2	5
IF_5	2	2	0.8	4.8

The result of this transformation is that

$$Iz' = \left(\frac{Ix + Iy + Iz}{\sqrt{3}}\right)$$

so if the points are co-planar, Iz' has the value of $N/1.732$. The parameters Ix' and Iy' can then be used to plot the relative shapes of molecules in 2-dimensional space.

For some limiting species, the shape is not uniquely defined by symmetry. For example in the square pyramid the L(axial)-M-L(equatorial) angle, denoted θ, is not uniquely defined, and an average value of 102° was used by Muetterties and Guggenberger. Holmes [36] showed that the L(axial)-M-L(equatorial) angle depends critically on the electronic structure of the central atom. This point will be discussed in more detail later in this section.

Table 12 summarises the calculated moments of inertia for hypothetical 5-coordinate molecules chosen to lie along either the Berry or the Turnstile rearrangement coordinates. The initial geometry in both cases is the trigonal bipyramid, and the transition state for the Berry coordinate is a square pyramid with θ = 100° or θ = 93°.

The Table illustrates that the moments of inertia vary smoothly as the coordinate is traversed, and that an approximately spherical distribution of ligands is maintained, since Ix + Iy + Iz deviates little from 5 for all intermediate structures. As the structures approach the transition states the sum of the moments of inertia deviates most from 5, which reflects the fact that the transition states are less symmetric than the initial trigonal bipyramid. Figure 15 illustrates the approximate coplanarity of the points.

Table 12. Variation of moments of inertia for possible rearrangement pathways for 5-coordinate molecules

	Ix	Iy	Iz	Ix + Iy + Iz
Berry	2.000	1.500	1.500	5.000
Coordinate	1.998	1.616	1.386	5.000
(100°)	1.990	1.719	1.288	4.997
	1.985	1.766	1.246	4.997
	1.978	1.809	1.208	4.995
	1.961	1.883	1.146	4.990
	1.940	1.940	1.105	4.985
Berry	2.000	1.500	1.500	5.000
Coordinate	1.999	1.863	1.108	4.970
(93°)	1.998	1.914	1.045	4.957
	1.994	1.994	0.907	4.895
Turnstile	2.000	1.500	1.500	5.000
Coordinate	2.008	1.631	1.360	4.999
	2.027	1.736	1.235	4.998
	2.045	1.816	1.135	4.996
	2.058	1.873	1.061	4.992
	2.059	1.875	1.058	4.993

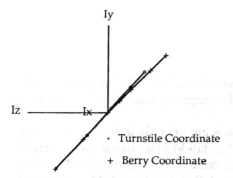

Fig. 15. 3-Dimensional plot of the variation in moments of inertia for rearrangement pathways for 5-coordinate molecules

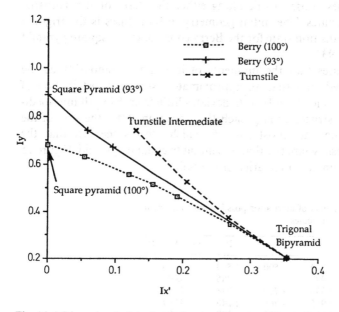

Fig. 16. 2-Dimensional plot of variation in moments of inertia for rearrangement pathways for 5-coordinate molecules

The projection of the points onto the plane $Ix + Iy + Iz = 5$ is displayed graphically in Fig. 16. The Figure shows a clear distinction between the alternative hypothetical rearrangement coordinates in the 2-dimensional space defined by the projection of the moments of inertia onto the plane $Ix + Iy + Iz = 5$. Therefore, if suitable fragments can be found which traverse a particular rearrangement pathway, then the moments of inertia of these fragments should provide a basis for distinguishing the alternative pathways.

Moments of Inertia in Cluster and Coordination Compounds

To illustrate the application of the methodology to real molecules, moments of inertia of 5-coordinate nickel compounds were calculated. 23 representative structures have been illustrated by Holmes in a review article [36], where the bond angles were given, together with the degree of distortion (as a percentage) from trigonal bipyramidal to square pyramidal geometry. These degrees of distortion were calculated using the dihedral angle method of Muetterties. The moments of inertia of the molecules described in this review have been calculated, and are given in Table 13, along with the degree of distortion and, where available, the spin state of the complex.

The structure of these compounds is illustrated in Appendix 1. For these experimentally determined structures the quantity $Ix + Iy + Iz$ is remarkably close to 5.000 predicted for spherical structures, although the individual structures appear asymmetric. It would appear that there is a conservation in the total moment of inertia for coordination compounds.

Holmes indicates that the value of θ in the limiting square pyramid varies with number of d electrons, and also with the spin state. The optimum values were found to be 100° for high spin d^8 complexes and 93° for low spin d^8 complexes, hence the variation of moments of inertia along pathways ending in square pyramids with θ = 100° and θ = 93°. Figure 17 illustrates the moments of inertia for the nickel complexes listed by Holmes, along with the theoretical Berry rearrangement pathways for limiting square pyramids of 93° and 100°. The turnstile pathway is also illustrated.

Table 13. Moments of inertia of representative 5-coordinate structures [36]

Number	Ix	Iy	Iz	Ix + Iy + Iz	% Distortion	Spin State
1	1.998	1.630	1.370	4.998	14.5	LS
2	1.998	1.602	1.399	4.999	15.0	LS
3	1.959	1.889	1.145	4.993	30.8	HS
4	1.975	1.649	1.373	4.997	18.9	LS
5	1.993	1.636	1.371	5.000	24.3	HS
6	1.987	1.690	1.323	5.000	32.2	LS
7	1.944	1.938	1.101	4.983	91.3	LS
7b	1.992	1.783	1.218	4.993	42.1	LS
8	2.122	1.969	0.891	4.982	53.3	HS
9	1.996	1.934	1.033	4.963	61.8	LS
10	1.948	1.774	1.278	5.000	56.6	LS
11	1.990	1.949	1.019	4.958	75.6	
12	1.998	1.977	0.947	4.922	79.4	LS
13	1.975	1.967	1.021	4.963	76.5	LS
14	1.958	1.907	1.124	4.989	82.0	HS
15	2.149	1.996	0.774	4.919	80.7	LS
16	1.939	1.937	1.109	4.985	91.5	
17	1.950	1.950	1.079	4.979	92.3	
18	2.123	1.951	0.905	4.979	78.6	HS
19	1.956	1.942	1.081	4.979	94.0	HS
20	2.003	1.984	0.944	4.931	93.4	LS
21	2.001	1.996	0.888	4.885	91.9	LS
22	1.950	1.947	1.083	4.980	97.0	HS

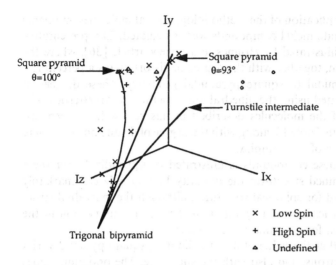

Fig. 17. 3-Dimensional plot of moments of inertia of representative 5-coordinate complexes on idealised proposed rearrangement pathways

Figure 18 illustrates the projection of these points onto the $Ix + Iy + Iz = 5$ plane. The Figures indicate that the majority of structures considered lie close to the lines for Berry pseudorotations leading to square pyramids with $\theta = 100°$ and $\theta = 93°$, rather than that for the turnstile coordinate. There is a good correlation between the position along the rearrangement pathway and the degree of distortion towards the square pyramid as defined by Holmes. Holmes plots the function $\Sigma_{ij}[\delta_{ij}(C) - \delta_{ij}(TBP)]$ versus $217.9° - \Sigma_{ij}[\delta_{ij}(C) - \delta_{ij}(SP)]$ (See Sect. 3.1) for these compounds to determine whether the compounds follow the Berry coordinate. The weakness of this approach is that initial assumptions must be made about the structure of the square pyramid. High and low spin complexes are distinguished in the Figure. The Figure confirms the general preference of low spin complexes for θ values approaching 93° in the square pyramid, and high spin complexes for θ values approaching 100°. A spectrum of θ values for the square pyramids, occurs, a point not revealed by the alternative method using dihedral angles.

The Figure also reveals that structures 8, 15 and 18, illustrated in Figs. 18 and 20 as open circles, do not lie close to any of the theoretical reaction coordinates. Their structures are illustrated in Fig. 19. Complexes 8 and 18 contain one bipyridyl ligand and one 4-membered ring. Both these ligands types exert strong steric requirements, as the 'bite' angles of the ligands are very small. These small chelating ligands force the stereochemistry away from the idealised C_{2v} Berry reaction coordinate. Structure 15 contains a 'tripod' type ligand, which is known to hinder the Berry type rotation [36]. Holmes described these structures as being 80.7%, 92.3% and 53.3% distorted along the Berry pathway.

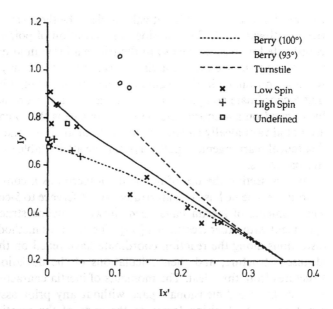

Fig. 18. Projection of Fig. 17 onto the Ix + Iy + Iz = 5 plane

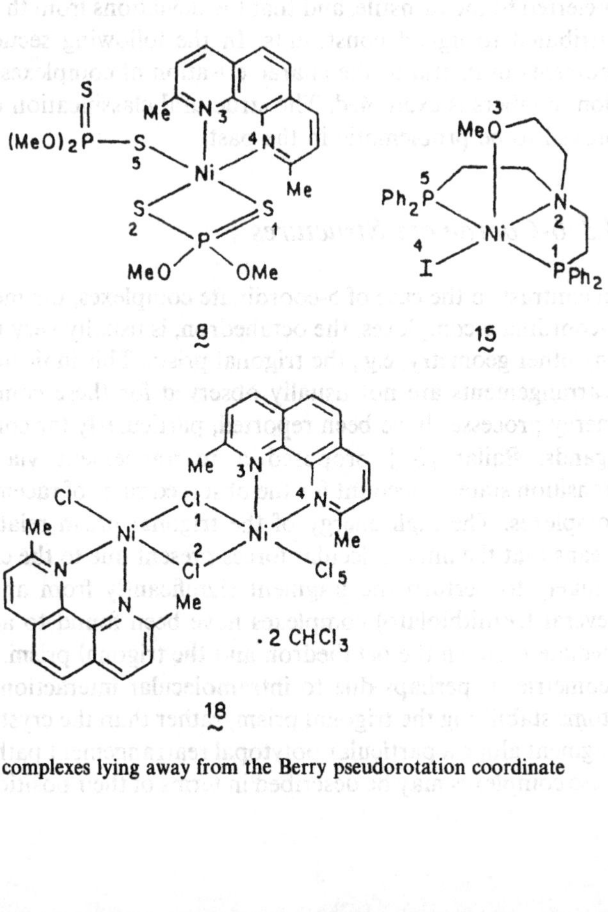

Fig. 19. Structures of nickel complexes lying away from the Berry pseudorotation coordinate

Figure 17 indicates that it is not valid to describe these structures in terms of the Berry pathway at all. The considerable deviation of points about the reaction coordinate for structures close to the trigonal bipyramid may indicate that the energy hypersurface does not rise sharply in a direction perpendicular to the reaction coordinate in this region of space. In addition, it has been shown [37] that 5-coordinate complexes may lie on a reaction coordinate involving attack by a ligand on a 4-coordinate centre. This means that 5-coordinate complexes are not always ideally suited for this form of analysis, and deviations from the theoretical rearrangement pathways may be attributable to alternative dissociative pathways.

In this section the use of moments of inertia as a convenient orthonormal coordinate scheme has been illustrated with reference to 5-coordinate structures. The moments of inertia have been shown to define structures in qualitative agreement with other treatments [36]. The previous methods of quantifying the distortions along the reaction coordinate have relied on the assumption of an idealised structure, and then calculations of the deviation of the individual structures from this ideal. The moments of inertia characterise the structure of a fragment in 3-dimensional space without any prior assumptions as to the structure of the limiting forms at the ends of the reaction coordinate. The Figures have also revealed that in the complexes studied, the Berry pathway is preferred to the turnstile, and that the deviations from the Berry pathway can be attributed to ligand constraints. In the following sections the application of moments of inertia to the characterisation of complexes with higher coordination numbers is examined. The structural classification of these complexes has proved to be problematic in the past.

4.3 6-Coordinate Structures

In contrast to the case of 5-coordinate complexes, the most stable geometry for 6-coordinate complexes, the octahedron, is usually very much more stable than any other geometry, e.g., the trigonal prism. This indicates that rapid polytopal rearrangements are not usually observed for these complexes, but some high energy processes have been reported, particularly for complexes with chelating ligands. Bailar [38] proposed a rearrangement via a trigonal prismatic transition state to account for the observed rates of racemisation of tris (chelate) complexes. The high energy of the trigonal prism relative to the octahedron means that the intermolecular forces present due to the crystal environment are unlikely to perturb the fragment significantly from an octahedral geometry. Several tris(dithiolato) complexes have been found to adopt geometries intermediate between the octahedron and the trigonal prism. The adoption of these geometries is perhaps due to intramolecular interactions between the sulphur atoms stabilising the trigonal prism, rather than the crystal lattice distorting the fragment along a particular polytopal rearrangement pathway. The structures of these complexes may be described in terms of their positions along an imaginary rearrangement coordinate between the octahedron and trigonal prism, in which 3-fold symmetry is maintained.

Various parameters have been used to describe the position of structures along this rearrangement coordinate. The twist angle, α, between the two parallel faces and the compession ratio, s/h, have been used [39], and are illustrated in Fig. 20.

The compression ratio is related to the angle formed between the polar axis an the atoms, illustrated in Fig. 21 by the expression.

$$\theta = \operatorname{Tan}^{-1}\left[\frac{2}{\sqrt{3}}\frac{s}{h}\right]$$

The moments of inertia for the octahedron and trigonal prism are dependent solely on θ, and are given by the expressions:

$$Iz = n\operatorname{Cos}^2\theta \quad Ix = Iy = n/2\operatorname{Sin}^2\theta$$

Therefore the compression ratio is directly related to the moments of inertia, by the equation

$$\frac{Ix}{Iz} = \frac{1}{2}\operatorname{Tan}^2\theta = \frac{2}{3}\frac{s^2}{h^2}$$

Distortions along a pathway which maintains D_3 symmetry can be divided into two main types [39].

i) A twist about the local C_3 axis, causing α to vary.
ii) A compression or elongation along the polar axis.

Where the ligand bite angle is variable, these two types of distortion can occur independently, so the two parameters, α and s/h (and therefore the moments of inertia), can define very different positions along the rearrangement coordinate for the same complex. If the ligand is rigid, then the parameters α and s/h are

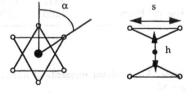

Fig. 20. Twist angle and compression ratio illustrated for an octahedron

Octahedron $\theta = 54.74°$

Trigonal Prism $\theta = 45°$

Fig. 21. Angle θ between the polar axis and the coordinating atoms

related, and so a variation in the moments of inertia will mirror a variation in α, and the parameters should define similar degrees of distortion along the rearrangement pathway.

The trigonal twist rearrangement from octahedral to trigonal prismatic geometry was modelled by varying the angle θ between the limits of an octahedron (54.74°) and trigonal prism (45°), whilst maintaining a constant ligand bite angle, χ, of 90°. The variation in moments of inertia as the rearrangement coordinate is traversed are summarised in Table 14. The sum of the moments of inertia remains close to 6, indicating that a spherical distribution of ligands is maintained. The variation in the moments of inertia reflects the transition from spherical to prolate geometries as θ decreases towards a value characteristic of the trigonal prism.

A change in the bite angle of the chelating ligands alters the geometry of the limiting octahedron, which can no longer show full O_h symmetry, and the complexes belong to the D_{3d} point group. For the trigonal prism the variation in the bite angle does not change the point group from D_{3h}, but the precise geometry of the trigonal prism is altered. The trigonal twist rearrangement coordinate has been modelled for ligand bite angles of 100° and 82°, along with the bite angle 90° already described. For a bite angle of 82°, which is characteristic of the commonly encountered dithiolato ligands, the limiting values of θ are 57.61° ('octahedron') and 49° (trigonal prism). Similarly for a bite angle of 100°, the values of θ are 51.28° ('octahedron') and 40° (trigonal prism). The calculated moments of inertia of intermediate structures as the rearrangement coordinate is traversed are summarised in Tables 15 and 16.

Table 14. Calculated moments of inertia for the bailar twist

	Ix	Iy	Iz	Ix + Iy + Iz
Octahedron	2.000	2.000	2.000	6.000
	2.403	1.798	1.798	5.999
	2.610	1.695	1.695	6.000
	2.819	1.591	1.591	6.001
Trigonal Prism	3.000	1.500	1.500	6.000

Table 15. Calculated moments of inertia for the trigonal twist with $\chi = 82°$

	Ix	Iy	Iz	Ix + Iy + Iz
Octahedron	2.139	2.139	1.722	6.000
	2.110	2.110	1.780	6.000
	2.013	2.013	1.974	6.000
	2.173	1.913	1.913	5.999
	2.376	1.812	1.811	6.000
Trigonal Prism	2.583	1.709	1.709	6.001

Moments of Inertia in Cluster and Coordination Compounds

Table 16. Calculated moments of inertia for the trigonal twist with $\chi = 100°$

	Ix	Iy	Iz	Ix + Iy + Iz
Octahedron	2.348	1.826	1.826	6.000
	2.686	1.657	1.657	6.000
	2.896	1.552	1.552	6.000
	3.104	1.448	1.448	6.000
	3.313	1.343	1.343	5.999
Trigonal Prism	3.520	1.240	1.240	6.000

Table 17. Summary of limiting θ and s/h values for different bite angles

	Octahedron		Trigonal Prism	
Bite Angle, χ	θ	s/h	θ	s/h
82°	57.61°	1.365	49°	0.996
90°	54.74°	1.225	45°	0.866
100°	51.28°	1.080	40°	0.727

It is once again remarkable that the sum of the moments of inertia remains close to 6.000.

The compression along the 3-fold axis of the octahedron associated with the reduction of bite angle from 90° to 82° causes the limiting 'octahedral' structure to be oblate, and the twist angle is reduced from 60° to 53.3°. Elongation caused by increasing the bite angle to 100° makes the limiting 'octahedral' geometry prolate. Table 17 summarises the values of θ and the compression ratios for limiting polyhedral forms for ligand bite angles of 82°, 90° and 100°. It is noteworthy that the ranges of θ associated with the individual bite angles overlap. Therefore, within the moments of inertia methodology it is important to compare structures with similar bite angles.

The variation in the calculated moments of inertia for the trigonal twist pathway for the three bite angles are illustrated individually as 2-dimensional projections onto the Ix + Iy + Iz = 6 plane in Figs. 22(a) to 22(c), and collectively in Fig. 22(d). For a bite angle of 82°, a spherical structure where Ix = Iy = Iz = 2 is reached at an intermediate point in the rearrangement process. This point marks a change between oblate and prolate structures, and accounts for the sharp turning point in Figure 22(b). For a bite angle of 100°, both the 'octahedron' and trigonal prism are prolate, and therefore a spherical structure is never generated. Figure 22(d) illustrates that the three separate rearrangement coordinates form a continuous series, because the values of θ overlap.

Table 18 summarises the molecular formulae, average bite angle, and calculated twist angle, α, of metal tris (dithiolate) complexes. The twist angle was

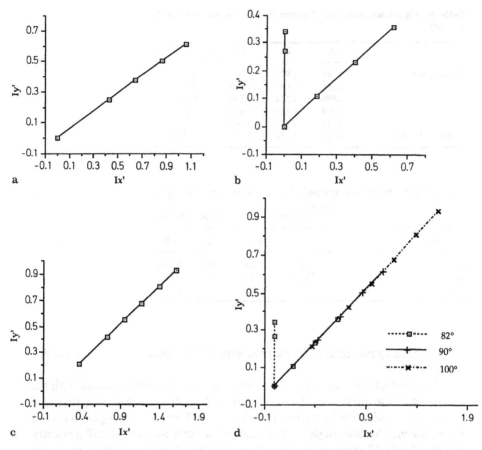

Fig. 22(a–d). (a) Bailar Twist for $\chi = 90°$; (b) Bailar Twist for $\chi = 82°$; (c) Bailar Twist for $\chi = 100°$; (d) Bailar Twists for $\chi = 82°$, $90°$ and $100°$

Table 18. Geometric properties of metal tris(dithiolato) complexes

	Av. χ	Ix	Iy	Iz	Ix + Iy + Iz	α	Ref
[Et$_4$N]$_3$[In(mnt)$_3$]	81.53°	2.082	2.031	1.887	6.000	47.3°	[40]
[PPh$_4$][Sb(tdt)$_3$]	83.97°	2.028	1.997	1.975	6.000	48.5°	[41]
[Et$_4$N]$_2$[Ti(edt)$_3$]	82.87°	2.231	1.911	1.858	6.000	38°	[42]
[AsPh$_4$][Ta(bdt)$_3$]	80.78°	2.361	1.845	1.794	6.000	30.8°	[43]
[AsPh$_4$]$_2$[W(mnt)$_3$]	82.04°	2.380	1.849	1.771	6.000	28°	[44]
[AsPh$_4$]$_2$[Mo(mnt)$_3$]	83.60°	2.397	1.858	1.745	6.000	27°	[44]
[PPh$_4$]$_2$[Mo(S$_2$C$_2$(CO$_2$Me)$_2$)$_3$]	80.51°	2.479	1.771	1.749	5.999	10.6°	[45]
V(S$_2$C$_2$Ph$_2$)$_3$	81.71°	2.574	1.730	1.696	6.000	4.3°	[46]
[AsPh$_4$][Nb(bdt)$_3$]	80.34°	2.499	1.759	1.742	6.000	1.6°	[47]
Mo(bdt)$_3$	82.12°	2.589	1.729	1.682	6.000	0°	[48]

mnt = maleonitrile dithiolate. tdt = toluene dithiolate. edt = ethylene dithiolate. bdt = benzene dithiolate

taken as the torsion angle defined by the two chelating S atoms and the centroids of the two parallel triangular faces.

Figure 23 illustrates the projection of the moments of inertia of the dithiolate ligands onto the Ix + Iy + Iz = 6 plane plotted on the rearrangement coordinate for the trigonal twist with $\chi = 82°$. The twist angle for each structure is also included.

Figure 23 shows that although most of the dithiolato complexes lie close to the idealised trigonal twist rearrangement pathway, they lie systematically within the 'V' defined by the idealised structures. The change in moments of inertia as the coordinate is traversed is followed by a change in the twist angle (structures closer to the octahedral limit are expected to have larger twist angles). Slight exceptions to this generalisation occur, for example the case of $[In(mnt)_3]^{3-}$ ($\alpha = 47.3°$) and $[Sb(bdt)_3]^-$ ($\alpha = 48.5°$), where the order of the twist angles is opposite to that expected. The difference is quite small, and the inversion of twist angles can be explained by the deviation of the bite angles from 82°. The distortions occurring in these compounds are not restricted to lie along the idealised trigonal twist pathway, and in many cases the D_3 symmetry is lost. This loss of D_3 symmetry is indicated by deviations away from the theoretical line in the Figure, which was calculated assuming D_3 symmetry. The $[Ta(bdt)_3]$-ion has two ligands which adopt a conformation characteristic of trigonal prismatic coordination, while the other adopts a conformation typical of octahedral coordination, and the point symmetry is reduced from D_3 to C_2. The quoted twist angle of 30.8° is in fact the average of 3 twist angles of 42.9°, 25°

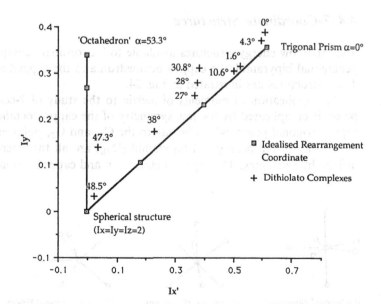

Fig. 23. Moments of inertia of dithiolato complexes plotted on the theoretical rearrangement coordinate for a trigonal twist ($\chi = 82°$)

and 24.9°, and the two opposite triangular S_3 faces are tilted away from the parallel by 12°. Similarly, in the case of $[Ti(edt)_3]^{2-}$ the quoted twist angle, 38°, is the average of three different angles, 36.6°, 37.6°, and 39.9°. Notably, $[Ta(bdt)_3]^-$ shows a very marked deviation away from the idealised geometry, reflecting the fact that the structure is not well described as either a distorted octahedron or distorted trigonal prism. The loss of D_3 symmetry is shown by a deviation from the theoretical rearrangement coordinate, and this deviation tends to occur such that the dithiolate complexes lie above the theoretical rearrangement coordinate, implying a more spherical structure is preferred. This may imply the existence of a rearrangement pathway of lower energy than the trigonal twist pathway, where the local D_3 symmetry is broken to reduce the energy of the intermediate structures.

Descriptions of the structure relative to limiting octahedral and trigonal prismatic forms are of limited use in cases where deviations from D_3 symmetry are very large. The two parameters α and s/h implicitly assume D_3 symmetry, and so are not strictly defined for complexes such as $[Ta(bdt)_3]^-$. If these parameters are used to describe structures such as that displayed by the $[Ta(bdt)_3]^-$, then incorrect conclusions about the structure may occur. The use of moments of inertia to describe the rearrangement pathway involves no prior assumption of point symmetry of any limiting polyhedral forms, and so, in addition to the relative position of fragments along any chosen idealised rearrangement coordinate, the deviation away from the rearrangement coordinate may be described in a quantitative manner.

4.4 7-Coordinate Structures

The three low energy structures available to 7-coordinate compounds are the pentagonal bipyramid, the capped octahedron and the capped trigonal prism. These structures are illustrated in Fig. 24.

The application of moments of inertia to the study of 7-coordinate complexes is complicated by the low symmetry of the capped octahedron and the capped trigonal prism, which belong to the C_{3v} and C_{2v} point groups respectively. Only in the pentagonal bipyramid (D_{5h}) are all the inter-ligand angles defined by symmetry. The capped octahedron and capped trigonal prism have

Pentagonal Bipyramid   Capped Octahedron   Capped Trigonal Prism

Fig. 24. Low energy structures of 7-coordinate complexes

a large number of symmetry independent angles, upon which the moments of inertia are strongly dependent, and so the moments of inertia of a capped octahedron and capped trigonal prism are not precisely defined. The idealised forms of these structures are illustrated in Fig. 24, but in the crystal environment it is found that the capped face expands, and the extent of this expansion is strongly dependent on the ligand type, reflecting the different distribution of electrons in different M − L bonds. It would therefore be preferable to study complexes of the type ML_7, where L is a similar ligand in all cases. Near the pentagonal bipyramid, where angles are defined by symmetry, the nature of the ligand is less important.

Few homoleptic 7-coordinate complexes have been well characterised, and so crystallographic data to compare with theoretical rearrangement pathways is scarce. The series of complexes $[M(CNR)_7]$ provides the most extensive set of data available for structures near the capped octahedron and capped trigonal prism [49], and so the moments of inertia of these structures, and those of the complexes $[M(CN)_7]$ have been calculated. The cyanide ions lie close to the pentagonal bipyramid, and so the differences in ligand character between CNR and CN do effect the resulting structures. The moments of inertia of the 3 limiting structures are summarised in Table 19. The geometry of the limiting capped trigonal prism and capped octahedron were chosen to represent the structures of the complexes found by inspection to most closely resemble the idealised structures. The bond angles were averaged where the structure deviated from idealised C_{2v} or C_{3v} symmetry, and the structures are illustrated in Figs 25 and 26.

As noted in [49], the capped octahedron and capped trigonal prism are structurally similar, and close in energy. Table 19 indicates that their moments of inertia are also very similar.

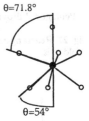

Fig. 25. Structure of a representative capped octahedron

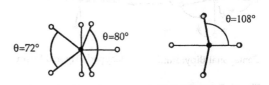

Fig. 26. Structure of a representative capped trigonal prism

Table 19. Moments of inertia of 7 coordinate complexes

	Ix	Iy	Iz	Ix + Iy + Iz
Pent. Bipyramid	2.500	2.500	2.000	7.000
Capped Oct.	2.336	2.335	2.325	6.996
Capped Trig.	2.396	2.310	2.295	7.001

Two possible rearrangement pathways have been proposed for the rearrangement from the pentagonal bipyramid to the capped trigonal prism [49], one of which retains a 2-fold axis during the rearrangement, defined by the bisector of the angle formed by atoms 5, 7 and the central atom, while the other retains a mirror plane, defined by atoms 1, 5 and 7.

The rearrangement which retains a mirror plane passes a geometry which is close to, but not identical to, the capped octahedron. This pathway can be separated into 2 separate sections,

Pentagonal Bipyramid → Capped Octahedron and

Capped Octahedron → Capped Trigonal Prism.

These separate pathways are illustrated in Fig. 29.

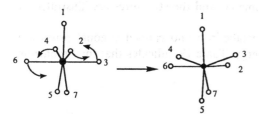

Fig. 27. Rearrangement from pentagonal bipyramid to capped trigonal prism with retention of a 2-fold axis

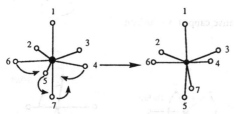

Fig. 28. Rearrangement from pentagonal bipyramid to capped trigonal prism with retention of a mirror plane

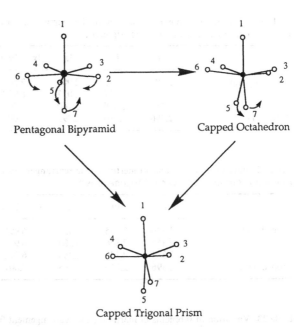

Fig. 29. Rearrangement from pentagonal bipyramid to capped trigonal prism with retention of a mirror plane, via a capped octahedron

Table 20. Variation in moments of inertia for the rearrangement from a pentagonal bipyramid to a capped octahedron

	Ix	Iy	Iz	Ix + Iy + Iz
Pent. Bipyramid	2.500	2.500	2.000	7.000
	2.485	2.475	2.037	6.997
	2.431	2.412	2.150	6.993
Capped Oct.	2.336	2.335	2.325	6.996

The variations in moments of inertia along these pathways are summarised in Tables 20, 21 and 22. Table 23 summarises the variation in moments of inertia along the pathway which retains a 2-fold axis. The projection of these moments of inertia onto the Ix + Iy + Iz = 7 plane is illustrated in Fig. 30.

The figure illustrates the close similarity between the capped octahedron (CO) and capped trigonal prism (CTP), both of which are well separated from the pentagonal bipyramid. The spatial separation of the two types of pathway is noticeable, with the variation in moments of inertia being much greater for the pathway where the 2-fold axis is retained.

Table 24 summarises the calculated moments of inertia of actual 7-coordinate complexes of the type ML_7.

Table 21. Variation in moments of inertia for the rearrangement from a pentagonal bipyramid to a capped trigonal prism.

	Ix	Iy	Iz	Ix + Iy + Iz
Pent. Bipyramid	2.500	2.500	2.000	7.000
	2.495	2.477	2.027	6.999
	2.453	2.405	2.140	6.998
Capped Trig.	2.396	2.310	2.295	7.001

Table 22. Variation in moments of inertia for the rearrangement from a capped octahedron to a capped trigonal prism.

	Ix	Iy	Iz	Ix + Iy + Iz
Capped Oct.	2.336	2.335	2.330	6.996
	2.365	2.346	2.284	6.995
	2.383	2.330	2.285	6.998
Capped Trig.	2.396	2.310	2.295	7.001

Table 23. Variation in moments of inertia for the rearrangement from a pentagonal bipyramid to a capped trigonal prism with retention of a 2-fold axis

	Ix	Iy	Iz	Ix + Iy + Iz
Pent. Bipyramid	2.500	2.500	2.000	7.000
	2.618	2.400	1.976	6.994
	2.682	2.326	1.972	6.980
	2.661	2.279	2.038	6.978
	2.558	2.272	2.157	6.987
	2.426	2.301	2.273	7.000
Capped Trig.	2.396	2.310	2.295	7.001

The projection of these points onto the $Ix + Iy + Iz = 7$ plane is plotted in Fig. 31, along with the theoretical rearrangement pathways. The Figure illustrates that the complexes cluster around the limiting geometries of the pentagonal bipyramid at one extreme, and the very similar capped octahedron and capped trigonal prism at the other. It should be emphasised that the structure of the theoretical capped trigonal prism was chosen by reference to the structures of $[W(CNBu)_7]^{2+}$ and $[Mo(CNBu)_7]^{2+}$. Structures intermediate between the pentagonal bipyramid and the capped trigonal prism are not observed, only ones exhibiting small distortions from one or other limiting form. This may be a consequence of a rapid increase in potential energy for small distortions.

The $[M(CN)_7]$ complexes, which lie close to the pentagonal bipyramid, appear to be slightly distorted along the pathway which retains 2-fold symmetry, while the $[M(CNR)_7]$ complexes, which lie close to the capped trigonal

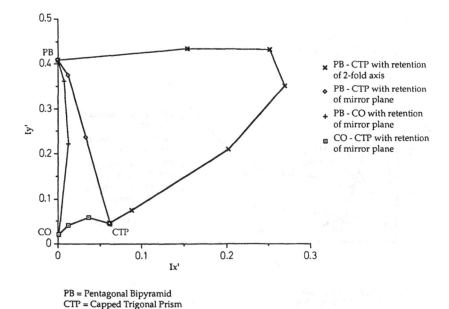

Fig. 30. Variation of moments of inertia along rearrangement pathways illustrated in Figs 27, 28 and 29

Table 24. Moments of inertia of 7-coordinate complexes

Compound	Ix	Iy	Iz	Ix + Iy + Iz	Ref.
[Mo(CNBu)$_7$][PF$_6$]$_2$	2.404	2.304	2.291	6.999	[50]
[Cr(CNBu)$_7$][PF$_6$]$_2$	2.417	2.357	2.226	7.000	[51]
[Mo(CNPh)$_7$][PF$_6$]$_2$	2.376	2.348	2.269	6.993	[52]
[Mo(CNMe)$_7$][BF$_4$]$_2$	2.421	2.348	2.231	7.000	[53]
[W(CNBu)$_7$][W$_6$O$_9$]	2.415	2.300	2.285	7.000	[54]
K$_4$[V(CN)$_7$](H$_2$O)$_2$	2.522	2.487	1.986	6.995	[55]
K$_5$[Mo(CN)$_7$](H$_2$O)	2.528	2.458	2.006	6.992	[56]
K$_4$[Re(CN)$_7$](H$_2$O)$_2$	2.543	2.481	1.982	7.006	[57]

prismatic extreme, are distorted along the pathway which retains a mirror plane. The complexes [Cr(CNBu)$_7$] and [Mo(CNMe)$_7$] lie along the pathway leading directly from the capped trigonal prism to the pentagonal bipyramid, illustrated in Fig. 28, and not towards the capped octahedron, while [Mo(CNPh)$_7$] lies much closer to the capped octahedron.

It should be emphasised that the low symmetry of the capped structures means that a higher degree of uncertainty is involved in defining the limting structure than is the case for other coordination numbers. Therefore quantitative conclusions as to the degree of distortion along a particular coordinate should be avoided, particularly in the region of the closely related capped octahedron and capped trigonal prism.

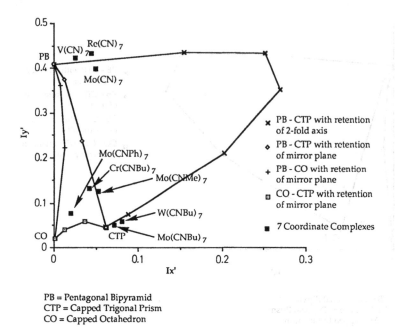

PB = Pentagonal Bipyramid
CTP = Capped Trigonal Prism
CO = Capped Octahedron

Fig. 31. 7-Coordinate complexes plotted onto the plane Ix + Iy + Iz = 7

4.5 8-Coordinate Complexes

8-coordinate complexes are another series of complexes where a number of structures are of similar energy, and so show a wide variety of structures in the solid state. In this section the moments of inertia are used to classify the shapes. Using pairwise repulsion energy calculations [49], Kepert has shown that there are three polyhedral geometries of low energy for octa-coordinated complexes, the square antiprism, dodecahedron and the cube, illustrated in Fig. 32.

Of these, the square antiprism is the most stable, with the dodecahedron of similar energy, and the cube at a higher energy.

The dodecahedron contains two distinct subsets of atoms, illustrated in Fig. 33.

The cone angle between the Z axis and the atoms a and b are calculated to be:

$$\phi a = 36.85° \quad \phi b = 69.46°$$

for a hard sphere repulsion model, and vary slightly depending on the form of the repulsion assumed to occur between the charge clouds.

In the cube and square antiprism all atoms are identical, and the cone angles between the Z axis and the atoms are:

Cube $\theta = 54.74°$ Square antiprism $\theta = 59.26°$.

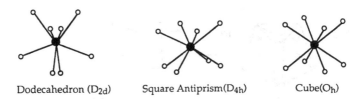

Fig. 32. Low energy structures of 8-coordinate complexes

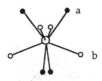

Fig. 33. Dodecahedron, showing the two symmetry distinct sets of atoms

In the case of the square antiprism this angle is not uniquely defined by symmetry, and the angle varies somewhat with the form of repulsive interaction assumed to occur between the ligands. This will be discussed in detail later.

Potential energy calculations [49] indicate that there is no potential energy barrier between the dodecahedron and the square antiprism, and so rapid rearrangement between the two geometries is to be expected, and structures intermediate between them may be found in the solid state. The regularity of the stereochemistry is not simple to describe, because the 8-coordinate species have 18 internal degrees of freedom, and so consideration of only a small subset of internal angles cannot fully describe the structure. The dodecahedron may be described as two intersecting orthogonal trapezoids [6], and the deviation from the perfect dodecahedron has been estimated by the non-planarity of the trapezoids. The variation of dihedral angles as two triangular faces become a square plane can also be used as a structural parameter to map the rearrangements involving a dodecahedron.

The rearrangements between all three species can be mapped using the variation of moments of inertia as the rearrangement coordinate is traversed. The proposed pathways for rearrangement are illustrated in Figs. 34, 35 and 36.

Figure 34 corresponds to elongation along the two a-a edges.

Figure 35 corresponds to elongation along the two b-b edges.

Figure 36 corresponds to a rotation of 45° about the Z axis, accompanied by an increase in the angle θ from 54.74° to 59.26°.

Table 25 summarises the moments of inertia for the three limiting polyhedra, calculated assuming unit bond length. There is a clear distinction between the shapes of the three limiting forms. The cube is spherical, while the square antiprism is oblate and the dodecahedron is prolate. The calculated moments of inertia for structures intermediate between the limiting structures, lying along the proposed reaction pathways proposed above, are shown in Tables 26, 27 and 28.

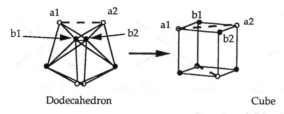

Fig. 34. Rearrangement from a dodecahedron to a cube

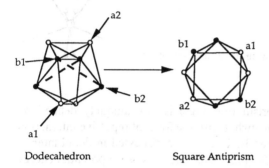

Fig. 35. Rearrangement from a dodecahedron to a square antiprism

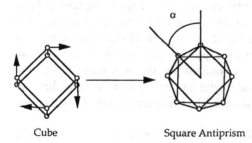

Fig. 36. Rearrangement from a cube to a square antiprism

In Fig. 37 the projection of these moments of inertia onto the $Ix + Iy + Iz = 8$ plane is illustrated. Table 29 describes the geometric properties of selected 8-coordinate complexes, along with the classification given in [49], were available. In Figure 38 the results are plotted, along with the theoretical rearrangement pathways derived from the moments of inertia analysis, as a projection onto the $Ix + Iy + Iz = 8$ plane. The complexes are identified on the figure. In the cases where more than one example of the same complex ion occurs, then the counter ion is also shown on the Figure.

Table 25. Moments of inertia of 8-coordinate polyhedra

	Ix	Iy	Iz	Ix + Iy + Iz
Square Antiprism	2.955	2.955	2.090	8.000
Dodecahedron	3.054	2.473	2.473	8.000
Cube	2.667	2.667	2.667	8.000

Table 26. Variation in moments of inertia for the rearrangement from a dodecahedron to a square antiprism

	Ix	Iy	Iz	Ix + Iy + Iz
Dodecahedron	3.054	2.473	2.473	8.000
	3.052	2.552	2.397	8.001
	3.043	2.634	2.323	8.000
	3.027	2.719	2.254	8.000
	3.005	2.807	2.188	8.000
	2.977	2.897	2.127	8.001
Square Antiprism	2.955	2.955	2.090	8.000

Table 27. Variation in moments of inertia for the rearrangement from a dodecahedron to a cube

	Ix	Iy	Iz	Ix + Iy + Iz
Dodecahedron	3.054	2.473	2.473	8.000
	2.985	2.508	2.508	8.001
	2.919	2.541	2.541	8.001
	2.858	2.571	2.571	8.000
	2.804	2.598	2.598	8.000
	2.755	2.623	2.623	8.001
	2.712	2.644	2.644	8.000
	2.676	2.662	2.662	8.000
Cube	2.667	2.667	2.667	8.001

Table 28. Variation in moments of inertia for the rearrangement from a cube to a square antiprism

	Ix	Iy	Iz	Ix + Iy + Iz
Cube	2.667	2.667	2.667	8.001
	2.732	2.732	2.535	7.999
	2.797	2.797	2.406	8.000
	2.860	2.860	2.279	7.999
	2.923	2.923	2.154	8.000
Square Antiprism	2.955	2.955	2.090	8.000

In discussing the rearrangement of the square antiprism to the cube (square prism) a similar problem is encountered to that which has been discussed above in 6-coordinate complexes. The variation in moments of inertia reflects only the change in cone angle, θ, and not the twist angle, α, and, for a square antiprism,

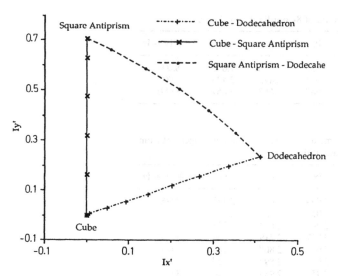

Fig. 37. Variation of moments of inertia along rearrangement pathways illustrated in Figs. 34, 35 and 36

Table 29. Geometric properties of 8-coordinate complexes

Molecular Formula	Ix	Iy	Iz	Ix + Iy + Iz	Class	Ref.
$[Cu_2ZrF_8](H_2O)_{12}$	2.980	2.676	2.349	8.005	SA	[58]
$Na_3[PaF_8]$	2.754	2.754	2.493	8.001	Cube	[59]
$(NO)_2[XeF_8]$	2.875	2.867	2.258	8.000		[60]
$Na_3[TaF_8]$	2.989	2.886	2.125	8.000	SA	[61]
$Li_4[UF_8]$	3.038	2.566	2.384	7.988		[62]
$Cd_2[Mo(CN)_8](N_2H_4)_2(H_2O)_4$	2.869	2.869	2.262	8.000	SA	[63]
$K_3[Mo(CN)_8](H_2O)_2$	2.962	2.530	2.507	7.999		[64]
$K_4[Mo(CN)_8](H_2O)_2$	3.058	2.482	2.454	7.994	Dod	[64]
$(Bu_4^n-N)_3[Mo(CN)_8]$	2.902	2.683	2.415	8.000	Dod	[65]
$(Et_3-NH)_2[Mo(CN)_8][H_3O]_2$	2.907	2.652	2.441	8.000	Int	[66]
$H_4[W(CN)_8](H_2O)_6$	2.822	2.780	2.396	7.999	SA	[67]
$(Et_4N)_4[U(NCS)_8]$	2.703	2.649	2.649	8.001	Cube	[68]
$Na_3[W(CN)_8](H_2O)_4$	2.983	2.872	2.144	7.999	SA	[69]
$H_4[W(CN)_8](HCL)_4(H_2O)_{12}$	2.835	2.667	2.498	8.000	SA	[70]
$Sr(H_2O)_8[AgI_2]_2$	2.899	2.899	2.202	8.000	SA	[71]
$Cs_4[U(NCS)_8]$	2.791	2.791	2.419	8.001	SA	[72]

SA = Square Antiprism   Dod = Dodecahedron   Int = Intermediate

the angle θ is not defined by symmetry. The complexes studied do not have chelating ligands, so the moments of inertia and twist angle are not related in a well defined manner. The considerable variation in θ for square antiprisms [49] means that the structure is not completely defined by the moments of inertia. Traversing the rearrangement coordinate towards the cube may merely imply that the limiting square antiprism becomes more elongated, rather than

Moments of Inertia in Cluster and Coordination Compounds

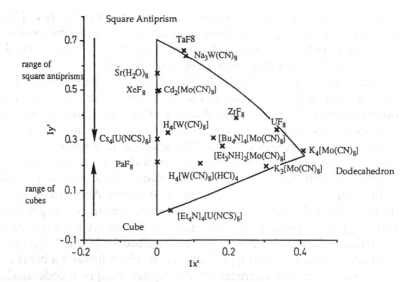

Fig. 38. 8-Coordinate complexes ploted onto the plane Ix + Iy + Iz = 8

reflecting any change towards cubic coordination. The twist angles of the complexes studied show that only $Cd_2[Mo(CN)_8](N_2H_4)_2(H_2O)_4$ shows appreciable intermediate character, with a twist angle of 38°. The others have twist angles of 45 ± 3° (square antiprism) or 0° (cube). There is no reason why the point labelled square antiprism in Fig. 38 should not be a prism derived from a cube by compression, such that $\theta = 59.26°$, while the point labelled cube could be an antiprism, elongated such that $\theta = 54.74°$. The calculations described in [49] indicate that the minimum value of θ attained by square antiprisms is approximately 57°, and so the energy of a square antiprism elongated so that $\theta = 54.74°$ would be high and similarly the energy of a highly compressed cube, where $\theta = 59.26°$ would be very high, and so it is reasonable to assume that near the limits defined by the moments of inertia of a cube and a square antiprism, the structure can be defined with a good degree of certainty, but in intermediate regions, no conclusion as to the degree of cubic character in a complex can be drawn without further information on the twist angle. This uncertainty is indicated in Fig. 38.

The square antiprisms can clearly be divided into two sets. Those which retain 4-fold symmetry lie along the pathway to the cube, illustrated in Fig. 36. This set includes the complexes $Sr(H_2O)_8[AgI_2]_2$, $(NO)_2[XeF_8]$, $Cd_2[Mo(CN)_8](N_2H_4)_2(H_2O)_4$, and $Cs_4[U(NCS)_8]$. In contrast $Na_3[TaF_8]$ and $Na_3[W(CN)_8](H_2O)_4$ clearly lie along rearrangement pathway illustrated in Fig. 35, leading to a dodecahedron. The structures of these two complexes show a pairwise deviation from equality of the cone angles of the atoms forming each 'square face'. A true square antiprism has θ equal for all 4 atoms forming a square face, but in $Na_3[TaF_8]$ two opposite ligands have $\theta = 59.72°$ and two have $\theta = 58.46°$. Similar distortions occur on the opposite face. $[Cu_2ZrF_8]$ $(H_2O)_{12}$ illustrates a greater degree of distortion along this pathway. One face has θ values of 62° (one pair) and 52°. The opposite face is much less distorted towards the dodecahedron, with θ values of 59° and 56°.

The complex $[Et_3-NH]_2[Mo(CN)_8][H_3O]_2$ has been described as intermediate [30], and the position in Fig. 38 confirms this classification. Several other complexes lie in the intermediate region of the Figure, indicating that the stereochemistry cannot be described adequately by any of the limiting geometries. The complex $H_4[W(CN)_8](H_2O)_6$ is slightly distorted from square antiprismatic coordination towards a dodecahedron, while $H_4[W(CN)_8](HCl)_4(H_2O)_{12}$ shows very strong distortions in this direction. Both complexes illustrate the pairing of θ angles described in $[Cu_2ZrF_8]$ $(H_2O)_{12}$. In both these cases, and in other intermediate complexes such as $Li_4[UF_8]$, the opposite faces of the square antiprism are not distorted towards the dodecahedron to the same extent, as described for $[Cu_2ZrF_8](H_2O)_{12}$. The common occurrence of this type of distortion, which involves a breaking of the 2-fold symmetry characteristic of the square antiprism-dodecahedron rearrangement process illustrated in Fig. 35, may indicate the existence of a rearrangement process of lower energy than that illustrated in Fig. 35.

Summary

The molecular moments of inertia provide a convenient framework for discussing the shapes and electronic structures of inorganic molecules and clusters. Since the moments of inertia and prolate and oblate perturbations of a spherical quantum mechanical solution may both be expressed in terms of spherical harmonic expansions they provide a common language for the discussion of the relationship between molecular and electronic structures. Alternative idealised co-ordination geometries are characterised by distinctive moments of inertia and therefore reaction coordinates connecting these alternative geometries may be expressed generally in terms of changes in moments of inertia in agreement with Bürgi and Dunitz's 'Structural Correlation' principle. The moments of inertia, however, only define the prolate and oblate distortions of a spherical shape and it may prove necessary to use higher moments to define more complex distortional modes.

Moments of Inertia in Cluster and Coordination Compounds

Appendix 1

Appendix 2

Acknowledgements: ALR thanks the British Council and the Association of Commonwealth Universities for a Commonwealth Scholarship. Thanks are also due to AFOSR for financial support. Prof. H.-B. Bürgi's valuable comments on the manuscript are also acknowledged.

5 References

1. Wade K (1976) Adv. Inorg. Chem. Rad. 18: 1; Mingos DMP (1972) Nature, Phys. Sci. 236: 99
2. Gillespie RJ (1972) Molecular Geometry, Van Norstrand Reinhold, London
3. Murray-Rust P, Bürgi H-B, Dunitz JD (1979) Act. Cryst. A 35: 703; (1978) Act. Cryst. B 34: 1787; 1793
4. Bürgi H-B (1975) Angew. Chem. Int. Ed. Engl. 14: 460
5. Bürgi H-B, Dunitz JD (1983) Acc. Chem. Res. 16: 153
6. Muetterties EL, Guggenberger LJ (1974) J. Am. Chem. Soc. 96: 1748
7. Porai-Koshits MA, Aslanov LA (1972) Zh. Struk. Kh. 13: 266
8. Dollase WA (1974) Act. Cryst. A30: 513
9. Brand JC (1975) Molecular structure-The physical approach, Edward Arnold, London
10. Gavezzotti AJ (1989) J. Am. Chem. Soc. 111: 1835
11. Schomaker V, Waser J, Marsh RE, Bergmann G (1959) Act. Cryst. 12: 600
12. Hermann A, Schumacher E, Wöste L (1978) J. Chem. Phys. 68: 2327
13. Foster PJ, Leckenby RE, Robbins EJ (1969) J. Phys. B 2: 478
14. Knight WD, Clemenger K, de Heer WA, Saunders W, Chou MY, Cohen ML (1984) Phys. Rev. Letters 52: 2141
15. Clemenger K (1985) Phys. Rev. B 32: 1359
16. Zhenyang L, Slee T, Mingos DMP (1989) Chem. Phys. 137: 15
17. Griffiths JS (1961) Theory of transition metal ions, Cambridge University Press, London
18. Chen J, Brink DM, Wille LT (1990) J. Phys. B 23: 885
19. Stone AJ (1980) Mol. Phys. 41: 1339
20. Boustani I, Pewestorf W, Fantucci P, Bonacic-Koutecky V, Koutecky J (1987) Phys. Rev. B 35: 9437
21. Bonacic-Koutecky V, Boustani I, Guest M, Koutecky J (1988) J. Chem. Phys. 89: 4861
22. Boustani I, Koutecky J (1988) J. Chem. Phys. 88: 5657
23. Fantucci P, Bonacic-Koutecky V, Koutecky J (1988) J. Phys. Rev. B 37: 4369
24. Kanters RPF (1990) Ph.D Thesis: Gold phosphine clusters, Katholieke Universiteit te Nijmegen
25. Kanters RPF, Steggerda JJ (1990) In Press
26. Murray-Rust P, Bürgi H-B, Dunitz JD (1979) Acta. Cryst. A35,703
27. Eyring H, Polyani M (1931) Z. Chem. Phys. B12: 279
28. Berry RS (1960) J. Chem. Phys. 32: 933
29. Rauk A, Allen LC, Mislow K (1972) J. Am. Chem. Soc. 94: 3035
30. Kepert DL (1973) Inorg. Chem. 12: 1938
31. Muetterties EL (1969) J. Am. Chem. Soc. 91: 1636
32. Ugi I, Marquarding D, Klusacek H, Gillespie P, Ramirez F (1971) Acc. Chem. Res. 4: 288
33. Meakin P, Muetterties EL, Jesson PJ (1972) J. Am. Chem. Soc. 94: 5271
34. Holmes RR, Deiters J (1977) J. Am. Chem. Soc. 99: 3318
35. Auf der Heyde TA, Nassimbini L (1984) Inorg. Chem. 23: 4525; Auf der Hyde TA, Bürgi H.-B, (1989) J. Am. Chem. Soc. 128: 3960; 3970; 3982
36. Holmes RR (1984) Prog. Inorg. Chem. 32: 119
37. Bürgi HB (1973) Inorg. Chem. 12: 2321
38. Bailar JC (1958) J. Inorg. Nucl. Chem 8: 165
39. Steifel EI, Brown GF (1972) Inorg. Chem. 11: 474
40. Einstein FWB, Jones RG (1971) J. Chem. Soc. A: 2762
41. Kisenyi JM, Willey GR, Drew MGB, Wandiga SO (1985) J. Chem. Soc. Dalton Trans.: 65
42. Dorfman JR, Rao CP, Holm RH (1986) Inorg. Chem. 25: 428
43. Martin JL, Takats J (1975) Inorg. Chem. 14: 1358
44. Brown GF, Steifel EI (1973) Inorg. Chem. 12: 2140
45. Draganjac M, Coucouvanis D (1983) J. Am. Chem. Soc. 105: 139
46. Eisenberg R, Gray HB (1967) Inorg. Chem. 6: 1844
47. Cowie M, Bennet MJ (1976) Inorg. Chem. 15: 1589

48. Cowie M, Bennet MJ (1976) Inorg. Chem. 15: 1584
49. Kepert DL (1987) in Comprehensive Coordination Chemistry, Vol. 1: 31, Pergamon,
50. Lewis DL, Lippard SJ (1975) J. Am. Chem. Soc. 97: 2697
51. Dewan JC, Mialki WS, Walton RA, Lippard SJ (1982) J. Am. Chem. Soc. 104: 133
52. Dewan JC, Lippart SJ (1982) Inorg. Chem. 21: 1682
53. Brant P, Cotton FA, Sekutowski JC, Wood TE, Walton RA (1979) J. Am. Chem. Soc. 101: 6588
54. LaRue WA, Liu AT, Filipo JS (1980) Inorg. Chem. 19: 315
55. Levenson RA, Towns RLR (1974) Inorg. Chem. 13: 105
56. Drew MGB, Mitchell PCH, Pygall CF (1977) J. Chem. Soc. Dalton Trans.: 1071
57. Manoli J-M, Potvin C, Bergeault J-M, Griffith WP (1980) J. Chem. Soc. Dalton Trans.: 192
58. Fischer J, Elchinger R, Weiss R (1973) Acta Cryst. B29: 1967
59. Brown D, Easey JF, Rickard CEF (1969) J. Chem. Soc. (A): 1161
60. Peterson SW, Holloway JH, Coyle BA, Williamson JM (1971), Science, 173: 1238
61. Hoard JL, Martin WJ, Smith ME, Whitney JF (1954) J. Am. Chem. Soc. 76: 3820
62. Brunton G (1967) J. Inorg. Nucl. Chem. 29: 1631
63. Chojnacki J, Grochowski J, Lebioda L, Oleksyn B, Stadnicka K (1969), Rocz. Chem. 43: 273
64. Hoard JL, Hamer TA, Glick MD (1968) J. Am. Chem. Soc. 90: 3177
65. Corden BJ, Cunningham JA, Eisenberg R (1970) Inorg. Chem. 9: 356
66. Leipoldt JG, Basson SS, Bok LDC (1980) Inorg. Chem. Acta 44: L99
67. Basson SS, Bok LDC, Leipoldt JG (1970) Act. Cryst. B 26: 1209
68. Countryman R, McDonald WS (1971) J. Inorg. Nucl. Chem. 33: 2213
69. Bok LDC, Leipoldt JG, Basson SS (1970) Act. Cryst. B 26: 684
70. Bok LDC, Leipoldt JG, Basson SS (1972) Z. Allg. Chem. 392: 302
71. Geller S, Dudley TO (1978) J. Solid State Chem. 26: 321
72. Bombieri G, Moseley PT, Brown D (1975) J. Chem. Soc. Dalton Trans.: 1520

Progress in Polymetallic Exchange-Coupled Systems; some Examples in Inorganic Chemistry

Marc Drillon[1] and Jacques Darriet[2]

[1] Groupe des Matériaux Inorganiques, IPCMS-EHICS, 1 rue B. Pascal, 67008 Strasbourg, France
[2] Laboratoire de Chimie du Solide, CNRS, 351 cours de la Libération, 33405 Talence, France

The aim of this paper is to review some basic concepts needed for describing polymetallic exchange-coupled systems and to treat problems of current interest in inorganic chemistry.
First, a brief survey of a priori conflicting descriptions of the exchange mechanisms is proposed. The overlap between the Heisenberg and Anderson models is demonstrated in the limiting case of highly distorted systems, in which the orbital degeneracy of interacting ions is removed. Then, the correlations between magnetic properties and structural features are discussed for some series of compounds characterized by specific arrangements of the metal ions such as dimers, trimers... The sign and magnitude of the exchange parameters are shown to depend on different factors such as the ground-state of the interacting ions, the geometry of the species, and the nature of the bridging ligands.

1 Introduction	56
2 Theoretical Basis for Exchange Coupling	57
2.1 Molecular Orbital Description	58
2.2 Localized-Electron Model	59
3 Interplay Between Heisenberg and Anderson Models	62
4 Polymetallic Systems and Related Structures	65
4.1 Hexagonal Perovskites and Related Systems	65
4.1.1 Strings of Increasing Size in $CsNiF_3$–$CsCdF_3$ Mixed Compounds	65
4.1.2 Binuclear Species in 2H Perovskite Derived Systems	68
4.1.3 Bi- and Trinuclear Species in Ruthenium Oxides with 2H, 9R and 12R Structure	76
4.2 Quasi-Isolated Pairs in Trirutile Compounds	81
4.2.1 Exchange Coupling in Chromium Dimers $(Cr_2O_{10})^{14-}$	81
4.2.2 Orbitally Degenerate Binuclear Units $(V_2O_{10})^{14-}$	84
4.3 Miscellaneous Structures	86
4.3.1 Pairs in the Layer Compound CsV_2O_5	86
4.3.2 Discrete Trimers in Copper(II) Phosphates	87
4.3.3 Tetranuclear Units in RuF_5 and Related Systems	91
4.3.4 Heterometallic Three-Centre System in $Ba_2CaCuFe_2F_{14}$	93
5 Concluding Remarks	97
6 References	99

Structure and Bonding 79
© Springer-Verlag Berlin Heidelberg 1992

1 Introduction

The chemical and physical properties of polynuclear systems always attract the attention of researchers concerned in inorganic and molecular chemistry, bioinorganics and solid state physics. In particular, we can mention the wide interest in the engineering of new systems with specific properties, the progress on the role of "active" sites in biological processes, and also the development of magneto-chemistry in the understanding of exchange mechanisms [1].

One characteristic receiving increasing interest is the relationships between properties and structural features, namely the key-role of the electronic structure of interacting ions and the way in which these are bound together.

Thus, the chemist is now able, through a planned synthesis or so-called "molecular engineering", to isolate materials in which the metal ions occur in some specific arrangement such as dimers, trimers... etc. The current interest in such systems results from their ability to be solved by standard techniques of quantum mechanics without using approximate statistical models [2].

The hexagonal perovskites, of general formula AMX_3, and related systems exhibit, in our view, one of the most fascinating series of polynuclear systems. They can be described on the basis of close-packing AX_3 layers (A is a large non-magnetic cation) which alternate with M cations in order to give corner- or face-sharing MX_6 octahedra. The Madelung energies of related systems are very similar so that a partial substitution of M by another cation, or high pressure effect, may induce a significant structural change. Thus, $CsNiF_3$ crystallizes with the 2H hexagonal structure in which face-sharing NiF_6 octahedra form infinite linear chains, and conversely adopts related structures when nickel(II) ions are partly substituted (in a given ratio) by cadmium(II) [3]. A very exciting series was, in this respect, isolated exhibiting dimers, trimers or tetramers. In these compounds, the most striking feature is that nickel(II) units are well-isolated magnetically and, in addition, constitute an increasing series of polynuclear units for studying magneto-structural correlations.

It should be noted that structural changes may equally be induced in perovskite derived structures by a high pressure effect. Thus, the proportion of cubic, close-packed layers may be increased by submitting the sample to high pressure, as observed for $BaRuO_3$. In this case, the sequence of structural packings 9R→4H→6H occurs [4], which is related to the increasing number of cubic layers. The same result can be obtained by substituting the A cation by a larger one, as in the mixed compound $Ba_{1-x}Sr_xRuO_3$ where strontium induces an internal pressure.

As shown in the above examples, several attractive situations, related to some typical combinations of the metal ions, may be investigated in particular from the viewpoint of the spin dimensionality and the nature of the exchange pathways [5-11]. Obviously, the binuclear systems present the simplest examples for studying exchange interactions, as reported in several review papers. They are commonly characterized by an antiferromagnetic coupling between the metal ions and an exchange constant varying, in relation with structural changes, in a large range of energy (a few wave-numbers to several hundreds) [12, 13]. In contrast to the profuse literature on antiferromagnetic systems, the examples where a ferromagnetic interaction occurs are very limited in number. This situation is realized for a near (or strict) orthogonality of magnetic orbitals involved in the exchange process as shown by Kahn et al. for $CuVO(fsa)_2$ en. CH_3OH [14]. In the same way, it corresponds to the intraionic coupling between d electrons, commonly referred to as Hund's rule.

Extensive theoretical research has been devoted to the understanding of exchange mechanisms and the structural factors which govern the sign and magnitude of the couplings [15–19]. The aim of this paper is to present a survey of the exchange models and to illustrate the current research in inorganic chemistry, by some specific examples. For an exhaustive coverage of exchange-coupled molecular systems, see for instance Ref. [1].

2 Theoretical Basis for Exchange Coupling

A rigorous description of exchange-coupled polynuclear systems requires the knowledge of crystal-field states of interacting ions. When the crystal-field ground state is orbitally non-degenerate, it is straightforward (for a small unit) to solve the eigenvalue problem by assuming a formal isotropic coupling between spin operators localized on sites i and j,

$$H_{ex} = -2 \sum_{ij} J_{ij} \vec{S}_i \vec{S}_j \tag{1}$$

where J_{ij} is the exchange constant for the pair (i, j).

Such an Hamiltonian is said to be of Heisenberg-like. In addition to this isotropic bilinear term, further contributions are sometimes added to improve the agreement with experiment. Note a biquadratic exchange term, $j(\vec{S}_i \vec{S}_j)^2$, an antisymmetric term $d(\vec{S}_i \times \vec{S}_j)$ tending to align the spins perpendicular to each other, and also an anisotropic coupling of pseudodipolar form [20–22]. Both the antisymmetric and anisotropic exchange contributions arise from spin-orbit coupling effects. When the orbital singlet ground-state of the ions under consideration is far away from the orbitally degenerate excited states, these extra-contributions are small perturbations. Thus, the isotropic bilinear interaction is the driving effect.

Using Eq. 1, the partition function and thermal variation of the magnetic susceptibility of limited metallic units may be derived, allowing one to deduce, after comparison with experiment, the exchange parameters. A large number of polynuclear systems have been thoroughly investigated in this way. Clearly, such a mathematical treatment does not provide any information above the ground exchange mechanism, in particular on the respective contributions of direct and superexchange (via bridging ligands) interactions. Attempts at a theoretical prediction were shown to be limited in scope because electronic levels, very close in energy, must be computed for a many electron system by considering electronic correlations. All the models derived so far use, for the description, Heitler-London wave-functions. They are referred to as molecular orbital models [23–25] or localized electron models (following Anderson's definition) [26–33], which, a priori, appear conflicting, but in fact describe an identical phenomenon from opposite approaches.

2.1 Molecular Orbital Description

Assuming a dimeric system with spin-1/2 interacting ions, the M.O. model developed by Hoffmann and coworkers [24] is based on the configuration interaction between orthogonalized molecular orbitals. If a_+ and a_- stand for the molecular orbitals built up from orthonormalized combinations of the magnetic orbitals (involving a partial delocalization toward the ligands) and E_+ and E_- are the respective energies, the singlet-triplet splitting is given by:

$$E_T - E_s = -2K_{ab} + (E_+ - E_-)^2/(J_{aa} - J_{ab}) \qquad (2)$$

In this expression, K_{ab} (always positive) is the exchange integral, J_{aa} and J_{ab} the one and two-center Coulomb integrals, respectively.

Clearly, the triplet is stabilized with respect to the singlet for $E_+ = E_-$. Conversely, when the energies of both symmetric and antisymmetric functions become different, an antiferromagnetic ground-state is usually observed. Using the extended Hückel model for determining the energy of molecular levels, the authors show that the change in J with structural parameters results from small energy variations of the molecular orbitals a_+ and a_-. Such a model was used to explain qualitatively the magnetic behavior of a series of Cu(II) complexes. However, we must keep in mind that this treatment involves severe approximations (it underestimates electronic correlations) so that its predictive character must be considered with great caution.

Kahn and Briat [25] suggested a quite similar approach by assuming, in the same way, Heitler-London wave-functions but without an admixture of polar states. Using now non-orthogonalized magnetic orbitals (called "natural") they state that both ferromagnetic and antiferromagnetic contributions to the exchange process result from the symmetry properties of the ground-configuration (with one electron per center); the configuration interaction with polar states is assumed to be negligible. In fact, the use of non-orthogonalized orbitals is equivalent to considering them in the limit of well-separated metal ions, so that the overlap between both models is clear for weakly interacting systems.

2.2 Localized Electron Model

It was shown previously that the spin Hamiltonian remains correct for describing systems without orbital degeneracy. Conversely, it cannot give a satisfactory result when orbital effects are important as it occurs with interacting ions of $^2T_{2g}$ or $^3T_{1g}$ ground-term. The spins are then strongly correlated to orbital moments so that a major deviation from Eq. (1) occurs.

The need to use a much more complete Hamiltonian was discussed in great detail by several authors (see [28] and references therein). In some cases, the problem includes a large number of independent parameters because all ion states are taken into account [29]. Actually, it seems always possible to reduce their number by some convenient approximations or local symmetry considerations. In other cases, the effective Hamiltonian is determined within the framework of the strong crystal-field scheme [30–32].

In a pioneering work on the exchange interaction, Anderson [26] stated that the excitation of an electron from one site to its nearest neighbor requires an energy U, which mainly results from Coulomb repulsion but also from pure one-electron terms, when there is no orbital degeneracy. The polar states so-created are connected to the ground-state through the transfer integral b, and thus contribute to lower the energy of the ground-configuration. The resulting Hamiltonian is written in second quantized notation as:

$$H = 2/U \sum_{m,n} \sum_{\sigma} b_{m'm} b_{n'n} C^*_{m'\sigma 1}(j) C_{n'\sigma 2}(j) C^*_{n\sigma 2}(i) C_{m\sigma 1}(i)$$

$$- 1/U \sum_{m,n} \sum_{\sigma} (b_{nm'm'm} C^*_{n\sigma}(i) C_{m\sigma}(i) + b_{m'm} b_{mn'} C^*_{m'\sigma}(j) C_{n'\sigma}(j)) \quad (3)$$

where $C^*_{m\sigma}$ and $C_{m\sigma}$ are anticommuting operators which, respectively, create and annihilate particles on an orthonormal set of states $|m\sigma\rangle$.

The last two terms are relative to the anisotropy effects on sites i and j, and so can be regarded as extra-contributions to the crystal-field; the first one is the exchange interaction term whose the parameter is usually written as:

$$J = -b^*_{mm'} b_{nn'}/U .$$

This way, we take into account the intraionic transfers between degenerate states as well as interionic transfers.

At this step, Drillon and Georges [30, 31] showed that any improvement of the exchange Hamiltonian must take into account spin-spin energy terms in the virtual excited states. This may be of major importance as clearly shown by considering two doubly degenerate ions, 1 and 2 (symmetry E_g), each one with one electron. Clearly, the e_g electron of 2 may be transferred to 1 whatever its spin, but the excitation energy will be lower if it is parallel to that of 1, that is when the intermediate state of 1 is high-spin (Fig. 1). This will result in an effective ferromagnetic exchange between the two ions, whose order of magnitude is given by:

$$S^{-2}(b^2/(U-KS/2) - b^2/(U+KS/2)) = Kb^2/S(U^2 - K^2S^2/4)$$

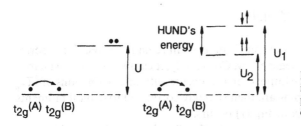

Fig. 1. Influence of the intraionic coupling in the polar states of a doublet degenerate dimeric system

where b is the usual transfer integral and $U \pm KS/2$ are the excitation energies for the high spin (−) and low spin (+) intermediate configurations. Such a contribution may not be neglected compared to the commonly encountered b^2/S^2U term, since K and U can be of the same order of magnitude. However, it is lost if the intermediate spin configurations are ignored. Note that this idea was reintroduced 4 years after by Torrance et al. [34] to explain the stabilization of ferromagnetism in organic complexes showing orbitally degenerate species.

A similar effect arises when a half filled orbital overlaps a completely filled one. In this case, one has to consider the intermediate spin configuration of the ion which emits the electron. The metal-metal interaction is then introduced from the Anderson Hamiltonian [26]:

$$H = \sum_m \sum_\sigma (b_{mm'} C^*_{m\sigma}(i) C_{m'\sigma}(j) + b_{m'm} C^*_{m'\sigma}(j) C_{m\sigma}(i))$$

$$H = H_{ij} + H_{ji} \tag{4}$$

where $H_{ij}(H_{ji})$ represents the part of H corresponding to the $i \to j$ ($j \to i$) electronic transfers. Let $|\varphi_e^\alpha\rangle$ (or $|\varphi_e^\beta\rangle$) be a state resulting from $i \to j$ (or $j \to i$) electron transfer, and U_e^α (or U_e^β) be the corresponding excitation energy. Clearly,

$$\langle \varphi_o | H | \varphi_e^\alpha \rangle = \langle \varphi_o | H_{ij} | \varphi_e^\alpha \rangle$$

and

$$\langle \varphi_o | H | \varphi_e^\beta \rangle = \langle \varphi_o | H_{ji} | \varphi_e^\beta \rangle \tag{5}$$

Since, within Anderson's framework, singly polarized states are involved together with unpolarized ones, the effective exchange Hamiltonian can be written as

$$H_{ex} = -\sum_{\substack{\alpha \\ \beta}} \frac{H_{ij} |\varphi_e^\alpha\rangle \langle \varphi_e^\alpha | H_{ji}}{U_e^\alpha} \frac{H_{ji} |\varphi_e^\beta\rangle \langle \varphi_e^\beta | H_{ij}}{U_e^\beta} \tag{6}$$

If H^s corresponds to the intraionic spin-spin coupling, the excitation energies are given by:

$$U_e^\alpha = U_e^a + \langle \varphi_e^\alpha | H^s | \varphi_e^\alpha \rangle$$
$$U_e^\beta = U_e^b + \langle \varphi_e^\beta | H^s | \varphi_e^\beta \rangle \tag{7}$$

Defining now the collective excited states from the total spin quantum numbers on each cation, the exchange Hamiltonian is given by:

$$H_{ex} = -\sum_{m,n} \sum_{\sigma_1,\sigma_2} \frac{b_{mm'}b_{n'n}[(C^*_{m\sigma 1}|S_iS_{iz}\rangle\langle S_iS_{iz}|C_{n\sigma 2}C_{m'\sigma 1}|S_jS_{jz}\rangle\langle S_jS_{jz}|C^*_{n'\sigma 2})}{(U^a_e + E^s(S_i, S_j))}$$

$$+ \frac{(C^*_{n'\sigma 2}|S_jS_{jz}\rangle\langle S_jS_{jz}|C_{m'\sigma 1}C_{n\sigma 2}|S_iS_{iz}\rangle\langle S_iS_{iz}|C^*_{m\sigma 1})]}{(U^b_e + E^s(S_i, S_j))} \quad (8)$$

where $E^s(S_i, S_j)$ refers to the eigenvalues of H^s. This Hamiltonian clearly shows its dependence on the various spin configurations in the excited states. Its expression gets simpler when some symmetries can be taken into account, and mainly when we are dealing with a pair of identical metal ions allowing us to ignore the distinction between U^a_e and U^b_e.

At first sight, the result of this approach strictly includes the kinetic contribution in terms of the Anderson model [26]. Any rigorous extension would have to take into account the self-energy of the overlap charges between partly occupied orbitals on adjacent atoms, namely the potential exchange. Actually, it may be shown that this does not modify basically the expression of the above Hamiltonian. This contribution, which is ferromagnetic by nature, is generally negligible but may be important when the orbitals located on adjacent sites are orthogonal as may be observed in hetero-bimetallic systems. Finally, before expanding the exchange Hamiltonian in some peculiar situations one can notice some simple results based on symmetry considerations.

At first, the general Hamiltonian can be described from products of spherical tensors expressed in terms of spin and orbital components. Since no spin-orbit coupling has been explicitly introduced in the general expression of H_{ex}, the spin operators must appear in a spherical way. As they occur linearly, the Hamiltonian will take the form:

$$H_{ex} = A(\vec{L}_i\vec{L}_j) + B(\vec{L}_i\vec{L}_j) \cdot \vec{S}_i\vec{S}_j \quad (9)$$

where A and B are polynomials involving L_i and L_j components. Due to the time reversal symmetry, they only involve terms with total even parity. Furthermore, we notice that orbital operators transform according to the D_0, D_1 and D_2 representations of the rotation group; as a result, the only components which are actually involved on each site are the following:

$$1 \quad \text{governed by } D_0$$

$$\left.\begin{array}{l} L_z \\ L^+ \\ L^- \end{array}\right\} \text{governed by } D_1$$

$$\left.\begin{array}{l} 3L_z^2 - L(L+1) \\ L^-L_z + L_zL^- \\ L^+L_z + L_zL^+ \\ L^{+2} \\ L^{-2} \end{array}\right\} \text{governed by } D_2$$

Table 1. Expressions of the exchange Hamiltonian in D_{nh} ($n=1, 2, 3$) symmetry and d^1 or d^2 electronic configurations

Interaction $t_{2g}^1 - t_{2g}^1$: $\quad J_1 = \dfrac{-b^2}{U_e + U^s(0,0)}, \quad J_2 = \dfrac{-b^2}{U_e + U^s(1,0)} \quad \begin{cases} J = \dfrac{J_1 + J_2}{2} \\ J' = \dfrac{J_2 - J_1}{2} \end{cases}$

D_{2h}: $H_{ex} = +\dfrac{J_1}{4}(4 - 3L_{1z}^2 - 3L_{2z}^2 - 2L_{1z}^2 L_{2z}^2)(1 - 4\vec{S}_1 \vec{S}_2) + \dfrac{J_2}{4}(L_{1z}^2 + L_{2z}^2 - 2L_{1z}^2 L_{2z}^2)(3 + 4\vec{S}_1 \vec{S}_2)$

D_{3h}: $H_{ex} = +\dfrac{J_1}{2}[1 - \vec{L}_1 \vec{L}_2 - (\vec{L}_1 \vec{L}_2)^2](1 + 4\vec{S}_1 \vec{S}_2) + \tfrac{1}{4}(J_2 - J_1)(1 - \vec{L}_1 \vec{L}_2)(2 + \vec{L}_1 \vec{L}_2)(3 + 4\vec{S}_1 \vec{S}_2)$

D_{4h}: $H_{ex} = -\tfrac{1}{4}(2L_{1z} L_{2z} + 2L_{1z}^2 L_{2z}^2 + L_1^{+2} L_2^{-2} + L_1^{-2} L_2^{+2})[2J' + J(1 + 4\vec{S}_1 \vec{S}_2)] + \tfrac{1}{2}(L_{1z}^2 + L_{2z}^2)[2J + J'(1 + 4\vec{S}_1 \vec{S}_2)]$

Interaction $t_{2g}^2 - t_{2g}^2$: $\quad J_1 = \dfrac{-2b^2}{U_e + U^s(1/2, 1/2)}, \quad J_2 = \dfrac{-2b^2}{U_e + U^s(1/2, 3/2)}$

D_{2h}: $H_{ex} = +\dfrac{J_1}{6}(L_{1z}^2 + L_{2z}^2 + L_{1z}^2 L_{2z}^2)(1 - \vec{S}_1 \vec{S}_2) + \dfrac{J_2}{6}(L_{1z}^2 + L_{2z}^2 - 2L_{1z}^2 L_{2z}^2)(2 + \vec{S}_1 \vec{S}_2)$

D_{3h}: $H_{ex} = +\dfrac{J_1}{6}[4 + \vec{L}_1 \vec{L}_2 + (\vec{L}_1 \vec{L}_2)^2](1 - \vec{S}_1 \vec{S}_2) + \dfrac{J_2}{3}[2 - \vec{L}_1 \vec{L}_2 - (\vec{L}_1 \vec{L}_2)^2](2 + \vec{S}_1 \vec{S}_2)$

D_{4h}: $H_{ex} = +\dfrac{J_1}{12}[12 + L_{1z} L_{2z} - 4L_{1z}^2 - 4L_{2z}^2 + L_{1z}^2 L_{1z}^2 + \tfrac{1}{2}(L_1^{+2} L_2^{-2} + L_1^{-2} L_1^{+2})](1 - \vec{S}_1 \vec{S}_2)$

$\qquad -\dfrac{J_2}{6}[L_{1z} L_{2z} - L_{1z}^2 - L_{2z}^2 + L_{1z}^2 L_{2z}^2 + \tfrac{1}{2}(L_1^{+2} L_2^{-2} + L_1^{-2} L_2^{+2})](2 + \vec{S}_1 \vec{S}_2)$

Rigorous calculations were carried out for systems involving $^2T_2$ or $^3T_1$ ground-term ions, and molecular symmetry obeying the D_{2h}, D_{3h} or D_{4h} (edge-, face- or corner-sharing octahedra, respectively) point group [30]. Related expressions of the Hamiltonian are given in Table 1.

3 Interplay Between Heisenberg and Anderson Models

Consider a dimeric unit built up from two ions in octahedral sites, with the overall symmetry D_{3h}. We assume on each site one unpaired electron corresponding to the $^2T_{2g}$ ground-term. Further, only exchange mechanisms involving similar t_{2g} orbitals for both ions are considered. Direct or superexchange-like interactions are not distinguished; both contribute to the effective coupling parameters, J_1 and J_2, defined as:

$$J_1 = -b^2/(U_e^a + U^s(0,0)), \qquad J_2 = -b^2/(U_e^a + U^s(1,0)) \qquad (10)$$

where b stands for the identical integrals $b_{xy,xy}$ $b_{yz,yz}$ and $b_{zx,zx}$, $U^s(0,0)$ and $U^s(1,0)$ refer, respectively to the $S=0$ and $S=1$ spin configurations in the t_{2g}^2 manifold.

A tedious though straightforward calculation thus leads to the expression of the exchange Hamiltonian:

$$H_{ex} = J_1/2(1 - \vec{L}_1\vec{L}_2 - (\vec{L}_1\vec{L}_2)^2)(1 + 4\vec{S}_1\vec{S}_2)$$
$$+ (J_2 - J_1)/4(1 - \vec{L}_1\vec{L}_2)(2 + \vec{L}_1\vec{L}_2)(3 + 4\vec{S}_1\vec{S}_2) \qquad (11)$$

The first term of this operator describes the exchange when one neglects the intra-ionic spin-spin coupling while the second one takes it into account precisely ("extended" Anderson model).

Obviously, this Hamiltonian shows drastic differences with the Heisenberg expression. To compare the results of this approach, for $J_1 = J_2$ and $J_1 \neq J_2$, to those of the Heisenberg Hamiltonian, we define a new set of parameters:

$$J = 1/2(J_1 + J_2) \qquad \text{and} \qquad J' = 1/2(J_2 - J_1) \qquad (12)$$

From the respective energy diagrams (Fig. 2), it can be emphasized that:

(1) the approximation $J' = 0$, and the Heisenberg model lead both to a two sublevel scheme with the energy gap $2|J|$, but the level multiplicities differ.
(2) the case $J' \neq 0$ is more interesting, since it describes real systems by taking into account the various excited spin configurations. It can be viewed that the low-lying level is now a 9-fold degenerate state $|L=1; S=1\rangle$ corresponding to a ferromagnetic spin-spin coupling. The first excited level (6 fold degenerate) $S=0$ is $4|J'|$ above the ground-level. This clearly emphasizes the importance of the distinction between J_1 and J_2.

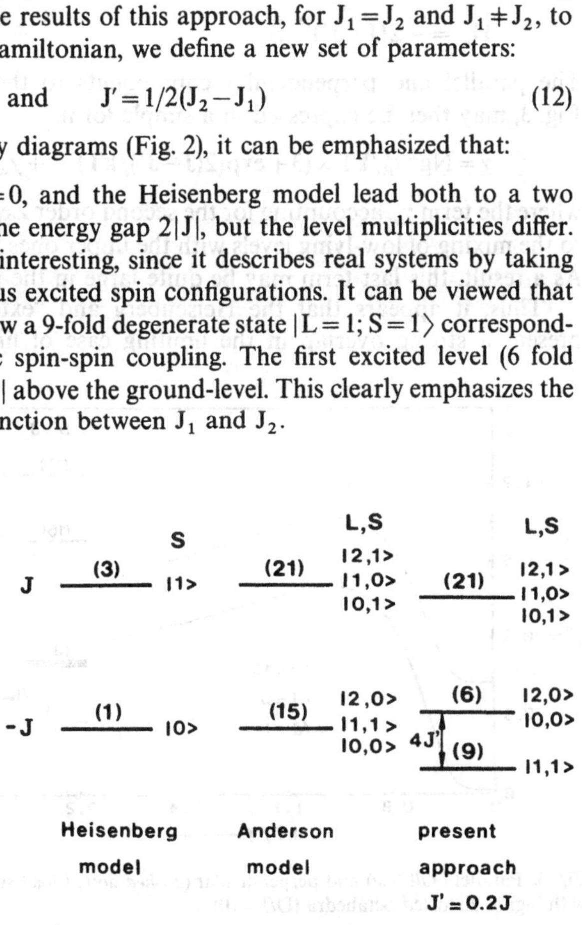

Fig. 2. Energy levels of an $t_{2g}^1 - t_{2g}^1$ dimeric species; comparison between different models

Heisenberg model   Anderson model   present approach $J' = 0.2 J$

Further investigations of this energy scheme would not be entirely realistic whether the effects of spin-orbit coupling and non-cubic components of the crystal-field was ignored. Then, the complete Hamiltonian must be written as:

$$H = H_{ex} - k \sum_i \vec{L}_i \vec{S}_i + D \sum_i L_{iz}^2 \tag{13}$$

in which all symbols have their usual meaning.

Consider now the case of a very large distortion parameter compared to the other ones and to the thermal energy available for the system. D is taken to be positive so that the ground-term of each isolated ion is the orbital singlet A_1, well-separated from the orbital doublet E. We are then dealing with a non-degenerate system with half-filled S-orbitals.

The remaining contributions of H act as perturbations and give an excited triplet at the energy $2|J|$ above the singlet ground-state. Distinguishing the two kinds of polarized states changes this gap into $2|J-J'|$. Finally, for highly distorted environments, we readily verify that the eigenvalues and degeneracies of the low-lying levels are just those of the isotropic Hamiltonian [31]:

$$H_{ex} = -2(J-J')\vec{S}_1\vec{S}_2 \tag{14}$$

The parallel and perpendicular components to the susceptibility plotted in Fig. 3, may then be expressed in a simple form:

$$\chi = Ng^2 u_B^2 / kT \times (3 + \exp(2(J-J')/kT))^{-1} + \chi_w \tag{15}$$

where the term χ_w accounting for the second order Zeeman contributions, is due to the mixing of low-lying levels with the upper ones located around D and 2D. As a result, this last term may be quite large in the present context.

Thus, it appears that the Heisenberg and "extended" Anderson models present a strong overlap in the limiting case of highly distorted octahedra.

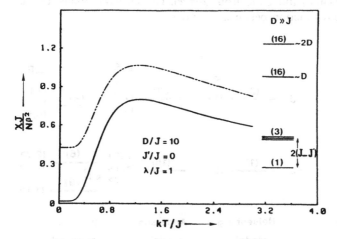

Fig. 3. Parallel (*full line*) and perpendicular (*dashed-dotted line*) susceptibilities of a t_{2g}^1–t_{2g}^1 species with highly distorted octahedra (D/J = 10)

Nevertheless, it is worth noticing that the latter allows us to describe all the situations, whatever the degree of orbital degeneracy of the interacting ions, and thus to predict the magnetic anisotropy of real systems.

4 Polymetallic Systems and Related Structures

Owing to the diversity of polymetallic exchange-coupled systems encountered in inorganic chemistry, it is a difficult task to classify them in a unique manner. Thus, while some research deals with the correlations between structure (local parameters, ligand effects ...) and magnetic properties, other research is more specifically concerned with exchange mechanisms in relation with local perturbations such as site distortions, spin-orbit coupling, etc.... Note that the first studies focus on chemical aspects while the second ones may be considered as a more physical point of view. In fact, this scheme appears to be arbitrary for some of the systems reported so far, so that we have chosen to classify them according to their structural types. The materials reported in the following belong to characteristic structural families of the inorganic chemistry, the main feature of which is to promote, at least potentially, "zero" dimensionality.

4.1 Hexagonal Perovskites and Related Systems

The structure of the hexagonal polytypes AMX_3 derives from the familiar perovskite network [35]. They can be described from the stacking of close-packing AX_3 layers in which M metals occupy only octahedral sites. Numerous structural types may be obtained theoretically depending on the sequence of AX_3 layers. Thus, they always consist in combinations of two types of packings: a complete hexagonal stacking of AX_3 layers resulting in a two-layer (2H) hexagonal unit cell, or a cubic AX_3 one yielding the familiar cubic perovskite structure (3C). The former corresponds to infinite chains of face-sharing MX_6 octahedra parallel to the c-axis (e.g. $CsNiF_3$), while the latter may be described as MX_6 octahedra sharing only corners (e.g. $CsCdF_3$). A detailed discussion of the sequences of close-packed stackings, with related symmetries and the descriptive Zhdanov notations, is provided in the International Tables for X-Ray Crystallography [36].

4.1.1 Strings of Increasing Size in the $CsNiF_3$-$CsCdF_3$ System

In the case of hexagonal perovskite-like fluorides, only the 2H, 6H, 10H, 9R and 12R arrangements have been found so far [36–38]. Thus, the system $CsNiF_3$–$CsCdF_3$ shows, according to the Ni/Cd ratio, 6H, 10H or 12R polytype

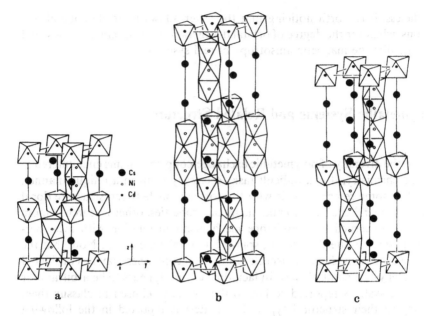

Fig. 4. Crystallographic order between Ni(II) and Cd(II) ions in the hexagonal structures of $Cs_3(NiCd)_3F_9$ compounds

structures (Fig. 4). These ones are characterized by the presence of a crystallographic order between Ni(II) and Cd(II) ions [38, 39]. Ni(II) ions are located in face-sharing octahedra and thus form finite strings of 2, 3 or 4 centres according to the composition. These units are isolated by CdF_6 octahedra and by coordination polyhedra of cesium. Notice that $Cs_3Ni_2CdF_9$ can only be prepared under high pressure.

The magnetic properties of $Cs_4Ni_3CdF_{12}$ (12R) and $Cs_5Ni_4CdF_{15}$ (10H) were shown to agree with a model of exchange-coupled units, on a large temperature scale [3]. The interactions are well-described by the Heisenberg Hamiltonian

$$H_{ex} = -2\sum_i J_i \vec{S}_i \vec{S}_{i+1} \qquad (16)$$

with a ferromagnetic exchange constant ranging between 7.5 and 9 K (Fig. 5). Similar couplings were reported for $RbNiF_3$ [40].

An analysis of the exchange mechanisms can only be formulated by considering the precise geometry of the molecular units. Clearly, the orientation of nickel(II) orbitals (containing unpaired electrons) with respect to bridging fluorines are very similar for both systems; they obey D_{3h} full symmetry. The metal orbitals are involved in symmetrical pathways (there is a mirror plane perpendicular to the metal-metal axis), so that an antiferromagnetic coupling should be observed, unlike experimental findings. In fact, it appears that orthogonal E' orbitals should be observed, unlike experimental findings. In fact,

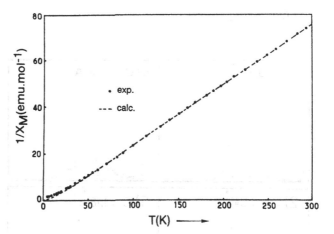

Fig. 5. Magnetic behavior of trimeric species in $Cs_4Ni_3CdF_{12}$ exhibiting a ferromagnetic exchange between Ni(II) ions

it appears that orthogonal E' orbitals are involved in a direct exchange process, which stabilize a net ferromagnetic interaction. A quantitative estimate is not available in this case.

Inelastic scattering of thermal neutrons provides an alternative way to determine the magnitude of the exchange couplings [41–43]. An example is given for $Cs_4Ni_3CdF_{12}$ in Fig. 6.

The inelastic peak appearing on the left of the elastic peak corresponds to the energy loss of neutrons. Its intensity decreases with increasing Q, as expected for a peak of magnetic origin. The energy of the transition $2J = 1.7 + 0.1$ meV agrees with that obtained from the magnetic investigation ($2J = 1.5$ meV).

Returning now to the magnetic properties, the plot of $\chi \cdot T$ vs T (Figs. 7–8) reveals, at very low temperatures, the influence of intercluster interactions and zero-field splitting for Ni(II) ions. Due to this last effect, the system is then anisotropic, and the complete Hamiltonian must be written:

$$H = H_{ex} + D\sum_i S_{iz}^2 - g\mu_B H(\sin\theta(S^+ + S^-)/2 + \cos\theta S_z) - 2zj S_\theta \langle S_\theta \rangle \qquad (17)$$

where θ is the angle between magnetic field and the Z axis. In this expression, $\langle S_\theta \rangle$ is defined as:

$$\langle S_\theta \rangle = kT/g\mu_B \frac{d}{dH}(\ln \text{Tr}(\exp - H/kT)) \qquad (18)$$

so that we have to solve an implicit equation in $\langle S_\theta \rangle$ with the density matrix technique.

This way, the exchange interactions between adjacent units are treated in the molecular field approximation. It appears that these interactions are antiferromagnetic and weak compared to intercluster ones. On the other hand, local distortions which stabilize the $S_z = 0$ component of Ni(II) ions strongly reinforce

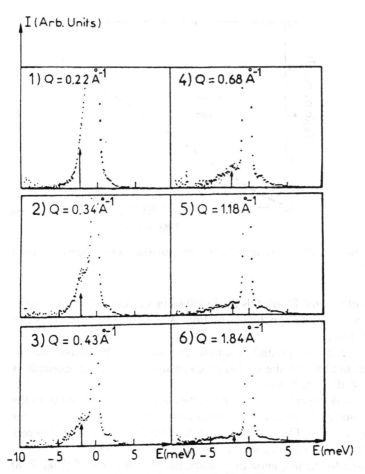

Fig. 6. Inelastic neutron scattering results for $Cs_4Ni_3CdF_{12}$

the low-dimensional character of these systems. Neutron diffraction measurements confirm the weak magnitude of intercluster couplings, since no 3d ordering occurs down to 4.2 K [38].

4.1.2 Binuclear Species in 2H Perovskite Derived Systems

Some of the binuclear systems reported so far derive from the 2H hexagonal-like structure described above, in which only two-thirds of the octahedral sites are occupied by metal ions.

The ordering between occupied and empty sites results in isolated binuclear units aligned along the c axis; each one may be schematized as two octahedra sharing a face, so that the site symmetry is C_{3v} and the molecular symmetry

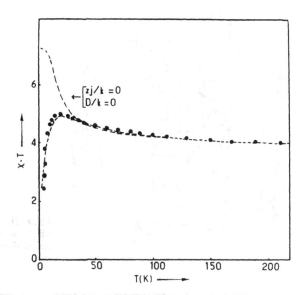

Fig. 7. Temperature dependence of the $\chi \cdot T$ product for $Cs_4Ni_3CdF_{12}$. Comparison between theory and experiment showing the influence of J and D parameters

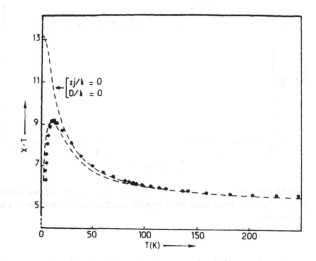

Fig. 8. Same as in Fig. 7, but for $Cs_5Ni_4CdF_{15}$

D_{3h} (Fig. 9). Thus, $Cs_3Fe_2F_9$, $Cs_3M_2Cl_9$ (with M = Ti, V), $Cs_3Gd_2Br_9$ and $Cs_3V_2O_2F_7$ where shown to present the same structural features [44, 48].

Owing to the metal-metal distances between adjacent dimers (> 4.8 Å), they are expected to behave as isolated binuclear species whose behavior may accurately be discussed from rigorous quantum analysis.

The simplest case corresponds to $Cs_3Fe_2F_9$ the behavior of which is plotted in Fig. 10, in the temperature range 2–40 K [49]. The increase of $\chi \cdot T$, upon

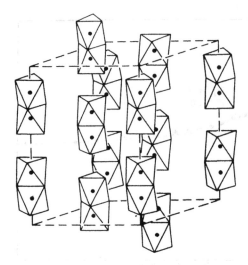

Fig. 9. 2H-perovskite derived structure in which two-thirds of the octahedral sites are occupied, leading to dimeric species

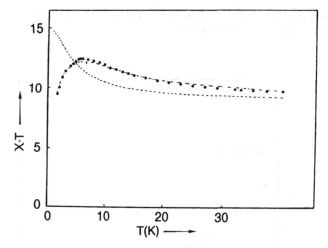

Fig. 10. Magnetic behavior of $Cs_3Fe_2F_9$ showing the ferromagnetic interaction within [FeFe] units. The very weak antiferromagnetic interaction between these units explains the $\chi \cdot T$ variation below 5K

cooling from 40 K down to 7 K, indicates that the coupling within [FeFe] units is ferromagnetic. Then, the ground-state should be given by $S = 5$, leading to a constant value, $\chi \cdot T = 15$, near absolute zero. In fact, a close analysis of the $\chi \cdot T$ variation at low temperature, shows clearly that intercluster interactions cannot be totally neglected, in the same way as for the Ni(II) systems. The result of the fits are listed in Table 2. They show that intercluster couplings are relatively weak, but their manifestation is obviously all the more significant as the spin-state of the ground-configuration becomes larger.

Table 2. Magnetic parameters in some hexagonal polytypes AMX_3

Compound	J (K)	d (Å)
$Cs_4Ni_3CdF_{12}$	9	2.74
$Cs_5Ni_4CdF_{15}$	7.5/9	
$Cs_5Ni_4CaF_{15}$	12/14	2.58/2.74
$Cs_3Fe_2F_9$	0.35(Zj = 0 K)	2.916
	0.98(Zj = −0.04 K)	
$Cs_3Gd_2Br_9$	−0.12	4.03
$Cs_3V_2O_2F_7$	−13.5	2.995

The system $Cs_3Gd_2Br_9$ differs from the previous one by the spin-state of interacting ions (S = 7/2), and further by the small value of intracluster exchange coupling [50]. Accordingly, dipolar interactions have been considered in the computation of energy levels, through the Hamiltonian:

$$H = -2J\vec{S}_1\vec{S}_2 + D_d(3S_{1z}S_{2z} - \vec{S}_1\vec{S}_2) - g\mu_B H(\sin\theta(S_{1x} + S_{2x}) + \cos\theta(S_{1z} + S_{2z})) \quad (19)$$

The dipolar magnetic interaction may roughly be estimated from the metal-metal distance; assuming isotropic magnetic moments, we obtain $D_d = -g^2\mu_B^2/r^3 = -0.04$ K.

By fitting the magnetization curve, at T = 2.3 K and 4.2 K, we show that the coupling is antiferromagnetic and close to −0.12 K (Fig. 11). It is to be noticed that the J value is significantly less than in [FeFe] units, as expected from the weak overlap between f orbitals.

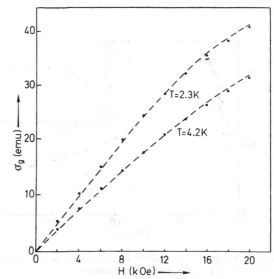

Fig. 11. Influence of temperature on the magnetization of $Cs_3Gd_2Br_9$ in the range 0–20 kG. Comparison between theory and experiment

Let us now consider dimeric systems in which the interacting ions exhibit an orbital degeneracy in their ground-state. The spins are then correlated to orbital moments, and the Heisenberg model becomes irrelevant. As an example, we will examine the behavior of vanadium(IV) pairs and titanium(III) pairs allowing us to compare the different contributions to the electronic structure.

By contrast with the other systems, $Cs_3V_2O_2F_7$ is characterized by a statistical distribution of oxygen and fluorine atoms over unbridging positions [48]. The bridging V–F bonds towards the common face are significantly longer than the outer V–(O, F) ones, giving a quite large V–V intradimer distance (d = 2.995 Å) for this kind of stacking. The magnetic data plotted in the range 4–120 K (Fig. 12) show a relatively broad maximum of susceptibility located around 16 K and a Curie-Weiss behavior at higher temperature (C = 0.76, θ = −9 K). Such a variation is well-known to feature the behavior of antiferromagnetic exchange-coupled dimers; the $\chi.T$ plot, given in the inset, continuously decreases upon cooling down and reaches a value very close to zero, confirming that the ground-state is non-magnetic.

In order to explain such a behavior, consider in a first step one unpaired electron located in degenerate t_{2g} orbitals ($^2T_{2g}$). Owing to molecular symmetry (roughly D_{3h}), it may be assumed that the t_{2g} orbitals are engaged in the exchange coupling with the same weight, although local symmetry removes the degeneracy of the orbital triplet (Fig. 13). The theoretical treatment was achieved on basis of the "extended" Anderson model, as developed in section III, and the complete energy matrix (36 × 36) solved for a set of selected (J_1, J_2), λ, D and k values [31].

Using this model, a least squares refinement of the data was performed, taking the free ion value for the spin-orbit constant (λ = 360 K) and k = 0.8 as orbital reduction parameter. We obtain $J_1 = -13.5$ K and D = 3000 K ± 200 K (Fig. 12).

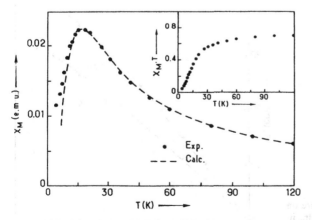

Fig. 12. Magnetic susceptibility of $Cs_3V_2O_2F_7$ in the temperature range 0–120 K. Comparison between theory (*dashed line*) and experiment

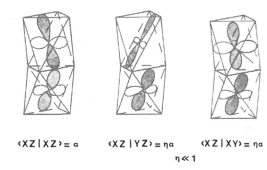

$\langle XZ|XZ\rangle = a$ $\langle XZ|YZ\rangle = \eta a$ $\langle XZ|XY\rangle = \eta a$
$\eta \ll 1$

Fig. 13. Available overlap integrals in dimeric species of symmetry D_{3h}

Due to the very large trigonal distortion, the system may nearly be described in neglecting the orbital constribution. This example confirms that this approach is not restricted to a limited range of J/D values; it may be applied to a large variety of systems involving competing effects without preconception above their relative magnitude.

Examine now the temperature dependence of the specific heat of $Cs_3V_2O_2F_7$. The magnetic contribution, plotted in the range 2–30 K in Fig. 14 corresponds to corrected values for the lattice contribution determined from the isostructural diamagnetic compound $Cs_3V_2O_4F_5$ [48].

The specific heat exhibits a well-defined schottky anomaly around T = 10 K related to the exchange coupling between metallic centres. Further, it is worth noticing the absence of λ-type anomaly characteristic of 3d ordering down to 2 K. Owing to magnetic measurement results, a singlet-triplet electronic structure may be assumed in the temperature range of interest, so that the specific heat can be expressed in closed form as:

$$C/R = 12(J/kT)^2 \exp(-2J/kT)/(1 + 3\exp(-2J/kT))^2 \qquad (20)$$

An approximate value of the exchange constant may easily be deduced from the position of the maximum close to 10 K. A refinement procedure over the whole temperature range leads to J = −13.8 K (Fig. 14).

For T > 25 K, the disagreement between calculated and experimental data results from the difficulty of estimating with accuracy the magnetic contribution of $Cs_3V_2O_2F_7$, in the range where the lattice contribution represents the major effect. In contrast, at a lower temperature where the magnetic term dominates, the agreement between theory and experiment is much more satisfactory; the deviation does not exceed 10%. Thus, the specific heat results fully confirm magnetic findings. Both agree with a simple electronic structure, at least for low-lying levels, since due to the strong distortion from O_h site symmetry, the problem is reduced to the interaction between two spin-doublets. Owing to structural features of the $(V_2O_2F_7)$ unit, such a result was, to a certain extent, predictable. Let us now examine the isoelectronic system $Cs_3Ti_2Cl_9$, which mainly differ by the nature of bridging ligands [31]. The metal-metal distance

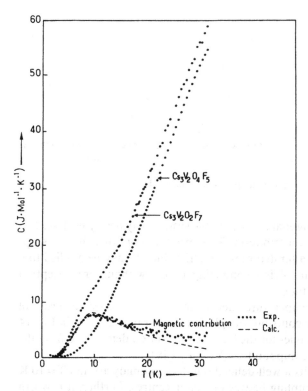

Fig. 14. Specific heat variation of $Cs_3V_2O_2F_7$. The lattice contribution is recorded for the diamagnetic compound $Cs_3V_2O_4F_5$

within (Ti_2Cl_9) units is 3.10 Å, while it is larger than 7.20 Å for adjacent units. Accordingly, the system can be considered as an assembly of isolated [Ti-Ti] pairs.

The susceptibility, measured by Briat et al. [51], is highly anisotropic (Fig. 15), the three-fold axis appearing to play a special role. At 4.2 K, the anisotropy $\Delta\chi$ is equal to 540×10^{-6} emu/mole. Using the above developed model, the analysis of experimental data, for the lower and upper limits of the orbital reduction factor k, leads to the parameters listed in Table 3.

On close examination of the results, it appears that:

(1) the agreement between theory and experiment is excellent since, whatever temperature, it corresponds to a mean divergence less than 2.5% and a very good description of the susceptibility anisotropy,
(2) 2(J-J') being the effective exchange parameter, we note a significant disagreement with previously reported results [50]. Actually, we do believe that the exchange Hamiltonian cannot be, in the present case, Heisenberg-like. Such an approximation may only be accounted for $D \gg J$ as shown in Fig. 3,

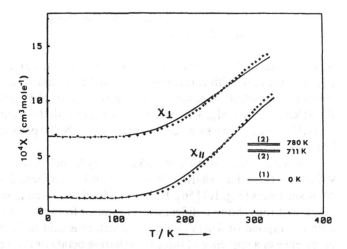

Fig. 15. Parallel and perpendicular molar susceptibilities of $Cs_3Ti_2Cl_9$. The comparison between theory (*full line*) and experiment is displayed for the set of parameters listed in Table 3

Table 3. Exchange parameters and local distortion in $Cs_3Ti_2Cl_9$

		k = 0.8	k = 0.9
J		549 K	520 K
J'		102 K	89 K
D		1200 K	1180 K
R	H//C	3.04×10^{-2}	2.90×10^{-2}
	H⊥C	2.52×10^{-2}	5.74×10^{-2}

(3) the low-lying level is a spin singlet well separated from two upper doublets located at 711 K and 780 K. This energy diagram is in close agreement with optical data [51] predicting a gap of 500 cm^{-1} (around 720 K).

(4) The exchange parameters are related by the relationship

$$J_2/J_1 = (U_e^a + U^s(0, 0))/(U_e^a + U^s(1, 0)) \tag{21}$$

so that they may be compared to expected values from the excitation energies.

U_e^a is obtained by comparing the ionization potentials for the configurations Ti(III)–Ti(III) and Ti(II)–Ti(IV), corrected for polarization effects [26] while spin-spin correlation energies in the excited configurations S = 0 and S = 1 are determined by using Racah parameters.

Finally, we get $J_2/J_1 = 1.3$, in good agreement with the experimental value, 1.5. This agreement could eventually be improved if we were able to take into account accurately the reduction effects of covalency on the Coulomb integrals involved in the calculation.

4.1.3 Bi- and Trinuclear Species in Ruthenium Oxides with 6H, 9R and 12R Structure

Several structural types were isolated in ruthenium oxides corresponding to polymetallic units with ruthenium in the oxidation state III, IV or V [52]. Thus, neutron diffraction measurements showed that the cationic ordering in $Ba_3MRu_2O_9$ (M = Mg, Ca, Sr or Cd) corresponds to isolated Ru(V)–Ru(V) pairs (Fig. 16) with the same geometry as Ni(II)–Ni(II) pairs previously reported [53–55].

Conversely, the structure of $BaRuO_3$ may be described from the stacking of strings of three face-sharing RuO_6 octahedra connected to each other by common apices (Fig. 16) [56]. Each outer ion is involved in metal-oxygen-metal bonds and thus shows a different environment from that of the central ion, but both correspond to Ru(IV) ions; the structure is said to be hexagonal 9R. The occurrence of a sequence of three face-sharing octahedra has equally been found in the system $Ba_4MRu_3O_{12}$ (with M = Nb, Ta) [57]. In fact, the structure of this last is in between those of $Ba_3MRu_2O_9$ (6H) and $BaRuO_3$ (9R). Here, the outer octahedra of trinuclear species share common apices with $(MO)_6$ octahedra (see Fig. 16); as a result, the system may be viewed as a set of well isolated trimers involving one Ru(III) and two Ru(IV) ions. The structure is referred to as 12R hexagonal perovskite.

Return now to $Ba_3MRu_2O_9$ (M = Mg, Ca, Sr or Cd). The intracluster metal distance is very short (2.65 Å for the calcium compound), thus entailing a strong exchange coupling. Non-magnetic reflections being observed down to 5 K, the magnetic properties are expected to agree with an isolated dimer model of spins $S_a = S_b = 3/2$, the eigenvalues of which are given by $E = -JS(S+1)$ with S running from 0 to 3. We thus obtain a satisfying description of experiments for J

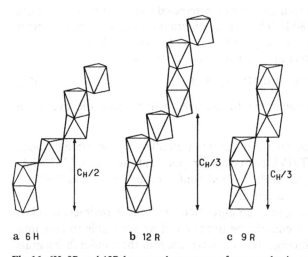

Fig. 16. 6H, 9R and 12R hexagonal structures of some ruthenium compounds

Table 4. Magnetic parameters for $Ba_3MRu_2O_9$ compounds

M	J (K)	g
Ca	−170	2
Cd	−173	2
Mg	−173	1.98
Sr	−138	1.88

close to −170 K, independently of the non-magnetic cation M (Table 4), except for the strontium compound [53]. The significant distortion of magnetic sites observed for this last leads to weaken the orbital overlap, and consequently the exchange constant (J = −138 K).

In fact, on close examination of the susceptibility variation, slight discrepancies appear at low temperatures, so that a model involving a biquadratic contribution was postulated in a second step [54]. In this assumption, the exchange Hamiltonian is written as:

$$H_{ex} = -2JS_1S_2 - j(S_1S_2)^2 \quad (22)$$

where the j value is positive and of the order of a few percent of J. The magnetic susceptibility, derived from the eigenvalues of this operator and Van Vleck's formula, is then given by:

$$\chi = Ng^2u_B^2/kT(\exp(2x-6.5y) + 5\exp(12x-3.5y) + 14\exp(12x-9y))/$$
$$(1 + 3\exp(2x-6,5y) + 5\exp(6x-13,5y) + 7\exp(12x-9y)) \quad (23)$$

where $x = J/kT$, and $y = j/kT$

The best fit to the experimental data, keeping g = 2 (determined by ESR) and letting J and j to vary, is shown in Fig. 17 for the final values J = −161 K and j = 6.6 K [54]. It may be seen that the comparison with experiment is significantly improved at low temperature. This model leads to a singlet-triplet splitting $\Delta E = 365$ K slightly larger than that obtained when neglecting the biquadratic exchange term ($\Delta E = 340$ K). Inelastic neutron scattering at various temperatures enables equally to find the energy splitting between the ground-state S = 0 and the first excited state, S = 1 [55]. The cross section $S(Q, \omega)$ versus energy gain is displayed in Fig. 18, at T = 6 K and 300 K, and for some values of Q. Phonons are present, but they can easily be identified. The magnetic peak at $\omega = 26$ meV corresponds to the spin singlet→spin triplet transition. The agreement between magnetic findings and inelastic scattering data confirms that a spin vector model is available in the present case, but does not allow us to conclude about the relative values of binuclear and biquadratic exchange terms.

Initially, $BaRuO_3$ was described as a 3d antiferromagnet. Actually, the magnetic susceptibility is low and rises with temperature to a broad maximum ($T_m = 440$ K) after which it decreases more slowly (Fig. 19) [57]. Mössbauer data show that this maximum does not correspond to a Néel temperature and that

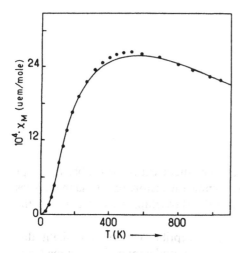

Fig. 17. Experimental and theoretical magnetic susceptibilities of $Ba_3CaRu_2O_9$ showing the influence of a biquadratic exchange coupling

this material is not magnetically ordered down to 4.2 K. However, strong antiferromagnetic couplings are to be considered between Ru(IV) ions within $(Ru_3O_{12})^{12-}$ units.

Referring to the short Ru-Ru distance, Gibb et al. considered the problem of the electronic structure from a molecular orbital viewpoint [58]. Such a treatment is a poor approximation in this case since it cannot explain the low temperature magnetic behavior, which mainly results from spin-orbit coupling effects. These ones prevail here as seen in other ruthenium(IV) compounds [57].

Then, the approach genuinely differs from that developed for $Ba_3MRu_2O_9$ [53] or Na_3RuO_4 [59] in which the ruthenium(V) polynuclear species are well isolated from each other and the magnetic ions orbitally non-degenerate (ground term $^4A_{2g}$). A realistic description of the observed magnetic susceptibility must take into account the major influence of spin-orbit coupling for Ru(IV) ions. This one removes the degeneracy of the orbital triplet $^3T_{1g}$, by giving three sub-levels, the lowest at -2λ (J=0), the others at $-\lambda$ (J=1) and λ (J=2). The non-magnetic ground-state of each ion introduces the fundamental problem of the interaction between the various sites at low temperature. The absence of long-range magnetic ordering shows that the excited states J=1 and J=2 are too far-away to induce a meaningful change of the magnetic behavior.

Conversely, the effect of the interactions cannot be neglected when temperature increases. The expression of H_{ex}, yielded in Table 1 for D_{3h} symmetry, enables us to determine the exact solutions of the overall Hamiltonian, but this necessitates a computer with a large storage capacity. Then, a qualitative approach may be achieved from the binuclear species $(Ru_2O_9)^{10-}$ which brings up the same problem, but in a more simple way. In this respect, we have plotted in Fig. 20 the theoretical curves of susceptibility for some values of the ratio J/λ (only one exchange constant was considered).

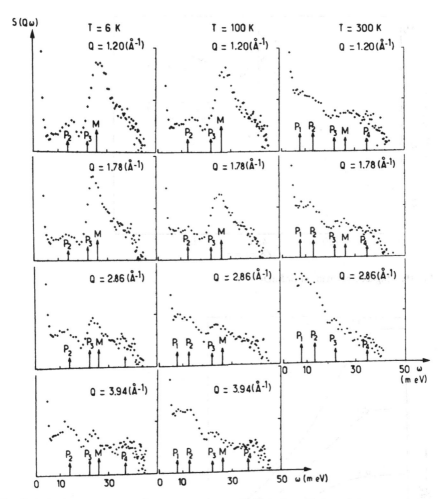

Fig. 18. Inelastic neutron scattering measurements exhibiting the variation of the cross section $S(Q, \omega)$ of $Ba_3SrRu_2O_9$

These curves suggest the following remarks:

(1) as T approaches zero, only the singlet-state $J = 0$ is appreciably populated so that the magnetic susceptibility arises from second order Zeeman contributions. Its value decreases with increasing values of the J/λ ratio,

(2) at intermediate temperature, there is a gradual increase in the $\chi(T)$ curves, which pass through a maximum when the interaction becomes quite large; such a maximum is observed experimentally. Finally, we can note that this model describes the experimental behavior in realistic terms as soon as the interaction becomes significant. It leads to good orders of magnitude, though an accurate comparison between theory and experiment is illusive on account of the suggested approximations.

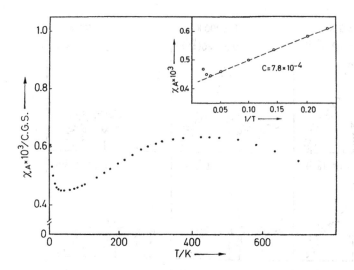

Fig. 19. Magnetic behavior of BaRuO$_3$

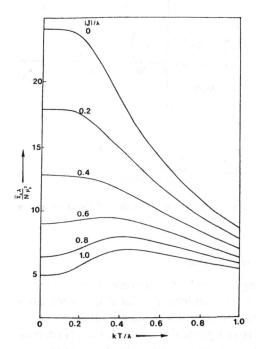

Fig. 20. Theoretical curves of susceptibility for a binuclear unit involving $^3T_{1g}$ sites. Influence of the ratio between the exchange constant and the spin-orbit coupling parameters

The trinuclear species of ruthenium occurring in Ba$_4$MRu$_3$O$_{12}$ (M = Nb, Ta) differ in that (1) they are well-isolated each other by corner-shared (MO$_6$) octahedra, (2) they contain ruthenium ions in mixed oxidation states [57]. Assuming that d electrons are localized at each site, two arrangements may be considered, namely Ru(IV)–Ru(III)–Ru(IV) which conserves the local symmetry, and Ru(IV)–Ru(IV)–Ru(III) which loses the inversion centre. An alternative way is that a substantial delocalization of electrons occurs, leading to charge averaging.

Using ^{99}Ru Mössbauer spectroscopy at 4.2 K for the above compounds provides evidence of the presence of Ru(III) and Ru(IV) ions in a ratio 1:2 [60]. However, it is not possible to draw more definite conclusions about the distribution of Ru(III) and Ru(IV) ions within the trinuclear units. In the assumption of a symmetrical stacking Ru(IV)–Ru(III)–Ru(IV), we can emphasize that, due to spin-orbit coupling effects, only the central atom presents a magnetic ground-state, $J = 1/2$. This enables to explain the absence of 3d long-range ordering upon cooling down to 4.2 K. Then, the variation of magnetic susceptibility is only related to the excitation within the Kramers doublet $J = 1/2$ (Fig. 21). The influence of the exchange coupling is weak (less than 1%) and thus does not modify noticeably the Landé factor, given by $g = (2 + 4k)/3$ where k is the usual orbital reduction factor.

Conversely, at higher temperatures, the upper levels $J = 3/2$ for Ru(III) and $J = 1$ and 2 for Ru(IV) become thermally populated; then we show that the interaction between nearest neighbors results in first order perturbations of the molecular levels, but this influence is very weak in the temperature range of interest.

4.2 Quasi Isolated Pairs in Trirutile Compounds

4.2.1 Exchange Coupling in Chromium Dimers $(Cr_2O_{10})^{14-}$

The MCr_2O_6 (M = Te, W) oxides show an ordered trirutile structure, first reported by Bayer [61] with space group $P4_2/mnm$. This structure can be described as $(Cr_2O_{10})^{14-}$ species built up from two octahedra sharing an edge;

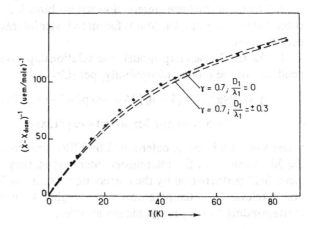

Fig. 21. Inverse magnetic susceptibility vs temperature of $Ba_4NbRu_3O_{12}$

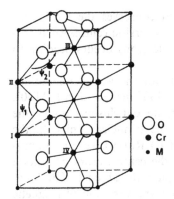

Fig. 22. Trirutile structure of MCr_2O_6 (M = Te, W) exhibiting the stacking of Cr(III) ions

these units are connected to each other by MO_6 octahedra along the c axis while they share a common corner in the other directions (Fig. 22). In this arrangement, each binuclear unit is surrounded by eight identical ones with superexchange paths consisting of Cr–O–Cr bonds, which are nearly the same for $TeCr_2O_6$ and WCr_2O_6. The Cr(III)–Cr(III) distances within binuclear species are 2.984 and 2.936 Å for Te and W compounds, respectively. Consequently, an increasing of the exchange interaction can be expected from $TeCr_2O_6$ to WCr_2O_6 [62].

The Cr(III) ions belonging to the adjacent entities are connected by much less favorable superexchange paths; they are separated by 4 Å through an oxygen atom and the bond angle is about 130°. Therefore, this coupling should be an order of magnitude smaller than the previous one. Neutron diffraction studies indicate that these compounds are magnetically ordered at 4.2 K, but with different magnetic structures [63]. Referring to Fig. 22, it appears that the interaction between nearest neighbors along the c axis (Cr_1–Cr_2) is always negative while it is either ferro- or antiferromagnetic for the others (Cr_2–Cr_3 and Cr_1–Cr_4) according to the non-magnetic cation (tungsten or tellurium, respectively). In this assumption, each magnetic site can be seen as interacting through two types of pathways (we only consider nearest neighbors). One of them (J) is related to the intracluster interaction within binuclear units and is the driving effect; the second one (j) accounts for intercluster interactions and acts as a small perturbation.

Using the Heisenberg model, the relationship giving the temperature dependence of the molar susceptibility, per $(Cr_2O_{10})^{14-}$ unit, is given by:

$$\chi = 2Ng^2\mu_B^2/k(T-\theta)\,[14 + 5\exp(6x) + \exp(10x)]$$
$$[7 + 5\exp(6x) + 3\exp(10x) + \exp(12x)] \qquad (24)$$

where $x = -J/kT$ and g, determined by ESR, is taken equal to 1.966. Owing to the 3d character of the intercluster interactions, they may be introduced as a mean field perturbation by the correcting term: $\theta = 8j\,S(S+1)/3k$. This procedure is relevant for temperatures much higher than the critical temperature corresponding to long-range magnetic order.

Table 5. Magnetic parameters for MCr_2O_6

Compound	J (intradimer)	j (interdimers)
$TeCr_2O_6$	−47	0
	−33.2	−4.5
WCr_2O_6	−37.8	0
	−44.5	1.4

Clearly, for $\theta = 0$, the obtained J values (Table 5) for both compounds disagree with the expected trend from the Cr(III)–Cr(III) distance. The agreement is greatly improved when considering intercluster couplings, as displayed in Fig. 23 for the best fit parameters listed in Table 5. Note that the calculated j parameters show a sign reversal which is fully consistent with neutron diffraction results.

It is of interest to compare the J values with that obtained in a similar entity isolated in $Cr_2Te_4O_{11}$ [63]. This example is the only one to our knowledge which shows binuclear units $(Cr_2O_{10})^{14-}$. The variation of the coupling with respect to the Cr(III)–Cr(III) distance (d) is plotted in Fig. 24 as $\log |J|$ against $\log d$. In the explored bond range, the data follow quite well a law of the type:

$$\log |J| = -\alpha \log d + \text{cste} \qquad (25)$$

where $\alpha = 16.5$ well agrees with the expected value [65]. This result supports the assumption that the contribution of t_{2g} orbitals is essential to describe the interaction mechanisms between the two Cr(III) ions. However, the various contributions (direct or superexchange through oxygen atoms) are difficult to separate in such a way.

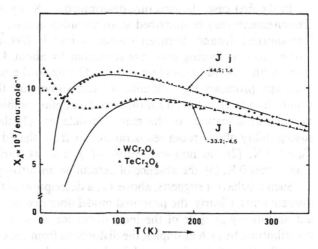

Fig. 23. Powder magnetic susceptibilities of $TeCr_2O_6$ and WCr_2O_6. The theoretical curves are given as *full lines*

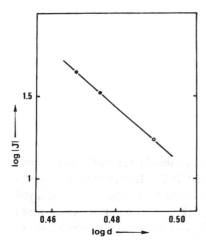

Fig. 24. Variation of the exchange interaction with the metal-metal distance in (Cr_2O_{10}) units

Finally, at low temperatures a discrepancy appears between experimental and theoretical curves. The influence of intercluster interaction being significant, one can expect a "crossover" from "0d" to 3d behaviors. A treatment with a mean field approximation then becomes questionable for giving a good picture of the system.

4.2.2 Orbitally Degenerate Binuclear Units $(V_2O_{10})^{14-}$

Depending on the thermal treatment of preparation, WV_2O_6 is either trirutile with a regular packing of the magnetic ions (vanadium(III)) and non-magnetic ones (tungsten(VI)) or rutile with a metal disorder [66, 67].

In the first case, the structure determined by X-ray and neutron diffraction measurements may be described as an assembly of discrete $(V_2O_{10})^{14-}$ species. The shortest distance between vanadium ions is 2.94 Å, while vanadium belonging to neighboring units are separated by about 4 Å through an oxygen atom, with a bond angle close to 130°. Owing to the orbitals involved in the exchange processes (t_{2g} orbitals), we can show that the magnetic behavior results, in a large extent, from the strong coupling within vanadium pairs. The most striking features of the magnetic data are (1) the broad maximum of susceptibility above room temperature, far from the 3d ordering temperature $T_N = 117$ K, (2) the non-zero value of χ and its variation as temperature approaches 0 K, (3) the absence of detectable anisotropy [67] (see Fig. 25).

Such a behavior suggests, above T_N, a description of the material in terms of discrete units. Clearly, the proposed model does explain the absence of significant anisotropy in spite of the local distortion at each site and the exchange contribution. In such examples, the distortions from the O_h symmetry constitute generally the driving force and lead to a high anisotropy of the magnetic properties. In fact, we have to describe here an unusual situation, in that the

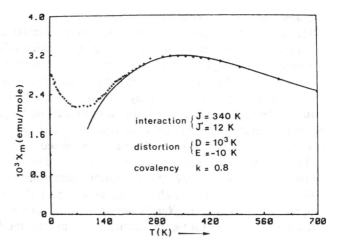

Fig. 25. Magnetic suceptibility of an orbitally degenerate binuclear species in WV$_2$O$_6$

relative coordinates (x, y, z) and consequently the C$_4$ axis of distortion of the octahedra (perpendicular to the pair axis) are different from a dimer to the adjacent one, thus averaging the anisotropy effect.

Let us now examine the influence of the exchange coupling between vanadium(III) ions (ground term $^3T_{1g}$). It can be stated, in a first analysis, that the direct coupling through the common edge is the only relevant contribution. In fact, a subsequent analysis of the magnetic data can only be made in the hypothesis of identical transfer integrals between t_{2g} orbitals. This is corroborated by a calculation of the respective direct and indirect contributions to the interaction [67].

Then, we show that the actual Hamiltonian may be expressed as:

$$H_{ex} = 1/6 \, J_1 (4 + \vec{L}_1 \vec{L}_2 + (\vec{L}_1 \vec{L}_2)^2)(1 - \vec{S}_1 \vec{S}_2) \\ + 1/3 \, J_2 (2 - \vec{L}_1 \vec{L}_2 - (\vec{L}_1 \vec{L}_2)^2)(2 + \vec{S}_1 \vec{S}_2) \quad (26)$$

where J_1 and J_2 have usual meanings.

Mention that the both terms of the Hamiltonian stabilize couplings with opposite signs, so that the resulting configuration will depend on the J_1/J_2 ratio [68]. Thus, the special case $J_1 = J_2$ reduces to the Anderson model [66] giving rise to an antiferromagnetic spin configuration in the ground-state. Taking into account the usual effects of spin-orbit coupling and local crystal-field distortions, with axial (D) and rhombic (E) terms, we are now able to compare theoretical predictions and experiment. The best fit values of the exchange and distortion parameters, found from a least squares refinement procedure, are:

$J_1 = -328$ K $D = 1000 \pm 50$ K

$J_2 = -352$ K $|E| < 10$ K

The agreement with experiment, shown in Fig. 25, is satisfying up to about 140 K; at lower temperatures, the intercluster interactions give rise to an additional contribution which cannot be easily introduced here, as usually done in orbitally non-degenerate systems. Further, we note that the large axial distortion of octahedra plays a significant role since it tends to remove the orbital degeneracy of the system by stabilizing an orbital singlet on each site. Actually, the system appears to be in between the totally quenched ($D \gg J$) and totally unquenched ($D=0$) limits, because of the strong exchange contribution allowing the upper levels to be thermally populated.

Finally, the relative value of the exchange constants agree well with the theoretical estimate. Using the ionization potentials of the configurations V(II)–V(IV) corrected by polarization effects, and the spin-spin correlation energies determined from Racah parameters, we obtain $b=0.36$ eV in two different ways thus ensuring the quality of the present model [67].

4.3 Miscellaneous Structures

4.3.1 Pairs in the Layer Compound CsV_2O_5

The crystal structure of CsV_2O_5 may be described in terms of $(V_2O_5)_n^{2-}$ layers held together by cesium atoms (Fig. 26) [69, 70]. Within layers, the vanadium(IV) ions are coupled together to form pairs of edge-sharing square pyramids. Note that similar units were observed in $(VO)_2P_2O_7$ [71].

Due to C_{4v} symmetry of vanadium sites, the ground-state is an orbital singlet $^2B_2$, well-separated from the doublet ^2E. Accordingly, the unpaired electron is mainly located in the d_{xy} orbital situated in the square base of the pyramid (in fact, it occurs a slight mixing of d_{xy} with out-of-plane orbitals due to the position of the metal ion). This entails a strong overlap of magnetic orbitals through the common edge of adjacent polyhedra, as may be observed from magnetic data. The typical variation of the magnetic susceptibility (broad maximum of χ around 100 K and drop to zero upon cooling down) is well-described, on the basis of the Heisenberg model with an isotropic coupling $J = -78$ K (Fig. 27). Such an agreement can only be explained by the strong stabilization of the

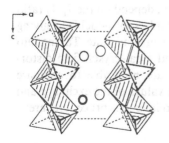 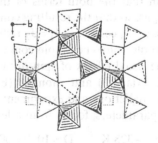

Fig. 26. Crystal structure of CsV_2O_5

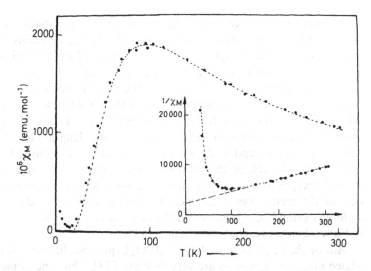

Fig. 27. Temperature dependence of the magnetic susceptibility of CsV_2O_5. The curve in the *dashed line* deals with a pair of exchange-coupled vanadium (IV) ions

orbital singlet ($^2B_2$) with respect to the doublet (^2E). Thus, we can note the analogy with $Cs_3V_2O_2F_7$, although the symmetry of the dimer, and accordingly the orbitals involved in the interaction, are different.

4.3.2 Discrete Trimers in Copper(II) Phosphates

The basic salts $Cu_3(PO_4)_2$ and $A_3Cu_3(PO_4)_4$ where $A=Ca$, Sr show similar structural species. Shoemaker et al. [72] first reported that $Cu_3(PO_4)_2$ is isostructural with the Stranskiite $Zn_2Cu(AsO_4)_2$. The crystal structure corresponds to infinite chains of $(Cu_3O_4)^{2-}$ trimers linked by bridging oxygen atoms along the b triclinic axis (Fig. 28). The middle copper(II) ion (on an inversion center) is in slightly distorted square planar coordination while the other two lie in irregular polyhydra of five oxygen atoms. Two trimers of adjacent chains are fastened by bridging oxygen atoms. Thus, copper(II) units are observed in the crystallographic plane (101) joined together by phosphate tetrahedra $(PO_4)^{3-}$. In these layers, the low dimensional character results, in a large extent, from the

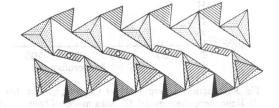

Fig. 28. Scheme showing the connections between CuO_4 and CuO_5 coordination polyedra in $Cu_3(PO_4)_2$

Jahn-Teller distortion of copper(II) polyhedra, which induces the localization of unpaired electrons in the basal plane of the square-pyramids. As a result, it can be assumed that the system reduces to a chain of alternatingly spaced metal ions with exchange interactions obeying the sequence J–J–j–J–J–j ... [73].

In $A_3Cu_3(PO_4)_4$ where A = Ca, Sr, the copper atoms show coordination polyhedra that are very similar to those of the above system [74]; one (Cu(1)) exhibits a square-planar surrounding, while the other two (Cu(2)) are in square-pyramidal surrounding (Fig. 29). The average bond lengths are 1.943 Å (square-planar polyhedron) and 2.029 Å (square-pyramid) as compared to 1.953 Å and 2.025 Å, respectively, in the orthophosphate. These polyhedra are connected through oxygen atoms to form infinite ribbons spreading along the b axis. In view of the metal-oxygen distances and bond angles, two different exchange pathways should be considered in the discussion, one within trimer units (J_1), and a weaker between adjacent trimers (J_2).

Although, $Cu_3(PO_4)_2$ and $A_3Cu_3(PO_4)_4$ present the same basic units, the related magnetic behaviors are very different [73]. Thus, the former compound exhibits a broad maximum of susceptibility around 30 K (Fig. 30) which is

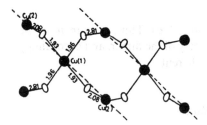

Fig. 29. Stacking of copper(II) ions in $A_3Cu_3(PO_4)_4$ with A = Ca, Sr

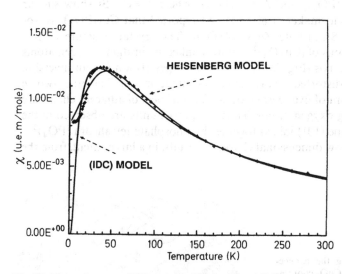

Fig. 30. Magnetic susceptibility of $Cu_3(PO_4)_2$. The *full lines* correspond to the best fits from (1) Heisenberg chain model, (2) Ising model of linearly connected trimers

characteristic of low-dimensional systems. This variation differs drastically from that expected for discrete [Cu$_3$] trimers whose susceptibility diverges as T→0 K. In fact, we show here that 1/2-spins are strongly correlated within trimer units (J = −125 K), but due to intercluster coupling (j) the non-compensated spins order antiparallely at absolute zero. The significant value of j thus entails a noticeable 1d character in a large temperature scale. The result of the fit by the regular Heisenberg chain is displayed in Fig. 30 for J = −68.8 K.

In contrast, the magnetic susceptibility of A$_3$Cu$_3$(PO$_4$)$_4$ (with A = Ca, Sr) may be described in terms of quasi-isolated [Cu$_3$] units, at least for T > 30 K (Fig. 31). Assuming that the interaction between nearest neighbors is isotropic, the exchange Hamiltonian is merely written as:

$$H_{ex} = -2J(\vec{S}_1\vec{S}_2 + \vec{S}_2\vec{S}_3) \tag{27}$$

where $S_1 = S_2 = S_3 = 1/2$.

The molar susceptibility, derived from the usual technique, was first compared to the experimental data by Boukhari et al., who found J values ranging between −100 and −110 K for the calcium and strontium compounds, respectively [74]. On close examination, it appears, in fact, that the agreement between theory and experiment is good, except in the low temperature region which cannot be merely described by such a model (Figs. 31–32). Actually, both systems show, at low temperature, the typical features of 1d ferrimagnets, namely a minimum of χ·T around 25 K and a strong divergence at lower temperature. Further measurements, performed in the very low temperature range, showed that Ca$_3$Cu$_3$(PO$_4$)$_4$ orders ferromagnetically at 0.8 K, while in

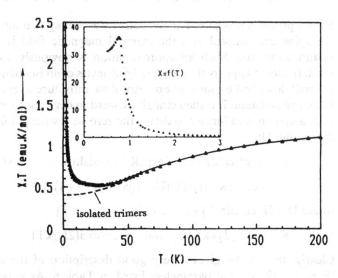

Fig. 31. Magnetic behavior of Ca$_3$Cu$_3$(PO$_4$)$_4$. The *dashed line* would correspond to isolated copper(II) trimers. 1d intercluster correlations give rise to the strong increase of χ·T observed at low temperature. The result of the fit is given as a *full line*

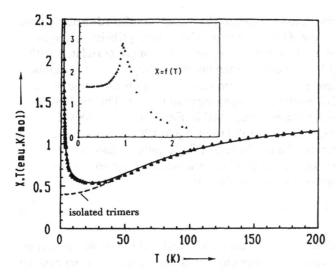

Fig. 32. Magnetic behavior of $Sr_3Cu_3(PO_4)_4$. The *full line* corresponds to the theoretical behavior for interconnected trimers within chains, the dashed line to isolated trimers

the case of $Sr_3Cu_3(PO_4)_4$ an antiferromagnetic ordering occurs below 0.9 K. This emphasizes that, even in the paramagnetic region, a model of interacting trimers (sketch in Fig. 29) must be considered according to the exchange Hamiltonian:

$$H_{ex} = -2\sum_i J_1 S_{3i}(S_{3i-1} + S_{3i+1}) + J_2 S_{3i}(S_{3i-2} + S_{3i+2}) \tag{28}$$

Such a problem was solved analytically by assuming that only Z components of the spins are coupled and the external magnetic field is applied along the quantization axis. Such an approximation is obviously questionable at first sight. In fact, it appears that the low-lying levels of an isolated Heisenberg trimer are well described by such an operator if we introduce a renormalization of the exchange constant. It is then straightforward to determine the transfer matrix of such a system, and further to derive the zero-field susceptibility from its largest eigenvalue U:

$$\chi = (Ng^2\mu_B^2/2kT[U(4\exp(K^+) + \cosh(K^+) + 1) - 8(\cosh(K^+) - \cosh(K^-))]/[U(U - 2(\cosh(K^+) + 1))] \tag{29}$$

where $U = 2(c\cosh(K^+) + c^{-1}\cosh(K^-) + c + c^{-1})$

$K^\pm = (J_1 \pm J_2)/kT$, and $c = \exp(2J_3/kT)$

Clearly, this model gives a very good description of the data down to 1 K (Figs. 31–32), for the parameters listed in Table 6. As anticipated above, the leading interaction (antiferromagnetic) occurs within $[Cu_3]$ trimers. Due to the relative stacking of these trimers, the non-compensated spins are parallel in the

Table 6. Best fit parameters for the compounds $M_3Cu_3(PO_4)_4$

Compound	J1 (intratrimer)	J2 (intertrimers)	g
$Ca_3Cu_3(PO_4)_4$	−68.8	−1.30	2.28
$Sr_3Cu_3(PO_4)_4$	−75.2	−1.25	2.25

ground-state, giving a ferromagnetic-like behavior below 10 K. Thus, although only one metal is involved and all the couplings are antiferromagnetic, the system is characterized by a high-spin ground-state.

4.3.3 Tetranuclear Units in RuF_5 and Related Systems

Transition metal pentafluorides fall into two main structural groups exhibiting either tetrameric rings or endless chain arrangements (Fig. 33). In MF_5 pentafluorides of the platinum-related metals (M = Ru, Os, Rh, Ir or Pt), the structure may be viewed as close-packed M_4F_{20} tetrameric units, in which (MF_6) octahedra share corners in *cis* positions [75–77]. A striking feature is the value of the bond angle M–F–M close to 135°, which induces a distortion of the clusters and a hexagonal close packing of fluorine anions. As expected, the

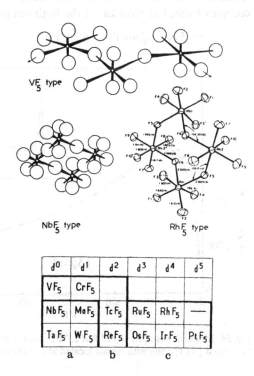

Fig. 33. Structures of MF_5 compounds

bridging fluorines are farther from the metal ions (=2.01 Å) than the non-bridging ones (=1.81 Å). Although pentafluorides of niobium, tantalum, molybdenum and tungsten also exhibit M_4F_{20} tetramers, they differ by the anionic packing resulting in linear M–F–M bond angles [78–80].

Let us examine the magnetic behavior of RuF_5 which may be solved quite easily since made of d^3 ions, characterized by a $^4A_{2g}$ ground-term. The magnetic susceptibility, plotted in the temperature range 4–300 K in Fig. 34, displays a rounded maximum around 40 K which cannot be ascribed to 3d long-range ordering. Neutron diffraction studies show that RuF_5 orders magnetically, but at much lower temperature ($T_N = 5$ K) [81]. Accordingly, the magnetic behavior may be related, in a large extent, to isolated [Ru_4] units, the moments of which are located at the apices of a rhombus [8].

The exchange Hamiltonian of such a unit is given by:

$$H_{ex} = -2J(\vec{S}_1\vec{S}_2 + \vec{S}_2\vec{S}_3 + \vec{S}_3\vec{S}_4 + \vec{S}_4\vec{S}_1) \tag{30}$$

whose eigenvalues are given by:

$$E(S, S', S'') = -J[S(S+1) - S'(S'+1) - S''(S''+1)]$$

S, S' and S'' refer to the total spin quantum number, $\vec{S}_1 + \vec{S}_3$ and $\vec{S}_2 + \vec{S}_4$, respectively.

The sequence of lowest energy levels is shown in Fig. 35, for a negative J value; the influence of applied magnetic field is also plotted. Owing to the number of spin states (256), the expression of the magnetic susceptibility takes a complex form, but reduces, in the high temperature limit, to:

$$\chi = 5Ng^2\mu_B^2/4kT(1 - 5J/kT)^{-1} \tag{31}$$

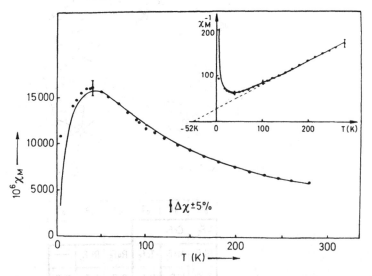

Fig. 34. Magnetic susceptibility of RuF_5. The maximum of susceptibility is described from a model of isolated [Ru_4] units with metals located at the apices of a rhombus

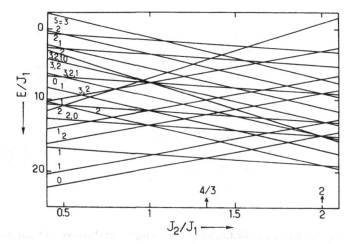

Fig. 35. Sketch of the low-lying energy levels for a tetranuclear unit with 3/2-spins located at the apices of a rhombus. The expression of the Hamiltonian is given by Eq. (30)

The comparison between the predicted Weiss temperature and the experimental one leads to an approximate value of the exchange constant ($J = -10$ K). Using the complete expression of the susceptibility, the agreement with experiment is shown to be very satisfying for $J = -8.3$ K and $g = 1.99$ (Fig. 34). We can note that, due to the weak coupling between Ru(V) ions, the energy levels are close to each other; so, high field magnetization could be used to improve the above model by considering the influence of next nearest neighbors, as developed for Na_3RuO_4 [82].

4.3.4 Heterometallic Three-Centre System in $Ba_2CaCuFe_2F_{14}$

We will now discuss the behavior of a new magnetic species, made of two iron(III) and one copper(II) exchange-coupled in the symmetrical lineal trimer (Fe–Cu–Fe). Such a system was isolated in the usovite-type fluoride $Ba_2CaCuFe_2F_{14}$ [83].

This compound can be described as the stacking of ($CaCuFe_2F_{14}$) layers separated in the spaces by barium atoms in twelve coordination (Fig. 36). In these layers, the Cu(II) and Fe(III) ions are octahedrally surrounded and share a coordination fluorine by forming closely interwining double chains spreading in the (010) direction; these ones are separated by calcium atoms. The cationic packing is such that Fe(III) and Cu(II) ions have respectively two and four nearest neighbors. Similar double chains were isolated in $Ba_2CaCoFe_2F_{14}$ and $M_3Cu_3(PO_4)_4$ (with M = Ca, Sr).

In fact, in the case under consideration, it may be undue to refer to as double chains, when examining the structural features. Thus, Cu(II) ions occupy axially distorted octahedra along Cu–F_4 bonds (2.31 Å) while the other bond lengths

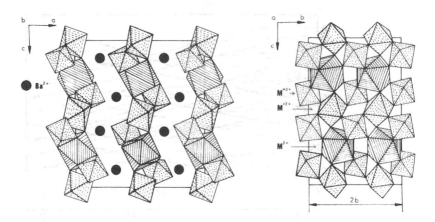

Fig. 36. Structure of $Ba_2CaCuFe_2F_{14}$ showing exotic chains in the b direction

are significantly shorter (Cu–F_7 = 1.91 Å and Cu–F_2 = 2.03 Å) (see Fig. 37). Such a distortion being related to the localization of the unpaired electron within the $d_{x^2-y^2}$ orbital (Jahn-Teller effect), it may be assumed that the coupling between Cu(II) and Fe(III) ions does mainly occur through the Cu–F_2–Fe exchange pathways. Conversely, the coupling via F_4 is expected to be very weak. Accordingly, the magnetic behavior of $Ba_2CaCuFe_2F_{14}$ may be described, to a large extent, in terms of quasi-isolated trimeric units (Fe(III)–Cu(II)–Fe(III)) [83].

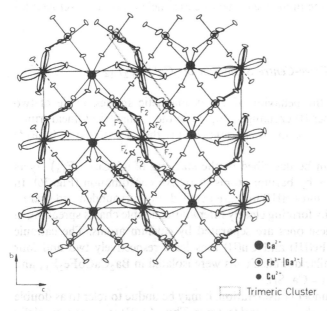

Fig. 37. Available exchange pathways between magnetic orbitals of Cu(II) and Fe(III) ions in $Ba_2CaCuFe_2F_{14}$

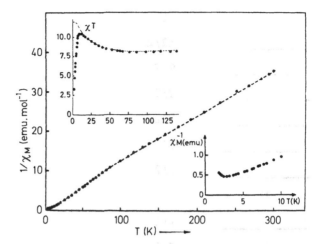

Fig. 38. Magnetic behavior of $Ba_2CaCuFe_2F_{14}$. The theoretical curve for a model of exchange-coupled trimers (5/2-1/2-5/2) is displayed as a *dashed line*. The decay of $\chi \cdot T$ observed at low temperature is provided by intercluster interactions

The thermal variation of the susceptibility, plotted in Fig. 38 as $\chi^{-1} = f(T)$, clearly illustrates the antiferromagnetic character of prevailing exchange interactions. When cooling down, the change of slope of $\chi^{-1} = f(T)$ around $T = 80$ K characterizes the increasing influence of the interactions within each trinuclear unit; finally, a minimum of χ^{-1} is observed around 2.5 K, which corresponds to the transition toward a 3d ordering.

Consider at first the trinuclear species $(S_1-S_2-S_3)$, where the spins $S_1 = S_3 = 5/2$ and $S_2 = 1/2$ are coupled by an isotropic interaction. According to the sign of this interaction, two situations may be encountered; let us focus on the antiferromagnetic exchange-coupled trimer where specific manifestations are expected.

The energy diagram given in Fig. 39 shows that the low-lying spin state is $S = 9/2$ and not the lowest spin configuration ($S = 1/2$) as generally observed in antiferromagnetic systems. On going up in energy, the spin configuration decreases from $S = 9/2$ to $S = 1/2$, then increases from $S = 1/2$ to $S = 11/2$. In the temperature range where only the lowest levels are thermally populated, the energy diagram is thus similar to that of a ferromagnetic coupled system. Accordingly, the $\chi \cdot T$ product does increase, upon cooling down, up to a constant value (close to 12.4) corresponding to a $S = 9/2$ ground-state. On the other hand, due to the spin multiplicity of the intermediate levels, $\chi \cdot T$ does exhibit a large minimum at intermediate temperature. Such a situation was actually observed in $Ba_2CaCuFe_2F_{14}$ and $(Mn(cth))_2Cu(pba)(CF_3SO_3)_2 \cdot 2H_2O$ [84] characterized by identical electronic structures.

The theoretical expression of the magnetic susceptibility has been established by assuming an isotropic g factor for Fe(III) and, due to the Jahn-Teller effect, an anisotropic g tensor for Cu(II). Then, the Hamiltonian to be solved is

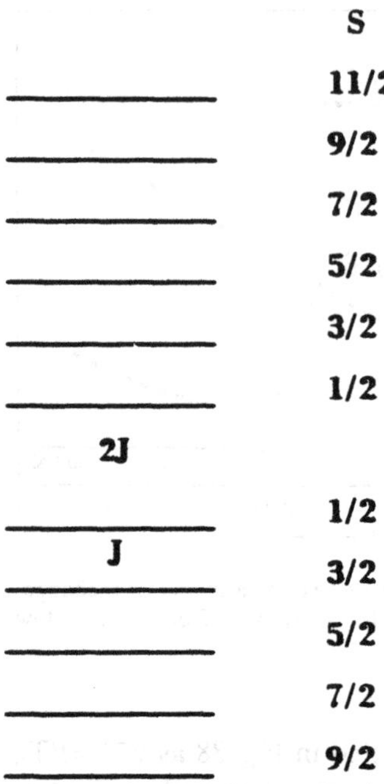

```
                S
_____        11/2

_____         9/2

_____         7/2

_____         5/2

_____         3/2

_____         1/2
  2J
_____         1/2
   J
_____         3/2

_____         5/2

_____         7/2

_____         9/2
```

Fig. 39. Energy level diagram of a [5/2-1/2-5/2] linear trimer

given by:

$$H = H_{ex} - g\mu_B H(\sin \Theta (S_1^+ + S_1^- + S_3^+ + S_3^-)/2 + \cos \Theta (S_{1z} + S_{3z}))$$
$$- \mu_B H(g_{2\parallel} \cos \Theta S_{2z} + g_{2\perp} \sin \Theta (S_2^+ + S_2^-)/2) \qquad (32)$$

where Θ is the angle between the applied magnetic field and the quantization axis.

The comparison between experimental and theoretical results for $Ba_2CaCuFe_2F_{14}$ is shown (Fig. 38) as $\chi \cdot T$ vs T. $g_{cu\parallel}$ and $g_{cu\perp}$ were both fixed, from ESR findings, to 2.507 and 2.091, respectively (average value 2.24). Then, the best agreement between theory and experiment corresponds to $J = -14.3$ K [83].

The proposed model appears to be satisfying, except for $T < 10$ K since instead of reaching the constant $\chi \cdot T$ value predicted by theory, a continuous decrease of $\chi \cdot T$ is observed. Such a deviation was explained by taking into account intercluster interactions which are drastically enhanced in the present case, because of the high spin-state of the ground-configuration. The correction for intercluster interactions was simply derived from the molecular-field approximation. The agreement between theory and experiment is then greatly improved for the intra- and inter-cluster interactions $J = -17.9$ K and $zj = -0.05$ K, respectively.

Although the ratio $zj/J = 3.10^{-3}$ agrees with quasi-isolated units, the manifestation of intercluster interactions appears unambiguously in the low temperature range. This example emphasizes the difficulty of obtaining well-isolated polymetallic systems exhibiting a high spin-degeneracy in their ground-state. While the interactions between antiferromagnetic pairs (with a S=0 ground-state) represent a second order effect, they act in turn to first order when assuming ferro- or ferrimagnetic units; then, their influence must be considered in all quantitative analysis of the magnetic properties. Finally, a striking feature to be emphasized is that the electronic scheme of this trimer is surprisingly similar to that of a mixed valence binuclear species (d^4-d^5) with one excess electron delocalized over two sites. In this last case, only the highest level S = 11/2 is to be ruled out. This remark could be of major importance in the future for stabilizing ferromagnetic species in an alternative way in order to obtain 3d ferromagnets.

5 Concluding Remarks

The experimental and theoretical works reported in the above sections show how isolated polynuclear systems remain an attractive research field in condensed matter. They constitute a first step to analysis of the behavior of low dimensional materials which are of current interest for both their transport and magnetic properties. Furthermore, it is clear that the study of small species enables a description of the exchange mechanisms at the microscopic scale, which is not available in the usual systems; accordingly, the results may be used to discuss phenomena observed in higher dimension, when these are related to the sign and magnitude of the exchange interactions.

The energy levels are usually determined by means of the isotropic Heisenberg Hamiltonian, whose solutions are trivial for 2 or 3 interacting sites. Additional contributions such as antisymmetric or anisotropic couplings are generally considered as small perturbations. It is worth noting that such a mathematical treatment does not provide any information on the respective contributions related to direct and superexchange couplings. In fact, on close examination of the symmetry of metallic and anionic orbitals involved in the exchange processes, an estimate of these contributions can be stated for a given series (same molecular symmetry for instance).

When considering different bridging ligands, a major deviation may be the difference of covalency of the metal-ligand bonds, so that it is a more difficult task to do realistic predictions. This influence is clearly emphasized in the isostructural compounds $Cs_3V_2O_2F_7$ and $Cs_3Ti_2Cl_9$ which present the same electronic configuration (d^1-d^1), but different bridging ligands (F^- and Cl^-, respectively). The vanadium atom is covalently linked to an outer oxygen atom, thus stabilizing a vanadyl unit (V=O); the exchange coupling through fluorine bridges is much weaker that it should be, compared to the titanium compound (J = −13.5 K against −360 K, while metal distances are 2.995 and 3.10 Å, respectively).

Obviously, the sign of the interaction may be predicted if we assume large trigonal distortion of the octahedra stabilizing the orbital singlet A, with respect to the doublet E. In fact even this assumption is questionable for $Cs_3V_2O_2F_7$ on account of the direction of the V=O bonds, which lowers the symmetry from D_{3h} to C_{2v}.

Conversely, the ferromagnetic coupling observed in the nickel(II) series, involving trimeric and tetrameric species, may be explained from orbital overlaps through linking fluorines of common octahedra. It results from couplings between orthogonal p_σ ligand orbitals and half-filled e_g metal ones, which stabilize a parallel configuration of magnetic moments.

In $Cs_3Fe_2F_9$, the value of the exchange parameter is of the same order (we consider the normalized energy $E = JS^2$) despite the significantly longer metal-metal distance within the (Fe_2F_9) unit. This may result from the competing effect of direct interactions between half-filled t_{2g} orbitals which stabilize an antiferromagnetic configuration. Indeed, we show that in D_{3h} symmetry, A_1 magnetic orbitals, pointing along the C_3 axis, present a strong direct overlap. This example demonstrates, if necessary, that the exchange constants are not uniquely related to the metal-metal distances. It is clear that we must be careful when comparing systems with different symmetries or electronic features.

Another factor emphasized in the present paper deals with the influence of the orbital contributions to the exchange process. These are worthwhile when considering single ions with an orbital triplet ground-state. Then, the resolution of the Hamiltonian becomes tricky even for a binuclear unit, and the eigenvalues are only available by computer calculation. Obviously, within the limit of strong local distortions stabilizing an orbital singlet, the solutions may be directly compared to those of the Heisenberg model. This is the case for $Cs_3V_2O_2F_7$ and CsV_2O_5 which both may be described on the basis of exchange-coupled spin pairs. This results from the significant overlap of d_{xy} orbitals with a single electron in the plane perpendicular to the V=O bond.

Conversely, the orbital component appears to be the dominant factor in the exchange process, when describing the magnetic behavior of $Cs_3Ti_2Cl_9$ and WV_2O_6. It provides, in particular for the former, an explanation on the origin of the strong magnetic anisotropy. Obviously, a comparison of the exchange parameters with previous systems is questionable, since several mechanisms are involved in the splitting between spin-states. They may even lead to some ambiguity about the sign of the interaction, since a spin-singlet may be observed experimentally while the true coupling is ferromagnetic.

Finally, we have shown that intercluster interactions have in some cases to be considered. Theoretically, these should be completely masked for antiferromagnetic species with a $S=0$ ground-state, well separated from the upper levels. It is clear that their influence may only be available at low temperature, whereas only the $S=0$ ground-state is thermally populated. In fact, it appears that, due to excited states, a 3d long range ordering may occur if the ratio zj/J (J and j being intra- and inter-cluster couplings) exceeds a critical value, even if it cannot be detected by susceptibility measurements. Conversely, for ferro- or ferrimagnetic species, the intercluster interactions represent a first order contribution. In the limit of strongly coupled species described at low temperature by a S molecular spin-state, the system reduces to a 3d network where only j exchange couplings are relevant.

6 References

1. Willett RD, Gatteschi D, Kahn O (1985) Magneto structural correlations in exchange coupled systems, NATO ASI Series, vol 140, Reidel, Dordrecht
2. See for instance Sinn E (1970) Coord Chem Rev 5:313
3. Darriet J, Dance JM, Tressaud A (1984) J Sol State Chem 54:29
4. Longo JM and Kafalas JA (1968) Mat Res Bull 3:687
5. Crawford WH, Richardson HW, Wasson JR, Hogson DJ, Hatfield WE (1976) Inorg Chem 15:2107
6. Bertrand JA, Ginsberg AP, Kirkwood RI, Martin RL, Sherwood RC (1971) Inorg Chem 10:240
7. Bonner JC, Kobayashi H, Tsujikawa I, Nakamura Y, Friedberg SA (1975) J Chem Phys 63:19
8. Drillon M, Darriet J, Georges R (1977) J Phys Chem Sol 38:411
9. Darriet J, Lozano L, Tressaud A (1979) Sol State Comm 32:493
10. Charlot MF, Kahn O, Drillon M (1982) Chem Phys 70:177
11. Leuenberger B, Briat B, Canit JC, Furrer A, Fischer P, Güdel HU (1986) Inorg Chem 25:2930
12. Julve M, Verdaguer M, Kahn O, Gleizes A, Philoche-Levisalles M (1983) Inorg Chem 22:368
13. Hodgson DJ, see Ref. [1], p 497
14. Kahn O, Galy J, Journaux Y, Jaud J, Morgenstern-Badarau I (1982) J Am Chem Soc 104:2165
15. Goodenough JB (1958) J Phys Chem Solids 6:287
16. Kanamori JJ (1959) J Phys Chem Solids 10:87
17. Forster LS, Ballhausen CJ (1962) Acta Chem Scand 16:1385
18. Ginsberg AP (1971) Inorg Chim Acta Rev 5:45
19. Van Kalkeren G, Schmidt WW, Block R (1979) Physica 97B:315
20. Drillon M (1977) Sol State Comm 21:425
21. Dzialoshinski I (1958) Phys Chem Solids 4:241
22. Stevens KWH (1976) Phys. Rep. 24:1
23. Dance IG (1973) Inorg Chem 12:2743
24. Hay PJ, Thibeault JC, Hoffmann R (1975) J Am Chem Soc 97:4884
25. Kahn O, Briat B (1976) J Chem Soc Faraday II 72:268
26. Anderson PW (1959) Phys Rev 115:2
27. Elliot RJ, Thorpe MF (1968) J Appl Phys 39:802
28. Wolf WP (1971) J Phys C32:26
29. Levy P (1969) Phys Rev 177:509
30. Drillon M, Georges R (1981) Phys Rev B24:1278
31. Drillon M, Georges R (1982) Phys Rev B26:3882
32. Leuenberger B, Gudel HU (1984) Mol Phys 51:1
33. De Loth P, Cassoux P, Daudey JP, Malrieu JP (1981) J Am Chem Soc 103:4007
34. Torrance JB, Oostra S, Nazzal A (1987) Synt Met 19:708
35. Katz L, Ward R (1964) Inorg Chem 3:205
36. International Tables for X Ray Crystallography (1959) vol II, p 342, Kynoch Press, Birmingham
37. Dance JM, Kerkouri N, Tressaud A (1979) Mat Res Bull 14:869
38. Dance JM, Kerkouri N, Soubeyroux JL, Darriet J, Tressaud A (1982) Mat Letters 1:49
39. Dance JM, Darriet J, Tressaud A, Hagenmuller P (1984) Z Anorg Allg Chem 508:93
40. Pickart SJ, Alperin HA (1968) J Appl Phys 39:1332; (1971) 42:1617
41. Gudel HU, Hauser U, Furrer A (1979) Inorg Chem 18:2730
42. Gudel HU, Furrer A (1977) Phys Rev Lett 39:657
43. Ferguson J, Gudel HU, Poza H (1973) Austr J Chem 26:513
44. Pausewang G, (1971) Z Anorg Allg Chem 381:189
45. Walterson K (1978) Crystal Struct Comm 7:507
46. Leuenberger B, Briat B, Canit JC, Furrer A, Fischer P, Gudel H (1986) Inorg Chem 25:2930
47. Meyer G, Schonemud A (1980) Mat Res Bull 15:89

48. Darriet J, Bonjour E, Beltran D, Drillon M (1984) J Magn Magn Mat 44:287
49. Dance JM, Mur J, Darriet J, Hagenmuller P, Massa W, Kummer S, Babel D (1986) J Sol State Chem 63:446
50. Darriet J, Georges R (1982) C.R. Acad Sc 295:347
51. Briat B, Kahn O, Morgenstern Badarau I, Rivoal JC (1981) Inorg Chem 20:4183
52. Callaghan A, Moeller CW, Ward R (1966) Inorg Chem 5:1572
53. Darriet J, Drillon M, Villeneuve G, Hagenmuller P (1976) J Sol State Chem 19:213
54. Drillon M (1977) Sol State Comm 21:425
55. Darriet J, Soubeyroux JL, Murani AP (1983) J Phys Chem Solids 44:269 (1983) J Magn Magn Mat 31:605
56. Denohue PC, Katz L, Ward R (1965) Inorg Chem 4:306
57. Drillon M, Darriet J, Hagenmuller P, Georges R (1980) J Phys Chem Solids 41:507
58. Gibb TC, Greatrex R, Greenwood NN, Kaspi P (1973) J Chem Soc Trans 1253
59. Drillon M, Darriet J, Georges R (1977) J Phys Chem Solids 38:411
60. Greatrex R, Greenwood NN (1980) J Sol State Chem 31:281
61. Bayer G (1960) J Am Ceram Soc 43:495
62. Drillon M, Padel L, Bernier JC (1979) Physica 97B:380 (1980) J Magn Magn Mat 15:317
63. Montmory MC, Newnham R (1968) Sol State Comm 6:323
64. Kahn O, Briat B, Galy J (1977) J Chem Soc Faraday II, 1453
65. Shrivastava KN, Jaccarino V (1976) Phys Rev B13:299
66. Drillon M, Padel L, Bernier JC (1980) Physica 100B:343
67. Pourroy G, Drillon M, Padel L, Bernier JC (1983) Physica 123B:21
68. Pourroy G, Drillon M (1983) Physica 123B:16
69. Mumme WG, Watts JA (1971) J Sol State Chem 3:319
70. Waltersson K, Forslund B (1977) Acta Cyst B33:789
71. Johnston DC, Johnston JW, Goshorn DP, Jacobson AJ (1987) Phys Rev B35:219
72. Shoemaker GL, Anderson JB, Kostiner E (1977) Acta Cryst B33:2569
73. Delhaes P, Drillon M (eds) (1987) Organic and inorganic low dimensional crystalline materials, NATO ASI Series, vol 168, Plenum, New York-London, p 421
74. Boukhari A, Moqine A, Flandrois S (1986) Mat Res Bull 21:395
75. Holloway JH, Peacock RD, Small RWH (1964) J Chem Soc 644
76. Mitchell SJ, Holloway JH (1971) J Chem Soc A 2789
77. Norell BK, Zalkin A, Tressaud A, Bartlett N (1973) Inorg. Chem 12:2640
78. Holloway JH, Rao PR, Bartlett N (1965) Chem Comm 393
79. Bartlett N, Rao PR (1965) Chem Comm 252
80. Bartlett N, Lohmann DH (1964) J Chem Soc 619
81. Darriet J, Soubeyroux JL, Touhara H, Tressaud A, Hagenmuller P (1982) Mat Res Bull 17:315
82. Drillon M, Darriet J, Georges R (1977) J Phys Chem Solids 38:411
83. Darriet J, Xu Q, Tressaud A, Hagenmuller P (1986) Mat Res Bull 21:1351
84. Pei Y, Journaux Y, Kahn O, Dei A, Gatteschi D (1986) J Chem Soc Chem Comm 1300

Stereochemical Aspects Associated with the Redox Behaviour of Heterometal Carbonyl Clusters

P. Zanello

Dipartimento di Chimica dell'Università di Siena, Piano dei Mantellini, 44-53100 Siena, Italy

Electrochemical techniques provide a clear picture of the ability of metal cluster compounds to undergo electron transfer processes, not only because they quantify the relative ease of such redox changes through the relevant standard electrode potential values, but also allow one to perceive the occurrence of chemical and/or stereochemical rearrangements within the starting molecular framework following these redox steps. An updated survey of the electrochemical behaviour of a wide series of heterometal carbonyl clusters of increasing nuclearity is presented. As much as possible, the stereochemical provisions, drawn on the basis of the electrochemical results, are supported by solid-state X-ray crystallographic data. When available, the correlation between *experimental* electrochemical parameters and *theoretical* molecular orbital analyses of the electron addition/removal processes is discussed.

1 Introduction 103
2 Trimetallic Assemblies 105
   2.1 Linear Frames 105
   2.2 Bent Frames 111
   2.3 Triangular Frames Without Triply-Bridging Capping Ligands 114
   2.4 Triangular Frames with One Triply-Bridging Capping Ligand 117
      2.4.1 Carbonyl Capping Unit 117
      2.4.2. Alkylidyne Capping Unit 119
      2.4.3 Phosphinidene Capping Unit 122
      2.4.4 Nitrosyl Capping Unit 124
      2.4.5 Sulfur Capping Unit 125
      2.4.6 Selenium Capping Unit 127
      2.4.7 Alkyne Capping Unit 128
   2.5 Triangular Frames Capped on Opposite Sides by Two Triply-Bridging Ligands .. 129
      2.5.1 Carbonyl Capping Units 129
   2.6 Trimetal Compounds Without Metal-Metal Bonds 132
3 Tetrametallic Assemblies 134
   3.1 Triangular Trimetallic Frames Connected to a Fourth Metal by Nonmetallic Bridge 135
   3.2 Spiked-Triangular Frames 138
   3.3 Planar Frames 141
      3.3.1 Square Planar Cores 141
      3.3.2 Triangulated Rhomboidal Cores 143
      3.3.3 Trigonal Planar Cores 148
   3.4 Butterfly Frames 149
   3.5 Tetrahedral Frames 155

4 Pentametallic Assemblies 166
   4.1 Spiked Triangulated Frames... 166
   4.2 Planar Frames 168
   4.3 Tetrahedral-like Frames 169
   4.4 Trigonal Bipyramidal Frames.. 171
   4.5 Star-like Frames 173

5 Hexametallic Assemblies... 174
   5.1 Raft-like Frames... 174
   5.2 Butterfly-Based Frames... 177
   5.3 Capped-Tetrahedron Frames 182
   5.4 Metallo-Ligated Square Pyramidal Frames 184
   5.5 Chair-like Frames.. 184
   5.6 Octahedral Frames.. 186

6 Heptametallic Assemblies 188
   6.1 Bow-Tie Frames... 188
   6.2 Uncommon Frames 192

7 Octametallic Assemblies 192
   7.1 Bicapped Octahedral Frames 192
   7.2 Metallo-Ligated Tricapped Tetrahedral Frames......................... 195
   7.3 Rhombohedral Frames.. 195

8 Decametallic Assemblies... 199
   8.1 Capped Square-Antiprismatic Frames 199

9 Undecametallic Assemblies... 200
   9.1 Pentagonal-Antiprismatic Frames...................................... 202

10 Duodecametallic Assemblies.. 202
   10.1 Tetracapped Tetragonal-Antiprismatic (or Tetracapped
        Triangulated-Dodecahedral) Frames................................... 202

11 Tridecametallic Assemblies.. 204
   11.1 Bicapped Pentagonal-Antiprismatic (Icosahedral) Frames 204

12 Pentadecametallic Assemblies.. 206
   12.1 Metallo-Ligated Icosahedral Frames................................... 206

13 Eneicosametallic Assemblies... 208
   13.1 21-Metal Atom Frames... 208

14 Summary 209

15 References.. 210

1 Introduction

The chemistry of heteronuclear cluster compounds is significantly less developed than that of homonuclear clusters. Since 1980, rational synthetic methods [1] and effective characterizational techniques have been developed [2–6]. Much of the interest provoked by these molecules stems from the fact that they have interesting chemical properties and potentially unique catalytic properties arising from the synergic effects of polar metal-metal bonds [6].

As far as the redox ability of clusters is concerned, two contrasting arguments are often invoked: (i) clusters act as electron sponges, which means that they have a high ability to add/lose electrons without framework destruction; (ii) the LUMO of metal-clusters is metal-metal antibonding, which means that addition of electrons tends to cause framework rearrangement. It is evident that generalizations cannot be drawn if the characteristics of their HOMO/LUMO levels are not known [7–13]. In connection with this, we have examined the stereodynamics of the redox properties of different classes of cluster compounds. Relationships between electrochemically induced electron-transfer processes and concomitant structural changes have been well established [13, 14]. These are mainly based on the rate of electron exchange between the electrode and the complex. In fact, a fast electron transfer (defined as *electrochemically reversible*) is indicative of a low reorganizational barrier experienced by the starting molecular assembly, so that one can foresee that geometrical changes will not be significant. A very slow electron transfer (defined as *electrochemically irreversible*) is indicative of a very high barrier to the structural change exhibited by the complex, so that it is likely that the electron transfer promotes such large stereochemical effects to cause destruction of the metal framework. An electron transfer neither too fast nor too slow (defined as *electrochemically quasireversible*) preludes to significant geometrical rearrangements within the coordination sphere of the redox-active site, but without breakage of the starting molecular frame.

One of the most simple methods for evaluating the extent of electrochemical reversibility is through cyclic voltammetry. As shown in Fig. 1, a totally reversible one-electron step is characterized by a peak current ratio (backward/forward) equal to one and a peak-to-peak separation, ΔE_p, of 59 mV; a quasireversible one-electron step again displays a peak current ratio of one, but a peak-to-peak separation higher than 59 mV; an irreversible step is indicated by the lack of the backward response [really some misunderstanding can occur in this case, in that an electrochemical process can appear as irreversible not because of its very low rate, but because of the presence of coupled fast chemical reactions, which consume the primarily electrogenerated species responsible for the backward response. This last mechanism is very informative about the redox pathway (much more than the occurrence of a simply slow electron exchange), although it is commonly defined as irreversible in organometallic electrochemistry].

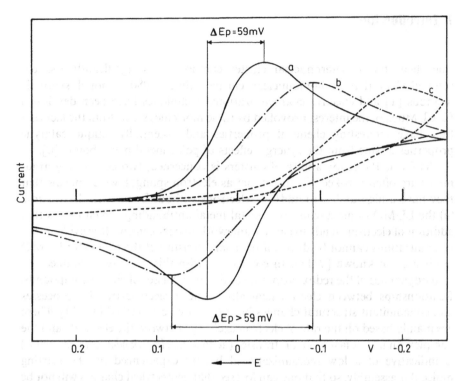

Fig. 1a–c. Schematic representation of the cyclic voltammetric response exhibited by the system Ox + e → Red, having a formal electrode potential $E^{o\prime} = 0.00$ V, which ideally proceeds through: **a)** a "reversible" electron transfer; **b)** a "quasireversible" electron transfer; **c)** an "irreversible" electron transfer

Finally the presence of homogeneous chemical complications coupled to the heterogeneous electron transfer is diagnosed by the ratio ip(backward)/ip(forward) which is lower (or less frequently higher) than one [15].

Having reviewed the electrochemistry of homo- [10] and heterometal-sulfur clusters [11], we have directed our attention to metal-carbonyl clusters. The stereochemical effects associated with the redox behaviour of homonuclear complexes have been recently reported [13]. In this review, those of heteronuclear compounds are discussed. Cluster compounds are classified according to their nuclearity as well as their geometrical assembly. Main-group elements at the borderline between metals and non-metals are treated as metallic elements. Occasionally, polynuclear compounds having no metal-metal bonds (and, as such, are not classifiable as clusters) are also included.

Throughout the paper, unless otherwise specified, the electrode potentials are referred to the saturated calomel electrode.

2 Trimetallic Assemblies

Scheme 1 summarizes the geometrical structures of the trinuclear clusters which have been electrochemically studied. The label of each assembly indicates the reference section.

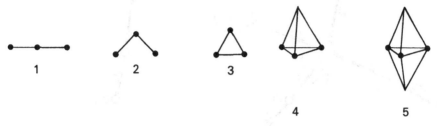

Scheme 1

2.1 Linear Frames

A number of heterotrimetallic clusters having a linear M-Hg-M assembly have been studied. Figure 2 illustrates the geometries of Hg[Co(CO)$_4$]$_2$ [16] and Hg[Mn(CO)$_5$]$_2$ [17], which are typical.

In the cobalt complex, each cobalt atom has a trigonal bipyramidal coordination geometry, whereas, in the manganese complex, each manganese atom is octahedrally coordinated. In both cases, the equatorial carbonyl groups are bent towards the central mercury atom. The structures of the related compounds

Table 1. Redox potentials (V) for the irreversible two-electron reduction of the linear M–Hg–M clusters

Complex	Ep	Solvent	Reference
Hg[Mn(CO)$_5$]$_2$	−0.34	DME[a]	21
Hg[Mn(CO)$_5$]$_2$	−0.74	PC[b]	25
Hg[Co(CO)$_4$]$_2$	−0.20	THF	23
Hg[Co(CO)$_4$]$_2$	−0.35	PC[b]	25
Hg[Co(CO)$_3${P(OPh)$_3$}]$_2$	−0.32	THF	23
Hg[Co(CO)$_3${P(OEt)$_3$}]$_2$	−0.96	THF	23
Hg[Co(CO)$_3$(PPh$_3$)]$_2$	−0.86	THF	23
Hg[Co(CO)$_3$(PMe$_2$Ph)]$_2$	−1.17	THF	23
Hg[Co(CO)$_3$(PMe$_3$)]$_2$	−1.45	THF	23
Hg[Co(CO)$_3$(PEt$_3$)]$_2$	−1.42	THF	23
Hg[Fe(CO)$_3$(NO)]$_2$	−0.20	MeCN	22
Hg[Fe(CO)$_2$(C$_5$H$_5$)]$_2$	−1.24	DME[a]	21
Hg[Cr(CO)$_3$(C$_5$H$_5$)]$_2$	−0.54	DME[a]	21
Hg[Mo(CO)$_3$(C$_5$H$_5$)]$_2$	−0.54	DME[a]	21

[a]DME = Dimethoxyethane; [b]PC = Propylene carbonate

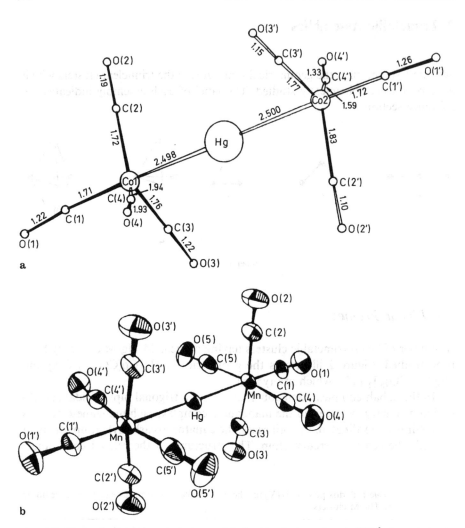

Fig. 2a, b. Molecular structure of: a) Hg[Co(CO)$_4$]$_2$; Hg–Co bond length, 2.50 Å (from Ref. 16); b) Hg[Mn(CO)$_5$]$_2$; Hg–Mn bond length, 2.61 Å [from Ref. 17]

[Hg{Fe(CO)$_4$}$_2$]$^{2-}$ [18, 19] and Hg[Fe(CO)$_2$(NO)(PEt$_3$)]$_2$ [20] have also been reported.

Generally, such clusters undergo irreversible cathodic reduction according to [21–25]:

$$M-Hg-M + 2e \rightarrow 2M^- + Hg$$

The electrode potentials for these reactions are summarized in Table 1. Because of the framework-destroying effects of such redox changes, they will not be discussed in detail. It is noteworthy that for the Hg[Co(CO)$_3$(L)]$_2$ series, the addition of electrons becomes considerably more difficult when the carbonyl groups are substituted by more electron-donating substituents.

Linear trimetallic clusters with the M–Au–M frame, e.g. $[Au\{Co(CO)_4\}_2]^-$ [26], have the geometry illustrated in Fig. 3.

As in the case of the isoelectronic $Hg[Co(CO)_4]_2$, each cobalt atom possesses a trigonal-bipyramidal geometry, with the equatorial carbonyl groups bent towards the central gold atom.

Those clusters studied electrochemically (namely, $[Au\{M(CO)_3(C_5H_5)\}_2]^-$ (M = Cr, Mo, W)) [24, 25] display no ability to support redox changes, and undergo irreversible one-electron reduction according to the following equation:

$$[M-Au-M]^- + e \rightarrow 2M^- + Au$$

As shown in Table 2, the relevant redox potentials (notably more negative than those of the M–Hg–M series) are indicative of an increased difficulty to add electrons.

Another family of linear trimetallic assemblies possess the frame M–Pt–M. Their structure can be representatively illustrated through those of $Pt(py)_2[Co(CO)_4]_2$ and $Pt(py)_2[Mn(CO)_5]_2$ shown in Fig. 4 [27]. Apart from the square-planar geometry of the central platinum atom, the geometrical assemblies of the lateral metal atoms parallel to those illustrated in Fig. 2, for the

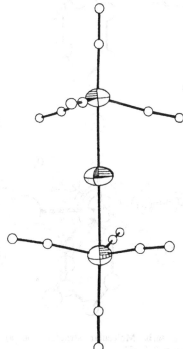

Fig. 3. Perspective view of the monoanion $[Au\{Co(CO)_4\}_2]^-$. Au–Co mean bond length, 2.51 Å [from Ref. 26]

Table 2. Electrode potential values (V) for the irreversible one-electron reduction of the linear species [M–Au–M]⁻ in Propylene carbonate solution [25]

Complex	Ep
[Au{Cr(CO)$_3$(C$_5$H$_5$)}$_2$]⁻	−2.02
[Au{Mo(CO)$_3$(C$_5$H$_5$)}$_2$]⁻	−2.05
[Au{W(CO)$_3$(C$_5$H$_5$)}$_2$]⁻	−2.10

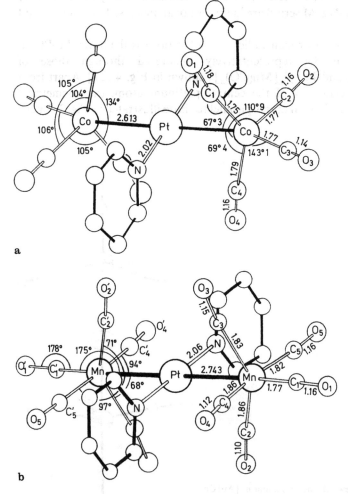

Fig. 4a, b. Molecular structure of: **a)** Pt(C$_5$H$_5$N)$_2$[Co(CO)$_4$]$_2$; **b)** Pt(C$_5$H$_5$N)$_2$[Mn(CO)$_5$]$_2$ [from Ref. 27].

Table 3. Redox potentials (V) for the irreversible one-electron reduction of the linear M–Pt–M clusters in Propylene carbonate solution [25]

Complex	Ep
Pt(cyclo-C$_6$H$_{11}$NC)$_2$[Co(CO)$_4$]$_2$	− 1.28
Pt(cyclo-C$_6$H$_{11}$NC)$_2$[Mn(CO)$_5$]$_2$	− 1.60
Pt(t-BuNC)$_2$[Fe(CO)$_3$(NO)]$_2$	− 1.18
Pt(t-BuNC)$_2$[Cr(CO)$_3$(C$_5$H$_5$)]$_2$	− 1.60
Pt(t-BuNC)$_2$[Mo(CO)$_3$(C$_5$H$_5$)]$_2$	− 1.62
Pt(t-BuNC)$_2$[W(CO)$_3$(C$_5$H$_5$)]$_2$	− 1.75

corresponding M–Hg–M complexes. Similar structures have been established by X-ray techniques for Pt(CO)$_2$[Mn(CO)$_5$]$_2$ [28] and Pt(CO)$_2$[Re(CO)$_5$]$_2$ [29].

Like the preceding linear trimetallic compounds, these platinum clusters seem unable to add/lose electrons without molecular destruction; in fact, in a series of related species (see Table 3), the only redox change is the irreversible one-electron reduction [24, 25]:

M–Pt(L)$_2$–M + e → [Pt(L)$_2$M]· + M$^-$

Fig. 5 shows the linear geometry of Pb[Mn(CO)$_2$(C$_5$H$_5$)]$_2$ [30]. The Pb–Mn bond length of 2.46 Å suggests that the Mn/Pb/Mn bonding network is best described by the cumulene-like double bonding system Mn=Pb=Mn.

Despite this unsaturation, such complexes apparently display the poor redox ability typical of the linear trimetallic frame, in that they undergo reduction to only transient congeners, as well as partaking in declustering oxidations (see Table 4) [30].

A recent report on the similar cumulenium species [As{Mn(CO)$_2$(C$_5$H$_4$Me)}$_2$]$^+$ [31] and [As{Mn(CO)$_2$(C$_5$Me$_5$)}$_2$]$^+$ [32, 33], the structure of which is illustrated in Fig. 6, throws some light on their propensity to add electrons.

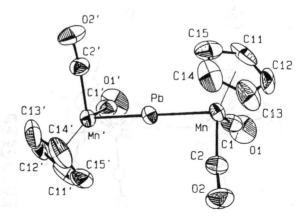

Fig. 5. Perspective view of Pb[Mn(CO)$_2$(C$_5$H$_5$)]$_2$ [from Ref. 30].

Table 4. Redox potentials (V) for the oxidation and reduction processes exhibited by the Mn=M=Mn (M=Pb,=Sn) clusters in Tetrahydrofuran solution [30]

Complex	E° reduction[a,b]	Ep oxidation[a]
Pb[Mn(CO)$_2$(C$_5$H$_5$)]$_2$	−1.22	+1.26
Sn[Mn(CO)$_2$(C$_5$Me$_5$)]$_2$	−1.54	+0.16

[a]Unknown number of electrons involved in the redox change;
[b]quasi-reversible process complicated by decomposition reactions

In Dichloromethane solution, [As{Mn(CO)$_2$(C$_5$H$_4$Me)}$_2$]$^+$ undergoes a one-electron reduction (Ep = −0.32 V), complicated by the fast generation of a species which, in turn, undergoes a reduction at more negative potential values (Ep = −1.28 V). There is good evidence that such a reaction product is simply the dimer [(C$_5$H$_4$Me)(CO)$_2$Mn]$_2$AsAs[Mn(CO)$_2$(C$_5$H$_4$Me)]$_2$, produced by the very reactive radical [As{Mn(CO)$_2$(C$_5$H$_4$Me)}$_2$]·, instantaneously generated in correspondence to the first reduction process. Interestingly, under the same experimental conditions, [As{Mn(CO)$_2$(C$_5$Me$_5$)}$_2$]$^+$ undergoes a substantially reversible one-electron reduction (E$^{°'}$ = −0.49 V). It has been argued that, in this case, the more encumbering pentamethylcyclopentadienyl ligand prevents the fast dimerization of the radical [As{Mn(CO)$_2$(C$_5$Me$_5$)}$_2$]· [31].

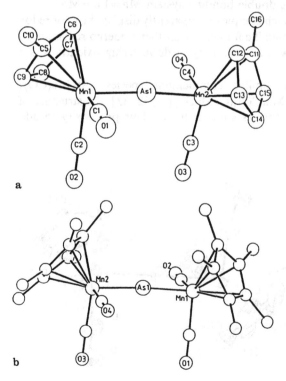

Fig. 6a, b. Molecular structure of: a) [As{Mn(CO)$_2$(C$_5$H$_4$Me)}$_2$]$^+$. Averaged As–Mn bond length, 2.12 Å; Mn–As–Mn, 175.3° (from Ref. 31); b) [As{Mn(CO)$_2$(C$_5$Me$_5$)}$_2$]$^+$. Averaged As–Mn bond length, 2.14 Å; Mn–As–Mn, 176.3° [from Ref. 32]

2.2 Bent Frames

The first example of bent trimetallic frame of known redox chemistry is Te[Mn(CO)$_2$(C$_5$Me$_5$)]$_2$ [2]. Its molecular structure is shown in Fig. 7 [34].

As in the case of the related, but linear, species M[Mn(CO)$_2$(C$_5$X$_5$)]$_2$ (M = Pb, Sn), the Mn–Te bond distance of 2.46 Å suggests the actual frame is better represented as Mn=Te=Mn. Nevertheless, in contrast with the latter complexes, the extended π-electron delocalization now makes the complex able to accept reversibly two electrons through two separated one-electron reductions. In N,N-Dimethylformamide solution, such steps occur at the formal electrode potential values of −0.82 V and −1.72 V, respectively [2]. The chemical reversibility of the [Te{Mn(CO)$_2$(C$_5$Me$_5$)}$_2$]$^{0/-}$ redox change testifies to the stability of the monoanion [Te{Mn(CO)$_2$(C$_5$Me$_5$)}]$^-$, which can be obtained using chemical reducing agents, such as cobaltocene or sodium naphtalenide. On the other hand, the relevant electrochemical reversibility allows one to foresee that the structure of the monoanion remains substantially unaltered.

The electrochemistry of a series of bent trimetallic clusters of general formula Ph$_2$SnMM′ [M = M′ = Fe(CO)$_2$(C$_5$H$_5$); M = M′ = Mn(CO)$_5$; M = M′ = Mo(CO)$_3$(C$_5$H$_5$); M = Mn(CO)$_5$, M′ = Fe(CO)$_2$(C$_5$H$_5$)] has been very recently reported [35]. Crystallographic details concerning Ph$_2$Sn[Mn(CO)$_5$]$_2$ [36] and Ph$_2$Sn[Mo(CO)$_3$(C$_5$H$_5$)]$_2$ [37] have been presented, but no diagrams of the relevant molecular structures are available. In this connection, Fig. 8 illustrates the structure of the related molecules Me$_2$Sn[Fe(CO)$_2$(C$_5$H$_5$)]$_2$ [38] and Cl$_2$Sn[Mn(CO)$_5$]$_2$ [39]. The tetrahedrally distorted coordination of the tin atom imparts a bent frame to these complexes.

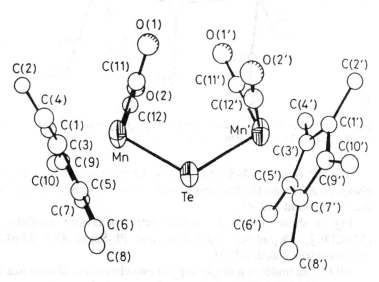

Fig. 7. Perspective view of Te[Mn(CO)$_2$(C$_5$H$_5$)]$_2$; Mn–Te–Mn, 123.8° [from Ref. 34]

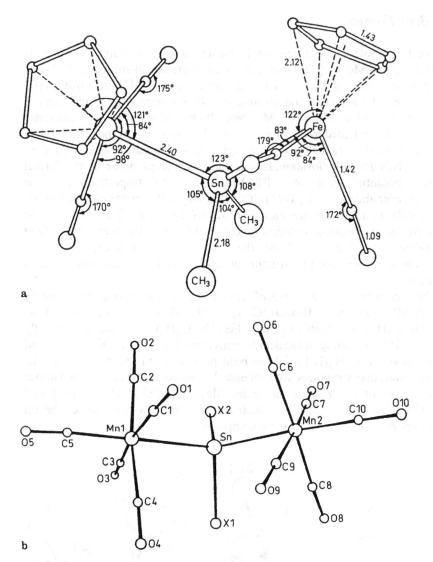

Fig. 8a, b. Perspective view of: **a)** Me$_2$Sn[Fe(CO)$_2$(C$_5$H$_5$)]$_2$ (from Ref. 38); **b)** Cl$_2$Sn[Mn(CO)$_5$]$_2$; Sn–Mn, 2.63 Å; Mn–Sn–Mn, 126.2° [from Ref. 39]

The fact that the M–Sn–M angle is always larger and the R–Sn–R angle is always smaller than the tetrahedral one, has been attributed to the s character of the Sn–M bond [40, 41].

Fig. 9 shows the cyclic voltammetric behaviour exhibited by Ph$_2$Sn[Mn(CO)$_5$]$_2$, Ph$_2$Sn[Fe(CO)$_2$(C$_5$H$_5$)]$_2$ and Ph$_2$Sn[Mn(CO)$_5$][Fe(CO)$_2$(C$_5$H$_5$)], in Acetonitrile solution [35].

All species undergo a single-stepped two-electron oxidation (see Table 5), the irreversibility of which testifies to the framework destruction operated by such

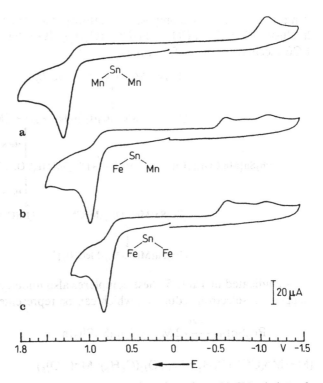

Fig. 9a–c. Cyclic voltammograms recorded at a glassy-carbon electrode on MeCN solutions of: a) Ph$_2$Sn[Mn(CO)$_5$]$_2$; b) Ph$_2$Sn[Mn(CO)$_5$][Fe(CO)$_2$(C$_5$H$_5$)]; c) Ph$_2$Sn[Fe(CO)$_2$(C$_5$H$_5$)]$_2$. Scan rate 0.2 Vs^{-1}. Ag/AgCl reference electrode [from Ref. 35 (Supplementary Material)]

electron removals. IR spectroelectrochemical measurements allowed the relevant redox pathways to be ascertained. For the hetero-dimetal complexes {M = Mo(CO)$_3$(C$_5$H$_5$), Mn(CO)$_5$, Fe(CO)$_2$(C$_5$H$_5$)} they may be represented by the following general equations [35]:

$$Ph_2SnM_2 \xrightarrow[MeCN]{-2e} [MSnPh_2(MeCN)_2]^+ + [M(MeCN)]^+$$

Table 5. Peak potential values (V) for the two-electron oxidation and two-electron reduction of the series Ph$_2$SnM$_2$, in Acetonitrile solution [35]

M$_2$	Ep$_{ox}^a$	Ep$_{red}^a$
[Mo(CO)$_3$(C$_5$H$_5$)]$_2$	+1.07	−1.54
[Mn(CO)$_5$]$_2$	+1.23	−1.87
[Fe(CO)$_2$(C$_5$H$_5$)]$_2$	+0.76	−2.04
[Mn(CO)$_5$][Fe(CO)$_2$(C$_5$H$_5$)]	+1.00	−1.84

[a]Measured at 0.25 Vs^{-1}

A more complicated decomposition pathway is assigned to the hetero-trimetallic cluster Ph$_2$Sn[Mn(CO)$_5$][Fe(CO)$_2$(C$_5$H$_5$)]. It probably proceeds through an ECE mechanism of the following type [35]:

$$[Ph_2Sn\{Fe(CO)_2(C_5H_5)\}(MeCN)_2]^+$$
$$\uparrow -e$$
$$[Ph_2Sn\{Fe(CO)_2(C_5H_5)\}(MeCN)_2]\cdot + [Mn(CO)_5(MeCN)]^+$$
$$\uparrow MeCN$$
$$Ph_2Sn\{Mn(CO)_5\}\{Fe(CO)_2(C_5H_5)\} \xrightarrow{-e} [Ph_2Sn\{Mn(CO)_5\}\{Fe(CO)_2(C_5H_5)\}]^+$$
$$\downarrow MeCN$$
$$[Ph_2Sn\{Mn(CO)_5\}(MeCN)_2]\cdot + [Fe(CO)_2(C_5H_5)(MeCN)]^+$$
$$\downarrow -e$$
$$[Ph_2Sn\{Mn(CO)_5\}(MeCN)_2]^+$$

As indicated in Table 5, these complexes also undergo, irreversibly, a single-stepped two-electron reduction, which can be represented by [35]:

$$Ph_2SnM_2 \xrightarrow{+2e} 2M^- + 1/n\,(SnPh_2)n$$

(M = Mo(CO)$_3$(C$_5$H$_5$); Fe(CO)$_2$(C$_5$H$_5$); Mn(CO)$_5$)

The redox data indicates that both the HOMO and LUMO levels of the present complexes are strongly tin-metal antibonding

2.3 Triangular Frames Without Triply-Bridging Capping Ligands

The present section is devoted to cluster compounds forming a closed trimetallic triangle not clasped by a triply bridging capping ligand.

Figure 10 shows the molecular structure of FeCO$_2$(μ-CO)(CO)$_7$(μ-PPh$_2$)$_2$ [42].

In the FeCo$_2$ core, each Fe-Co bond (averaged length, 2.57 Å) is bridged, by opposite sides with respect to the metallic plane, by one PPh$_2$ ligand, whereas the Co-Co edge (2.64 Å) is asymmetrically bridged by one carbonyl group. The remaining seven terminal carbonyls are bound: two to the iron atom; two and three to the two cobalt atoms, respectively.

As shown in Fig. 11, the cluster undergoes a series of one-electron reductions, the first of which (E°′ = − 0.64 V) only has the features of chemical and electrochemical reversibility. The second cathodic step (Ep = − 1.20 V), likely generating the corresponding dianion, is complicated by successive reactions, which afford a new, unidentified species responsible for the third reduction process (Ep = − 1.45 V) [42].

The cluster also exhibits an irreversible oxidation step at Ep = + 0.75 V.

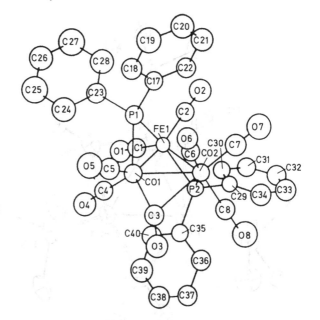

Fig. 10. Perspective view of FeCo$_2$(CO)$_8$(PPh$_2$)$_2$ [from Ref. 42]

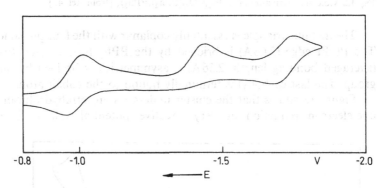

Fig. 11. Cyclic voltammogram exhibited, at a mercury electrode, by FeCo$_2$(CO)$_8$(PPh$_2$)$_2$ in MeCN solution. Scan rate 0.2 Vs^{-1}. Potential values refer to an Ag/AgNO$_3$ reference electrode [from Ref. 42]

Chemical reduction by sodium amalgam (which does not seem the best choice to obtain [FeCo$_2$(μ-CO)(CO)$_7$(PPh$_2$)$_2$]$^-$, because of its much too negative reduction potential) afforded a species not yet characterized, but able to restore the starting precursor upon protonation with H$_3$PO$_4$. Nevertheless, it can be foreseen that the monoanion is stable and possesses a geometry quite similar to that of the neutral parent.

A somewhat related complex is CoPt$_2$(μ-CO)$_2$(CO)(PPh$_3$)$_3$(μ-PPh$_2$), the structure of which is illustrated in Fig. 12 [43].

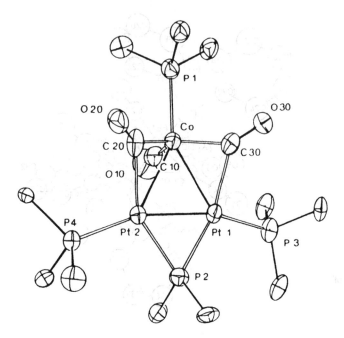

Fig. 12. Molecular structure of $CoPt_2(CO)_3(PPh_2)(PPh_3)_3$ [from Ref. 43]

The metallic triangle is essentially coplanar with the four phosphorus atoms. The Pt–Pt edge (2.66 Å) is bridged by the PPh_2 ligand. Each Pt–Co vector (averaged bonding length, 2.56 Å) is asymmetrically bridged by one carbonyl group. The last carbonyl is terminally bound to the cobalt atom.

Figure 13 shows that the cluster undergoes, in tetrahydrofuran solution, a one-electron reduction at very negative potential values ($E^{o'} = -2.0$ V),

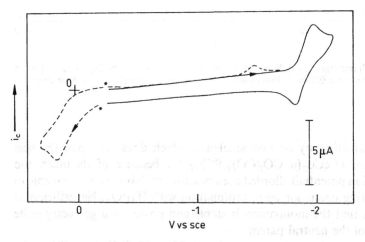

Fig. 13. Cyclic voltammetric behaviour of $CoPt_2(CO)_3(PPh_3)_3(PPh_2)$ in THF solution. Platinum working electrode. Scan rate 0.4 Vs^{-1} [from Ref. 43]

as well as an easy irreversible oxidation (Ep = + 0.14 V). In spite of the apparent chemical reversibility of the reduction step, the monoanion [CoPt$_2$(μ-CO)$_2$(CO)(PPh$_3$)$_3$(μ-PPh$_2$)]$^-$ is not long-lived, decomposing to [Co(CO)$_3$(PPh$_3$)]$^-$ and other unidentified products [43].

The last members of the group in this section are the pure carbonyl molecules Fe$_2$Ru(CO)$_{12}$ and FeRu$_2$(CO)$_{12}$. In fact, regardless of their precise structure, (which can be similar either to that of Fe$_3$(CO)$_{12}$, with one Fe-Fe edge of the trimetallic triangle doubly bridged by two CO groups [44], or to that of Ru$_3$(CO)$_{12}$, with all the carbonyls in terminal positions [45]) they are likely constituted by a closed trimetallic triangle and they do not possess triply bridging capping units. Even if electrochemical details are lacking, it has been briefly reported that electrolytic reduction, at low temperature, affords the corresponding monoanions [Fe$_2$Ru(CO)$_{12}$]$^-$ and [FeRu$_2$(CO)$_{12}$]$^-$, which are very short-lived [46].

2.4 Triangular Frames with One Triply-Bridging Capping Ligand

A series of triangular trimetallic frames with different capping ligands have been studied.

2.4.1 Carbonyl Capping Unit

PtCo$_2$(μ$_3$-CO)(CO)$_6$(PPh$_3$)(CO) and PtCo$_2$(μ$_3$-CO)(CO)$_6$(Ph$_2$PCH$_2$CH$_2$PPh$_2$) are constituted by an heterometallic triangle capped by a carbonyl group. Figure 14 shows the structure of the former cluster [47].

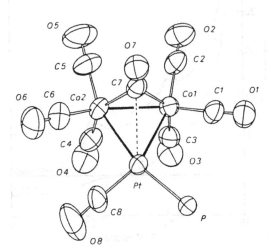

Fig. 14. Perspective view of PtCo$_2$(CO)$_8$(PPh$_3$) [from Ref. 47]

The almost equilateral metallic triangle (Co–Co, 2.51 Å; averaged Pt–Co, 2.52 Å) can be considered asymmetrically capped by one carbonyl carbon atom (C7) (averaged Co–C7, 1.91 Å; Pt–C7, 2.57 Å). Each cobalt atom bears three terminal carbonyls, whereas the platinum atom binds both one terminal carbonyl and the phosphine ligand. A similar structure is assigned to PtCo$_2$(CO)$_7$(Ph$_2$PCH$_2$CH$_2$PPh$_2$), where the CO and PPh$_3$ ligands of the Platinum atom are substituted by the bidentate phosphine Ph$_2$PCH$_2$CH$_2$PPh$_2$ [47]. On the other hand, such an assembly has been confirmed for

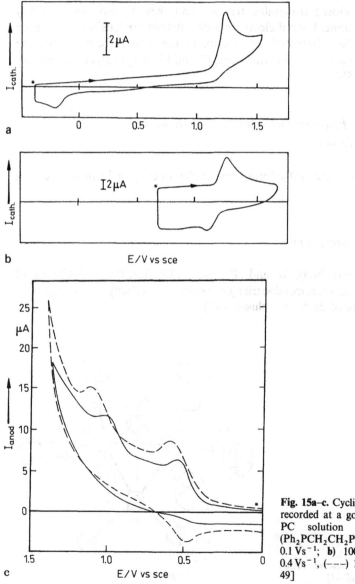

Fig. 15a–c. Cyclic voltammograms recorded at a gold electrode on a PC solution of PtCO$_2$(CO)$_7$(Ph$_2$PCH$_2$CH$_2$PPh$_2$). Scan rate: a) 0.1 Vs^{-1}; b) 100 Vs^{-1}; c) (——) 0.4 Vs^{-1}, (– – –) 1 Vs^{-1} [from Ref. 49]

Table 6. Electrode potential values (V) for the redox changes exhibited by carbonyl-capped PtCo$_2$ clusters in Propylene carbonate solution [49]

Complex	$E^{o/a}_{0/-}$	$E^{o/a}_{+/0}$	$E_{p2+/+}$
PtCo$_2$(CO)$_8$(PPh$_3$)	− 0.75	+ 0.71	+ 0.95
PtCo$_2$(CO)$_7$(Ph$_2$PCH$_2$CH$_2$PPh$_2$)	− 1.20	+ 0.47	+ 0.85

[a] Complicated by subsequent chemical reactions

PtCo$_2$(CO)$_7$(1, 5–Cyclooctadiene) by single crystal X-ray studies [48]. Fig. 15 illustrates the redox propensity of PtCo$_2$(CO)$_7$(Ph$_2$PCH$_2$CH$_2$PPh$_2$) in propylene carbonate solution [49].

The comparison between the responses a and b indicates that the cluster undergoes a (one-electron) reduction followed by decomposition of the initially electrogenerated monoanion. The increase in the scan rate prevents such chemical complications. It can be roughly estimated that the monoanion [PtCo$_2$(CO)$_7$(Ph$_2$PCH$_2$CH$_2$PPh$_2$)]$^-$ has a lifetime of about 50 msec. The two voltammetric profiles shown in Fig. c suggest that chemical complications are associated also with the anodic generation of the monocation [PtCo$_2$(CO)$_7$(Ph$_2$PCH$_2$CH$_2$PPh$_2$)]$^+$, which is relatively more long-lived (about 1 sec) than the monoanion. An analogous pattern is displayed by PtCo$_2$(CO)$_8$(PPh$_3$). Table 6 summarizes the relevant redox potentials.

It is interesting to note how the capping unit in the present triangular PtCo$_2$ assembly does not improve the stability of redox congeners with respect to that of the linear PtCo$_2$ assembly in Pt(RNC)$_2$[Co(CO)$_4$]$_2$ discussed in Sect. 2.1.

2.4.2 Alkylidyne Capping Unit

Detailed electrochemical characterization has been presented for a series of alkylidyne-capped triangular clusters of general formula MCo$_2$(μ$_3$–CR)(CO)$_8$(C$_5$H$_5$), (M = Mo, Cr, W; R = Ph, Me), [50, 51]. As an example, Fig. 16 shows the molecular structure of MoCo$_2$(C–Ph)(CO)$_8$(C$_5$H$_5$) [52].

The isosceles metallic triangle (Mo–Co, 2.68 Å; Co–Co, 2.48 Å) is asymmetrically capped by the carbon phenylidyne atom (Mo–C3, 2.10 Å; Co–C3, 1.93 Å). All the carbonyl groups are terminal.

A similar structure is displayed by WCo$_2$(C–p–Tol)(CO)$_8$(C$_5$H$_5$) [53].

The redox propensity of these species is rather complicated [50, 51]. As shown in Fig. 17, they undergo two consecutive, one-electron reduction processes; the first one is irreversible at low scan rates, whereas the second one exhibits features associated with chemical reversibility.

Either increasing the scan rate or decreasing the temperature make the first step more and more chemically reversible. In addition, cyclic voltammograms run under CO atmosphere, show that the two cathodic steps to become closer each other, coalescing in a single two-electron step for the CrCo$_2$ cluster. All

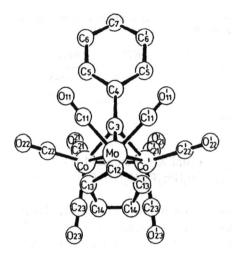

Fig. 16. Perspective view of MoCo$_2$(C-Ph)(CO)$_8$(C$_5$H$_5$) [from Ref. 52]

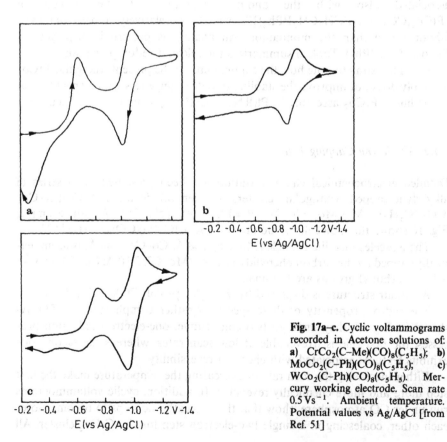

Fig. 17a–c. Cyclic voltammograms recorded in Acetone solutions of: a) CrCo$_2$(C-Me)(CO)$_8$(C$_5$H$_5$); b) MoCo$_2$(C-Ph)(CO)$_8$(C$_5$H$_5$); c) WCo$_2$(C-Ph)(CO)$_8$(C$_5$H$_5$). Mercury working electrode. Scan rate 0.5 Vs^{-1}. Ambient temperature. Potential values vs Ag/AgCl [from Ref. 51]

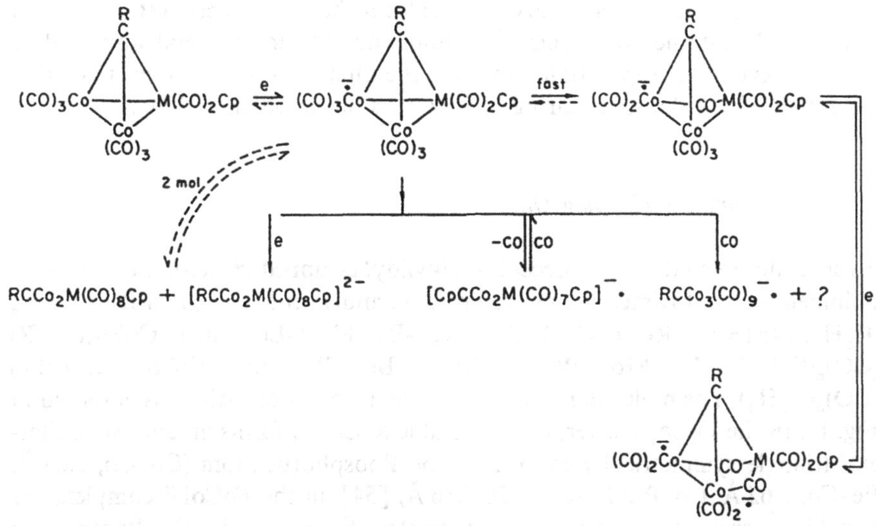

Scheme 2

these data are diagnostic of a series of chemical complications involving the initially electrogenerated monoanion $[MCo_2(C-R)(CO)_8(C_5H_5)]^-$, which can be accounted for by the sequence illustrated in Scheme 2 [51].

For the purposes of the present review, the most interesting aspect is the relative instability of the monoanion, which tends to break a Co–M edge of the starting closed metallic triangle to assume an open triangular geometry. In turn, the subsequent one-electron addition causes breakage of another Co–M edge. In both events, one carbonyl group becomes edge-bridging.

Table 7. Redox potentials (V) for the two reduction processes exhibited by the clusters $MCo_2(C-R)(CO)_8(C_5H_5)$ [50, 51]

M	R	$Ep_{0/-}$	$E°_{-/2-}$	Solvent
Cr	Me	− 0.81	− 1.26	CH_2Cl_2
		− 0.62	− 1.08	Me_2CO
Cr	Ph	− 0.78	− 1.22	CH_2Cl_2
Mo	Me	− 0.92	− 1.23	CH_2Cl_2
		− 1.19	− 1.31	$C_2H_4Cl_2$
Mo	Ph	− 0.84	− 1.16	CH_2Cl_2
		− 0.66	− 0.98	Me_2CO
		− 1.10	− 1.27	$C_2H_4Cl_2$
W	Me	− 0.91	− 1.23	CH_2Cl_2
		− 1.25	− 1.46	$C_2H_4Cl_2$
W	Ph	− 0.89	− 1.19	CH_2Cl_2
		− 0.71	− 1.01	Me_2CO
		− 1.17	− 1.43	$C_2H_4Cl_2$

Table 7 summarizes the redox potentials for the two reduction steps. Since it is assumed that the two electron addition fills the metal-metal antibonding LUMO level, the separation between the first and the second reduction potential reflects the spin-pairing energy. It is of the order of 0.3 eV.

2.4.3 Phosphinidene Capping Unit

Structurally related to the preceding alkylidyne-capped clusters are the phosphinidene-capped trimetallic molecules of formula: (i) MCo$_2$ (μ_3-PR) (CO)$_{9-n}$ (C$_5$H$_5$)$_n$, (M = Fe, Ru; n = 0, 1; R = Me, t-Bu, Ph, NEt$_2$); (ii) FeCoM(μ_3-PR) (CO)$_8$(C$_5$H$_5$), (M = Mo, W; R = Me, t-Bu; Ph); (iii) FeCoNi(μ_3-PBu^t) (CO)$_6$(C$_5$H$_5$). The molecular structure of one member of each class is shown in Fig. 18. In the FeCo$_2$ cluster, the trimetallic assembly forms an almost equilateral triangle symmetrically capped by the Phosphorus atom (Co–Co, 2.63 Å; Fe–Co, 2.62 Å; Co–P 2.12 Å; Fe–P, 2.16 Å) [54]; in the FeCoNi complex, the metallic isosceles triangle is again symmetrically capped by the Phosphorus atom (Co–Ni, 2.51 Å; Ni–Fe, 2.50 Å; Fe–Co, 2.64 Å; Ni–P, 2.10 Å; Co–P = Fe–P, 2.12 Å) [56]; the FeCoMo scalene triangle is asymmetrically capped by the phosphorus atom (Fe–Co, 2.64 Å; Co–Mo, 2.80 Å; Mo–Fe, 2.87 Å; Fe–P, 2.12 Å; Co–P, 2.13 Å; Mo–P, 2.38 Å) [55]. All the carbonyl groups are terminal.

The redox activity of these molecules is a function of the number of heteroatoms. In fact, the MCo$_2$ cores display at least one reversible one-electron reduction, whereas the FeCoM species exhibit only irreversible reduction steps. This means that the MCo$_2$ monoanions are relatively stable, even more long-lived than the similar carbon-capped species. In contrast, the FeCoM monoanions are completely unstable [50, 51]. The relevant redox potentials are summarized in Table 8. The FeCoM complexes are significantly more difficult

Table 8. Electrode potentials (V) and peak-to-peak separation (in millivolts) for the reduction processes exhibited by Phosphinidene trimetallic clusters [50, 51]

Complex	Ep$^{0/-}_0$	ΔEp[a]	Ep$_{-/2-}$	Solvent
RuCo$_2$(P–Ph)(CO)$_9$	−0.88	78	—	C$_2$H$_4$Cl$_2$
FeCo$_2$(P–Ph)(CO)$_9$	−0.83	60	−1.23	C$_2$H$_4$Cl$_2$
FeCo$_2$(P–Ph)(CO)$_8$(C$_5$H$_5$)	−0.58	60	−0.98	CH$_2$Cl$_2$
FeCo$_2$(P–Me)(CO)$_8$(C$_5$H$_5$)	−0.62	120	—	CH$_2$Cl$_2$
FeCo$_2$(P–t–Bu)(CO)$_8$(C$_5$H$_5$)	−0.61	72	—	CH$_2$Cl$_2$
FeCo$_2$(P–NEt$_2$)(CO)$_8$(C$_5$H$_5$)	−0.63	75	—	CH$_2$Cl$_2$
FeCoMo(P–Ph)(CO)$_8$(C$_5$H$_5$)	−1.13[b]	—	—	C$_2$H$_4$Cl$_2$
	−1.13[b]	—	—	CH$_2$Cl$_2$
FeCoMo(P–Me)(CO)$_8$(C$_5$H$_5$)	−1.17[b]	—	—	C$_2$H$_4$Cl$_2$
	−1.17[b]	—	—	CH$_2$Cl$_2$
FeCoMo(P–t–Bu)(CO)$_8$(C$_5$H$_5$)	−1.15[b]	—	—	CH$_2$Cl$_2$
FeCoW(P–Ph)(CO)$_8$(C$_5$H$_5$)	−1.25[b]	—	—	C$_2$H$_4$Cl$_2$
FeCoW(P–Me)(CO)$_8$(C$_5$H$_5$)	−1.43[b]	—	—	C$_2$H$_4$Cl$_2$
FeCoNi(P–t–Bu)(CO)$_6$(C$_5$H$_5$)	−1.19[b]	—	—	CH$_2$Cl$_2$

[a] Measured at 0.02 Vs^{-1}; [b] peak potential value for irreversible process

Redox Behaviour of Heterometal Carbonyl Clusters

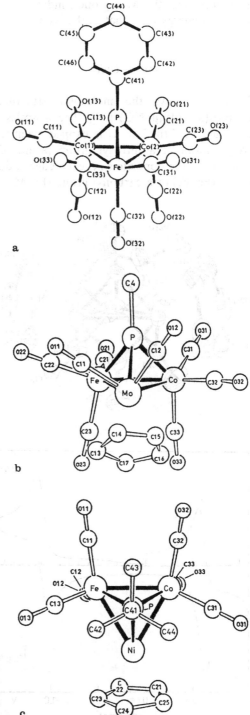

Fig. 18a–c. Perspective view of:
a) FeCo$_2$(P–Ph)(CO)$_9$ [from Ref. 54];
b) FeCoMo(P–Me)(CO)$_8$(C$_5$H$_5$) [from Ref. 55]; c) FeCoNi(P–t–Bu)(CO)$_6$(C$_5$H$_5$) [from Ref. 56]

to reduce than the MCo$_2$ ones, indicating that the LUMO of the former is higher in energy than that of the latter.

2.4.4 Nitrosyl Capping Unit

Figure 19 shows the molecular structure of MnFe$_2$(μ_3–NO)(C$_5$H$_4$Me)(C$_5$H$_5$)$_2$(μ–CO)$_2$(μ–NO) [57]. The isosceles metallic triangle (Fe–Fe, 2.44 Å; averaged Fe–Mn, 2.56 Å) is capped symmetrically by the triply-bridging nitrogen atom of a nitrosyl group (averaged metal-N bond length, 1.89 Å), whereas, at the opposite side, each edge is doubly-bridged by either a nitrosyl group (Fe–Fe) or a carbonyl group (Fe–Mn).

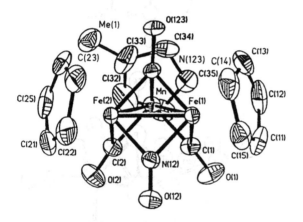

Fig. 19. Molecular structure of MnFe$_2$(μ_3–NO)(C$_5$H$_4$Me)(C$_5$H$_5$)$_2$(CO)$_2$(NO) [from Ref. 57]

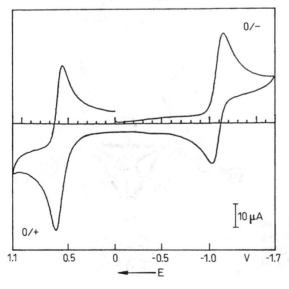

Fig. 20. Cyclic voltammogram recorded at a platinum electrode on a CH$_2$Cl$_2$ solution of MnFe$_2$(μ_3–NO)(C$_5$H$_4$Me)(C$_5$H$_5$)$_2$(CO)$_2$(NO). Scan rate 0.2 Vs^{-1} [from Ref. 57]

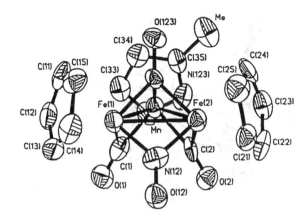

Fig. 21. Perspective view of the monoanion [MnFe$_2$(μ_3-NO)(C$_5$H$_4$Me)(C$_5$H$_5$)$_2$(CO)$_2$(NO)]$^-$ [from Ref. 57]

Figure 20 indicates that, in dichloromethane solution, the cluster is able to add, as well to lose, one electron through chemically reversible, but electrochemically quasireversible, steps ($E^{\circ\prime}_{(+/0)}$ = +0.60 V, ΔE_p = 80 mV; $E^{\circ\prime}_{(0/-)}$ = −1.08 V, ΔE_p = 130 mV) [57].

Chemical reduction by potassium-benzophenone afforded the monoanion [MnFe$_2$(μ_3-NO)(C$_5$H$_4$Me)(C$_5$H$_5$)$_2$(μ-CO)$_2$(μ-NO)]$^-$, the molecular structure of which is illustrated in Fig. 21 [57].

In agreement with the quasireversibility of the relevant redox change, which preludes to a significant geometrical reorganization, the cluster, although maintaining the structural assembly of the neutral parent, undergoes some structural modification. In fact, the Fe–Fe bond length markedly lengthens by 0.16 Å (2.60 Å), whereas the Fe–Mn distances slightly shorten by about 0.01 Å (2.55 Å). In addition, the metal-N$_{(capping)}$ distances become unequal (averaged Fe–N, 1.92 Å; Mn–N, 1.85 Å). This variation suggests that the added electron enters a LUMO level which is antibonding with respect to the Fe/Fe interaction.

In spite of the evidence for the stability of the monocation [MnFe$_2$(μ_3-NO)(C$_5$H$_4$Me)(C$_5$H$_5$)$_2$(CO)$_2$(NO)]$^+$, chemical oxidation by AgPF$_6$ afforded the monocation [MnFe$_2$(μ_3-NH)(C$_5$H$_4$Me)(C$_5$H$_5$)$_2$(μ-CO)$_2$(μ-NO)]$^+$. The transformation of the capping nitrosyl into an imido ligand is attributed to an electrophilic attack of Ag$^+$ on the μ_3-NO unit [57].

2.4.5 Sulfur Capping Unit

FeCo$_2$(μ_3-S)(CO)$_9$ is the first sulfur-capped heterotrimetallic cluster structurally characterized (see Fig. 22) [58].

The almost equilateral metallic triangle (averaged M–M distance, 2.55 Å) is symmetrically triply-bridged by the sulfur atom (averaged M–S, 2.16 Å). All the carbonyls are terminal.

It is interesting to note that this cluster is isoelectronic with the monocation [Co$_3$(μ_3-S)(CO)$_9$]$^+$. Since in the isostructural neutral paramagnetic precursor

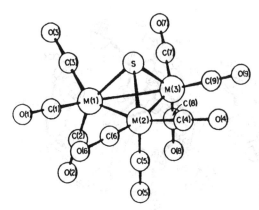

Fig. 22. Molecular structure of FeCo$_2$(S)(CO)$_9$ [from Ref. 58]

Co$_3$(S)(CO)$_9$, the metal-metal distance is on average 0.09 Å longer (2.64 Å) and the Co–S distance is almost unchanged (2.14 Å), it was argued that the extra electron in Co$_3$(S)(CO)$_9$ resides in a trimetal-based antibonding orbital [58]. Indeed, the strict correlation between FeCo$_2$(S)(CO)$_9$ and [Co$_3$(S)(CO)$_9$]$^+$ does not seem so straightforward since the ESR spectrum of the monoanion [FeCo$_2$(S)(CO)$_9$]$^-$, which is equivalent to Co$_3$(S)(CO)$_9$, indicates that the majority of the spin density is localized on the Co–Co edge [59].

The latter conclusion is based on the redox ability of the sulfur capped species MM$_2'$(S)(CO)$_{9-n}$(C$_5$H$_5$)$_n$, MM'M''(S)(CO)$_{9-n}$(C$_5$H$_5$)$_n$(n = 0 – 2). Pertinent X-ray structures are those relevant to FeW$_2$(S)(CO)$_7$(C$_5$H$_5$)$_2$ [60], shown in Fig. 23, and FeCoM(S)(CO)$_7$(PMePrPh)(C$_5$H$_5$)(M = Mo, W), homologues of FeCoM(S)(CO)$_8$(C$_5$H$_5$) [61].

These molecules generally undergo an initial one-electron reduction, quasi-reversible in character, which generates the corresponding monoanions

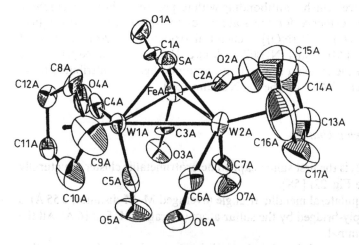

Fig. 23. Molecular structure of FeW$_2$(S)(CO)$_7$(C$_5$H$_5$)$_2$. W–W, 3.04 Å; averaged W–Fe, 2.80 Å; W–S, 2.36 Å; Fe–S, 2.20 Å [from Ref. 60]

Table 9. Redox potentials (V) and peak-to-peak separation (in millivolts) for the one-electron oxidation and reduction of sulfur-capped trimetallic species

Complex	$Ep_{+/0}$	$E^{o'}_{0/-}$	ΔEp^a	Solvent	Ref.
FeCo$_2$(S)(CO)$_9$	+ 1.4	− 0.65	96	C$_2$H$_4$Cl$_2$	50
	—	− 0.28	—	Me$_2$CO	59
FeCo$_2$(S)(CO)$_8$[P(OPh)$_3$]	—	− 0.90	—	Me$_2$CO	59
FeCo$_2$(S)(CO)$_8$(C$_6$H$_5$NC)	+ 1.13	− 0.92	74	C$_2$H$_4$Cl$_2$	50
FeCo$_2$(S)(CO)$_8$(p-MeC$_6$H$_4$NC)	+ 1.04	− 0.94	76	C$_2$H$_4$Cl$_2$	50
FeCo$_2$(S)(CO)$_8$(p-NO$_2$C$_6$H$_4$NC)	+ 1.25	− 0.87	66	C$_2$H$_4$Cl$_2$	50
FeCo$_2$(S)(CO)$_7$(C$_6$H$_4$NC)$_2$	+ 1.4	− 1.12	71	C$_2$H$_4$Cl$_2$	50
FeCo$_2$(S)(CO)$_7$(p-MeC$_6$H$_4$NC)$_2$	+ 1.2	− 1.15	79	C$_2$H$_4$Cl$_2$	50
FeCo$_2$(S)(CO)$_7$(p-NO$_2$C$_6$H$_4$NC)$_2$	+ 1.5	− 1.06	70	C$_2$H$_4$Cl$_2$	50
FeCo$_2$(S)(CO)$_6$(C$_6$H$_5$NC)$_3$	+ 1.6	− 1.37	81	C$_2$H$_4$Cl$_2$	50
FeCo$_2$(S)(CO)$_6$(p-MeC$_6$H$_4$NC)$_3$	+ 1.6	− 1.42	78	C$_2$H$_4$Cl$_2$	50
RuCo$_2$(S)(CO)$_9$	+ 1.2	− 0.80	98	C$_2$H$_4$Cl$_2$	50
[CoFe$_2$(S)(CO)$_9$]$^-$	+ 0.57	− 1.71	118	C$_2$H$_4$Cl$_2$	50
FeW$_2$(S)(CO)$_7$(C$_5$H$_5$)$_2$	+ 0.89	− 1.27^b	—	MeCN	60
FeCoMo(S)(CO)$_8$(C$_5$H$_5$)	+ 1.1	− 1.20^b	—	C$_2$H$_4$Cl$_2$	50
FeCoW(S)(CO)$_8$(C$_5$H$_5$)	+ 1.0	− 1.34^b	—	C$_2$H$_4$Cl$_2$	50
RuCoMo(S)(CO)$_8$(C$_5$H$_5$)	+ 0.9	− 1.35^b	—	C$_2$H$_4$Cl$_2$	50

^{a}Measured at 0.02 Vs^{-1}; ^{b}followed by fast decomposition of the monoanion

[50, 59, 60]. Such monoanions are not fully stable, in that the appearance of further irreversible cathodic steps has been attributed to the reduction of their fragmentation products.

The sulfur-capped clusters also undergo irreversible one-electron removal. The redox potentials of these steps are summarized in Table 9.

The substitution of CO groups for electron donating ligands makes the electron addition more and more difficult. The same result is obtained by substituting one metal atom with an heavier heteroatom; in this case, the lifetime of the monoanion is also significantly reduced, suggesting an increased antibonding character of the LUMO level.

2.4.6 Selenium Capping Unit

The structure of FeCo$_2$(μ_3-Se)(CO)$_9$ is quite similar to that of FeCo$_2$(μ_3-S)(CO)$_9$. The averaged M–M distance is 2.58 Å; the averaged M–Se is 2.28 Å [62].

Table 10. Redox potentials (V) and peak-to-peak separation (mV) for the redox changes exhibited by some selenium-capped trimetallic clusters in 1,2-Dichloroethane solution [50]

Complex	$Ep_{+/0}$	$E^{o'}_{0/-}$	ΔEp^a
FeCo$_2$(Se)(CO)$_9$	+ 1.3	− 0.58	99
RuCo$_2$(Se)(CO)$_9$	+ 0.1	− 0.73	85
FeCoMo(Se)(CO)$_8$(C$_5$H$_5$)	+ 1.0	− 1.08^b	—

^{a}Measured at 0.02 Vs^{-1}; ^{b}peak potential value for irreversible process

FeCo$_2$(Se)(CO)$_9$, RuCo$_2$(Se)(CO)$_9$ and FeCoMo(Se)(CO)$_8$(C$_5$H$_5$) parallel the redox behaviour of the corresponding sulfur-capped species, in that they undergo a one-electron reduction, with features of significant chemical reversibility, and a declustering one-electron oxidation [50]. As shown in Table 10, the redox potentials are only slightly different from those of the sulfur-capped homologues.

2.4.7 Alkyne Capping Unit

The M$_3$C$_2$ cluster cores resulting from the interaction of alkynes with trimetallic compounds may be viewed in terms of two geometries: (i) square-based pyramidal, if the alkyne is parallel to one metal-metal edge; (ii) trigonal-bipyramidal, if the alkyne is perpendicular to one metal-metal edge [63]. As a matter of fact, the orientation of the alkyne is dictated by the electronic situation of the complex [64]. FeCo$_2$(μ_3-η^2-∥-EtC$_2$Et)(CO)$_9$, formally being a saturated 48-electron species, belongs to the first class, Fig. 24 [65].

The capping diethylacetylene molecule together with the parallel Co–Co edge form the square base; the apical Fe atom completes the geometry.

Fe$_3$(EtC$_2$Et)(CO)$_9$, which has two fewer electrons, adds reversibly and stepwisely two electrons. The present heteronuclear cluster in dichloromethane solution, undergoes first a one-electron reduction (E$^{\circ\prime}$ = $-$ 0.86 V), generating the corresponding, short-lived (t$_{1/2} \approx$ 0.3 s) monoanion. Then, it is presumed that this monoanion congener reorganizes to a new (unknown) monoanion, which, in turn, undergoes a further one-electron reduction to its relatively stable dianion (E$^{\circ\prime}$ = $-$ 1.09 V) [66].

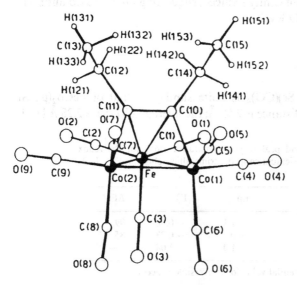

Fig. 24. Molecular structure of FeCo$_2$(EtC$_2$Et)(CO)$_9$. Bond lengths: Co–Co, 2.58 Å; averaged Fe–Co, 2.48 Å; Co1–C10. Co2–C11, 1.96 Å; Fe–C10, FeC11, 2.04 Å [from Ref. 65]

2.5 Triangular Frames Capped on Opposite Sides by Two Triply-Bridging Ligands

2.5.1 Carbonyl Capping Units

Fig. 25 shows the geometry of PtRh$_2$(μ_3–CO)$_2$(PPh$_3$)(CO)(C$_5$Me$_5$)$_2$ [67]. The almost equilateral triangle (averaged metal-metal bond length, 2.65 Å) is capped, at opposite sides, by two carbonyl groups, which bridge the Rh–Rh edge (mean Rh–C$_{(capping)}$ distance, 2.00 Å), but are semibridging with respect to the platinum atom (mean Pt–C$_{(capping)}$, 2.52 Å).

It has been briefly reported that the cluster in dichloromethane solution, undergoes first a reversible one-electron oxidation ($E^{o'} = +0.3$ V), followed by second an irreversible oxidation (Ep = $+0.8$ V). As expected, chemical oxidation by ferrocenium salt afforded the monocation [PtRh$_2$(μ_3–CO)$_2$(PPh$_3$)(CO)(C$_5$Me$_5$)$_2$]$^+$, which has not been characterized by X-ray [67].

Another molecule, the monocation [CoPd$_2$(μ_3–CO)$_2$(Ph$_2$PCH$_2$PPh$_2$)$_2$(CO)$_2$]$^+$, exhibits some metal-carbon bonds of the capping units which are semibridging, Fig. 26 [68].

The metallic isosceles triangle (Pd1–Pd2, Pd2–Co1, 2.55 Å; Pd1–Co1, 2.62 Å) is capped at opposite sides by carbonyl ligands. At each side, one of the Pd–C$_{(capping)}$ distances results in semibridging carbonyls (Pd1–C51, 2.36 Å; Pd2–

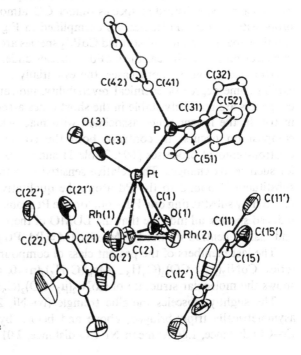

Fig. 25. Perspective view of PtRh$_2$(CO)$_2$(PPh$_3$)(CO)(C$_5$Me$_5$)$_2$ [from Ref. 67]

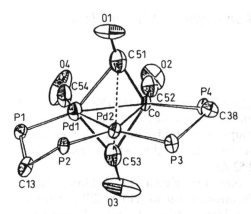

Fig. 26. Molecular structure of [CoPd$_2$(μ_3–CO)$_2$(Ph$_2$PCH$_2$PPh$_2$)$_2$(CO)$_2$]$^+$ [from Ref. 68]

C51, 2.47 Å; Co–C51, 1.67 Å; Pd1–C53, 2.21 Å; Pd2–C53, 2.47 Å; Co–C53, 2.08 Å).

This complex belongs to a series of cluster compounds displaying a metallic triangle bridged by two bidentate phosphine ligands, the redox behaviour of which has been studied (69, 70). These complexes have general formulae: (i) [CoPd$_2$(μ_3– CO)$_2$(Ph$_2$PCH$_2$PPh$_2$)$_2$(CO)(X)]$^{n+}$ (n=0, X=I; n=1, X=CO, PPhMe$_2$); (ii) [CoPdPt(μ_3– CO)(Ph$_2$PCH$_2$PPh$_2$)$_2$(CO)(X)]$^{n+}$ (n=0, X=I; n=1, X=CO, PPh$_3$). In polar solvents, the ligand X is labilized, particularly when X=I, CO, giving rise to the corresponding solvate species. Nevertheless, the electrochemical behaviour of either the monophosphine-substituted species or the carbonyl-substituted species (under CO atmosphere) reflects the redox propensity of these molecules. As exemplified in Fig. 27, in Dimethylsulfoxide solution, the phosphine-substituted CoPd$_2$ species are reduced in two successive one-electron steps, whereas the CoPdPt species undergoes a single stepped two-electron reduction. In both cases, the essentially reversible electron transfers exhibit some degree of chemical reversibility, indicating that the isostructural congeners are relatively stable in the short times associated with cyclic voltammetry. In the longer times associated with macroelectrolysis, irreversible decomposition occurs. In contrast, both the complexes undergo irreversible electron-removal processes [69]. Table 11 summarizes the electrode potentials for such redox changes. The relative sensitivity to the electronic effects of the substituent X bound to the Pd atom, the qualitative and quantitative effects induced by substituting one Pd atom for one Pt atom, the transient nature of the reduced species, all indicate that the LUMO of these complexes is antibonding and mainly centred on the Pd–M edge (M=Pd, Pt) [70].

The last members of the present class of compounds are constituted by the series CoNi$_2$(μ_3– CO)$_2$(C$_5$H$_{5-n}$Me$_n$)(C$_5$H$_5$)$_2$(n=0, 1, 5) [71, 72]. Figure 28 shows the molecular structure of CoNi$_2$(μ_3–CO)$_2$(C$_5$Me$_5$)(C$_5$H$_5$)$_2$.

The slightly isosceles metallic triangle (Ni–Ni, 2.33 Å; Ni–Co, 2.37 Å) is asymmetrically triply-bridged, above and below, by carbonyl groups (mean Co–CO distance, 1.86 Å; mean Ni–CO distance, 2.01 Å).

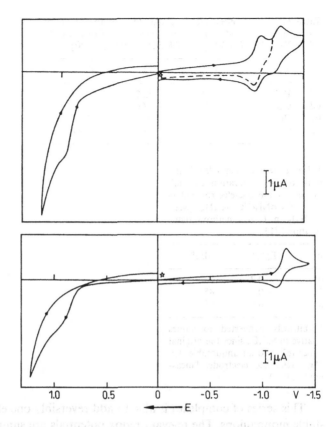

Fig. 27a, b. Cyclic voltammetric responses recorded at a platinum electrode on DMSO solutions of:
a) [CoPd$_2$(CO)$_2$(Ph$_2$PCH$_2$PPh$_2$)$_2$(CO)(PPhMe$_2$)](PF$_6$); b) [CoPdPt(CO)$_2$(Ph$_2$PCH$_2$PPh$_2$)$_2$(CO)(PPh$_3$)](PF$_6$). Scan rate 0.1 Vs^{-1} [from Ref. 69]

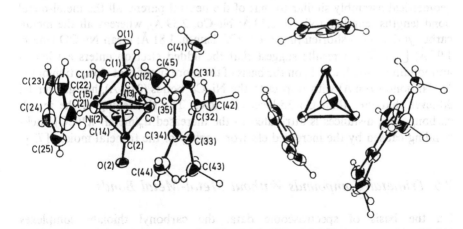

Fig. 28. Perspective view of CoNi$_2$(CO)$_2$(C$_5$Me$_5$)(C$_5$H$_5$)$_2$ [from Ref. 71]

Table 11. Formal electrode potentials (V) for the reduction processes exhibited by the clusters [CoPdM(CO)$_2$(Ph$_2$PCH$_2$PPh$_2$)$_2$(CO)(X)]$^+$ in DMSO solution [69, 70]

M	X	$E^{o'}_{+/0}$	$E^{o'}_{0/-}$	$E^{o'}_{+/-}$
Pd	PPhMe$_2$	−1.04	−1.27	—
Pd	CO	−0.77	−1.00	—
Pt	PPh$_3$	—	—	−1.24
Pt	CO	—	—	−1.00

Table 12. Redox potentials (V) and peak-to-peak separation (in millivolts) for the one-electron reduction of CoNi$_2$(CO)$_2$(C$_5$H$_{5-n}$Me$_n$)(C$_5$H$_5$)$_2$ in 1, 2-Dimethoxyethane solution [71]

n	$E^{o'}_{0/-}$ [a]	ΔE_p [b]
0	−0.77	70
1	−0.80	85
5	−0.93	85

[a] Arbitrarily converted to values relative to S.C.E., since the original paper refers to an unquotable Ag/Ag$^+$ reference electrode; [b] measured at 0.1 Vs^{-1}

This series of complexes is able to add reversibly one electron to afford the stable monoanions. The relevant redox potentials are summarized in Table 12.

The ΔE_p parameter suggests the quasireversible character of the electron-transfer. This preludes to the occurrence of some structural reorganization as a consequence of the one-electron addition. The structure of the paramagnetic monoanion [CoNi$_2$(μ$_3$- CO)(C$_5$Me$_5$)(C$_5$H$_5$)$_2$]$^-$ reveals that, within a gross geometrical assembly similar to that of the neutral parent, all the metal-metal bond lengths lengthen (Ni–Ni, 2.39 Å; Ni–Co, 2.39 Å), whereas all the metal-carbonyl distances shorten (mean Co–CO length, 1.81 Å; mean Ni–CO length, 1.97 Å) [72]. These results suggest that the added electron enters a trimetal antibonding orbital, which, on the basis of the more significant elongation of the Ni–Ni bond (0.06 Å) with respect to the Ni–Co bonds (0.02 Å), has greater Ni 3d-level character than Co 3d-level character. The shortening of the metal-carbon$_{(capping)}$ distances is attributed to the increased d_π(metal)–π*(CO) back-bonding caused by the increased electron density of the trimetal moiety [72].

2.6 Trimetal Compounds Without Metal-Metal Bonds

On the basis of spectroscopic data, the carbonyl thiolate complexes [M(SR)$_4${Mo(CO)$_4$}$_2$]$^{2-}$ (M = Fe, Co; R = Ph, Bz) have assigned the open trimetallic structure shown in Scheme 3 [73]. Strictly speaking, the lack of any direct metal-metal interaction does not permit their classification as cluster compounds.

The central metal atom is thought to be M(II), while the outer molybdenum atoms are classified as Mo(0). Chemical oxidation by iodine affords a neutral molecule formulated as depicted in Scheme 4. The change in oxidation state involves the oxidation of both the outer molybdenum atoms to Mo(I).

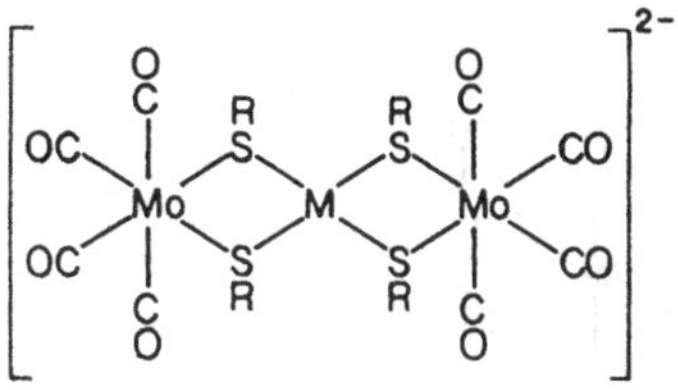

Scheme 3 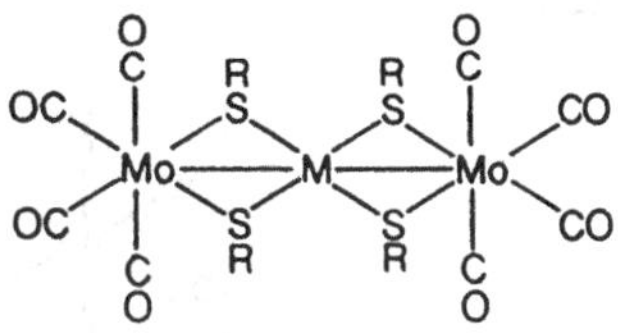   Scheme 4

Interestingly, the simultaneous two-electron removal should be accompanied by the formation of metal-metal bonding.

In agreement with the chemical pathway, the starting dianions undergo a single-stepped two-electron oxidation, followed, however, by slow chemical decomposition of the electrogenerated neutral species. As shown in Fig. 29, which refers to $[Co(SPh)_4\{Mo(CO)_4\}_2]^{2-}$, the occurence of such a decomposition may be significantly quenched by working at low temperatures [73]. Table 13 reports the redox potentials of such oxidation steps. Comparison of the relevant peak-to-peak separations with the theoretical value of 29 mV expected for a reversible two-electron transfer (and hence involving no significant structural reorganization) may account for the large geometrical changes supposed.

A somewhat related compound is the carbonyl sulfide dianion $[W(S)_4\{Mo(CO)_4\}_2]^{2-}$. Tentatively it has been assigned the structure illustra-

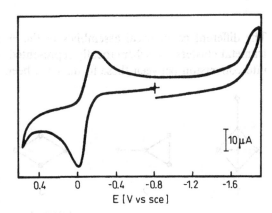

Fig. 29. Cyclic voltammogram recorded at a platinum electrode on a MeCN solution of $[Co(SPh)_4\{Mo(CO)_4\}_2]^{2-}$. T = −40°C. Scan rate 0.1 Vs^{-1} [from Ref. 73]

Table 13. Redox potentials (V) and peak-to-peak separation (in millivolts) for the two-electron oxidation of the complexes [M(SR)$_4${Mo(CO)$_4$}$_2$]$^{2-}$ [73]

M	R	E$^{o'}_{0/2-}$	ΔEp[a]	Solvent	T (°C)
Fe	Ph	0.00	200	MeCN	−40
Fe	Bz	−0.20	130	DMF	−37
Co	Ph	−0.10	200	MeCN	−40
Co	Bz	−0.19	120	DMF	−50

[a] Measured at 0.1 Vs^{-1}

Scheme 5

ted in Scheme 5, in which a central W(VI) atom is connected through sulfide bridges, to two Mo(0) centres [74].

In acetonitrile solution, it undergoes a quasireversible one-electron reduction (E$^{o'}$ = −1.99 V, at −40 °C) as well as an irreversible one-electron oxidation (Ep = +0.18 V, at −40 °C). Such redox changes are attributed to the reduction of the central tungsten atom and to the oxidation of one outer molybdenum atom, respectively. At variance with the previously discussed thiolate bridged trimetal clusters (which undergo a chemically reversible two-electron oxidation), the occurrence of an irreversible one-electron oxidation is attributed to the inability of the central W(VI) to form metal-metal bonds with the outer Mo(I) moieties [74].

3 Tetrametallic Assemblies

The different geometrical assemblies of the electrochemically investigated tetrametal clusters are schematically represented in Scheme 6. Pathways causing interconversion between these forms have been discussed [75–77].

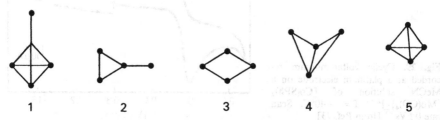

Scheme 6

3.1 Triangular Trimetallic Frames Connected to a Fourth Metal by a Nonmetallic Bridge

A series of heteronuclear clusters have been synthesized by incorporating a heteroatom into the sulfur-capped or the alkylidyne-capped trinuclear clusters previously discussed. The first example is constituted by FeCo$_2$[μ_3-S-Cr(CO)$_5$](CO)$_9$, the structure of which is shown in Fig. 30 [78].

The FeCo$_2$ core is substantially similar to that of the precursor FeCo$_2$(S)(CO)$_9$, except for a slight shortening of the metal-S$_{(capping)}$ distances (less than 0.02 Å).

In the sulfur-capped trimetallic clusters, the antibonding LUMO is centred on the metallic triangle (in particular, the FeCo$_2$ core, it is mainly localized on the Co-Co edge). It has been briefly reported that the present chromium adduct undergoes a quasireversible one electron reduction just at the same potential value (E$^{o'}$ = −0.28 V, in Acetone solution) as that exhibited by FeCo$_2$(S)(CO)$_9$ [59].

Another series of tetranuclear compounds has been basically obtained by adding a ferrocenyl moiety to the tricobalt methylidyne cluster Co$_3$(μ_3-CH)(CO)$_9$ [79, 80]. Fig. 31, which refers to Co$_3$(C{(C$_5$H$_4$)-Fe(C$_5$H$_5$)}(CO)$_9$, illustrates the main structural features of this class of compounds [80]. With respect to the symmetrical assembly of Co$_3$(CH)(CO)$_9$ [81], the presence of the sterically encumbering ferrocenyl substituent induces deformation from C$_{3v}$ symmetry displacing the capping-carbon atom towards one cobalt vertex.

This molecule has potentially two redox active centres: (i) the ferrocenyl moiety is able to undergo one-electron oxidation; (ii) the carbon-capped tricobalt fragment undergoes one-electron reduction [13]. Fig. 32 shows that it

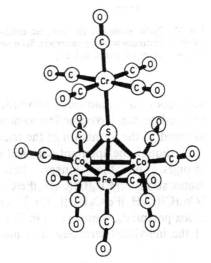

Fig. 30. Perspective view of FeCo$_2$[S-Cr(CO)$_5$](CO)$_9$. Metal-metal mean bond length, 2.56 Å; metal-S$_{(capping)}$, 2.14 Å, Cr-S, 2.35 Å) [from Ref. 78]

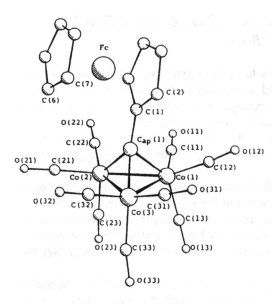

Fig. 31. Perspective view $Co_3[\mu_3-C\{(C_5H_4)Fe(C_5H_5)\}](CO)_9$. Co–Co averaged bond length, 2.47 Å; Co1–Cap, 1.97 Å; Co2–Cap, 1.93 Å; Co3–Cap, 1.87 Å; averaged Fe–C (cyclopentadienyl rings) distance, 2.05 Å; C1–Cap, 1.45 Å [from Ref. 80]

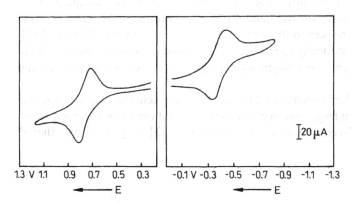

Fig. 32. Cyclic voltammetric response exhibited by $Co_3[C\{(C_5H_4)Fe(C_5H_5)\}](CO)_9$ in Acetone solution. Platinum working electrode. Scan rate 0.2 Vs^{-1}. Potential values are referred to Ag/AgCl reference electrode [from Ref. 82]

undergoes, in nonaqueous solvents, both a one-electron oxidation, just attributed to the ferrocene/ferrocenium couple, and a one-electron reduction, assigned to the formation of the tricobalt carbon monoanion [82].

Although the apparent chemical reversibility of these quasireversible redox changes, in the longer times of macroelectrolysis, either the monocation $[Co_3(C\{(C_5H_4)Fe(C_5H_5)\})(CO)_9]^+$ or the monoanion $[Co_3(C\{(C_5H_4)Fe(C_5H_5)\}(CO)_9]^-$ undergo slow decomposition. The relevant redox potentials, summarized in Table 14, are only slightly different from those of the individual ferrocene and methylidine-tricobalt cluster (see caption of

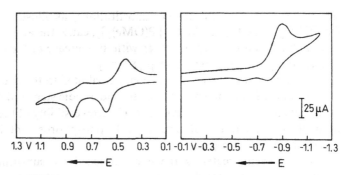

Fig. 33. Cyclic voltammogram recorded at a platinum electrode on a Me₂CO solution of Co₃[C{(C₅H₄)Fe(C₅H₅)}](CO)₇[P(OMe)₃]₂. Scan rate 0.2 Vs⁻¹ [from Ref. 82]

Table 14). This suggests the existence of a weak conjugation between the two redox centres and indicates that, in the mixed-valent monocation and monoanion ions, the charge is localized on the ferrocenium and tricobalt centres, respectively.

The effect of substituents both in the ferrocenyl fragment and in the tricobalt carbon moiety (by substituting CO groups for phosphines or phosphites) has been evaluated. While the presence of substituents in the cyclopentadyenyl rings of the ferrocenyl group affects the redox potentials in a predictable fashion (electron-donating groups make the oxidation easier; electron-withdrawing groups make the oxidation more difficult) (see Table 14), the more significant variation in the redox propensity is induced by the progressive substitution of one carbonyl group of each cobalt atom for phosphorus-containing ligands. In fact, if the substitution of the first carbonyl only makes transient the

Table 14. Formal electrode potentials (V) and peak-to-peak separation (mV) for the first one-electron oxidation and the one-electron reduction exhibited by Co₃[C{C₅H₅R)Fe(C₅H₄R′)}](CO)$_{9-n}$L$_n$ clusters. By way of comparison, one can take into account that, in Acetone solution, ferrocene undergoes one-electron oxidation at E°′ = + 0.51 V, whereas Co₃(CH)(CO)₉ undergoes one-electron reduction at − 0.45 V

R	R′	n	L	$E^{o\prime}_{+/0}$	ΔE_p^a	$E^{o\prime}_{0/-}$	ΔE_p^a	Solvent	Reference
H	H	0	—	+ 0.63	90	− 0.53	110	Me₂CO	82
				+ 0.63	150	− 0.77	160	CH₂Cl₂	79
H	1′ − Ac	0	—	+ 0.73	150	− 0.51	150	Me₂CO	82
2 − Ac	H	0	—	+ 0.73	100	− 0.62	120	Me₂CO	82
H	H	1	PPh₃	+ 0.63	120	− 0.75	80	Me₂CO	82
H	H	1	P(C₆H₁₁)₃	+ 0.67	100	− 0.85	280	Me₂CO	82
H	H	1	P(OMe)₃	+ 0.47	80	− 0.87	130	Me₂CO	82
H	H	2	P(OMe)₃	+ 0.40	140	− 1.16^b	220	Me₂CO	82
H	H	2	P(OPh)₃	+ 0.50	80	− 0.84^b	290	Me₂CO	82
				+ 0.43	—	− 1.40^b	—	CH₂Cl₂	82
H	H	3	P(OMe)₃	+ 0.31	120	− 1.42^b	160	Me₂CO	83
H	H	3	P(OPh)₃	+ 0.42	100	− 1.12^b	110	Me₂CO	83

^{a}Measured at 0.2 Vs⁻¹; ^{b}complicated by following chemical reaction

relevant monoanion, the further substitutions, as illustrated in Fig. 33 for $Co_3[C\{(C_5H_4)Fe(C_5H_5)\}](CO)_7[P(OMe)_3]_2$, cause the appearance of a second irreversible oxidation step, together with an increase of the extent of chemical irreversibility of the reduction step [82, 83].

The first oxidation step is once again attributed to the one-electron removal centred on the ferrocenyl moiety; the second anodic step, and the reduction step are assigned to the tricobalt fragment. The irreversibility of the reduction step is thought to be due to the labilization of the phosphorus-containing ligand upon one-electron addition [82].

Finally, even greater is the effect induced by substitution of two CO groups of one cobalt atom for one nitrosyl group, e.g. in $[Co_3[C\{(C_5H_4)Fe(C_5H_5)\}](CO)_7(NO)]^-$. In this case, the irreversible one-electron oxidation of the tricobalt carbon fragment (Ep = + 0.09 V, in Acetone solution) precedes the one-electron oxidation of the ferrocenyl group (Ep = + 0.65 V), making the latter also irreversible [84].

3.2 Spiked-Triangular Frames

The so-called "spiked-triangular" geometry of tetranuclear clusters (or "metallo-ligated" clusters) is here represented by the structure of $Pd_2Co_2(\mu_3-CO)_2(Ph_2PCH_2PPh_2)_2(CO)_5$, Fig. 34 (68, 85].

The assembly is quite reminiscent of that of $[Pd_2Co(\mu_3-CO)_2 (Ph_2PCH_2PPh_2)_2(CO)_2]^+$ (illustrated in Sect. 2.5.1, Fig. 26). An isosceles metallic triangle (mean Pd1–Pd2, Pd1–Co1 distance, 2.60 Å; Pd2–Co1, 2.51 Å) capped, at opposite sides, by two triply semibridging carbonyl groups, and having two edges (Pd1–Pd2, Pd2–Co1) bridged by the bidentate phosphine. In turn, the triangle is connected to an exocyclic $Co(CO)_4$ group (Pd1–Co2, 2.73 Å), which replaces the terminal CO group present in the trinuclear species.

Such a structure is also attributed to $PdPtCo_2(\mu_3-CO)_2 (Ph_2PCH_2PPh_2)_2 (CO)_2$.

The redox behaviour of these complexes is illustrated in Fig. 35 [70].

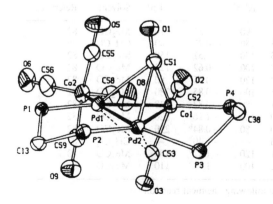

Fig. 34. Perspective view of $Pd_2Co_2 (CO)_7(Ph_2PCH_2PPh_2)_2$ [from Ref. 68]

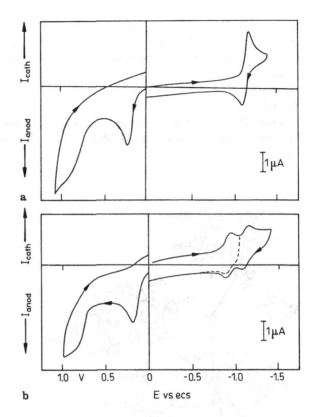

Fig. 35a, b. Cyclic voltammetric responses exhibited by a) PdPtCo$_2$(CO)$_7$(Ph$_2$PCH$_2$PPh$_2$)$_2$ and b) Pd$_2$Co$_2$(CO)$_7$(Ph$_2$PCH$_2$PPh$_2$) in DMSO solution. Platinum working electrode. Scan rate 0.1 Vs^{-1} [from Ref. 70]

In agreement with the discussion in Section 2.5.1, in Dimethyl–sulfoxide solution, the species containing the CoPd$_2$ core undergoes two successive one-electron reductions (E$^{\circ\prime}$ = $-$0.90 V and $-$1.15 V, respectively), whereas the species containing the CoPtPd core undergoes a single-stepped two-electron

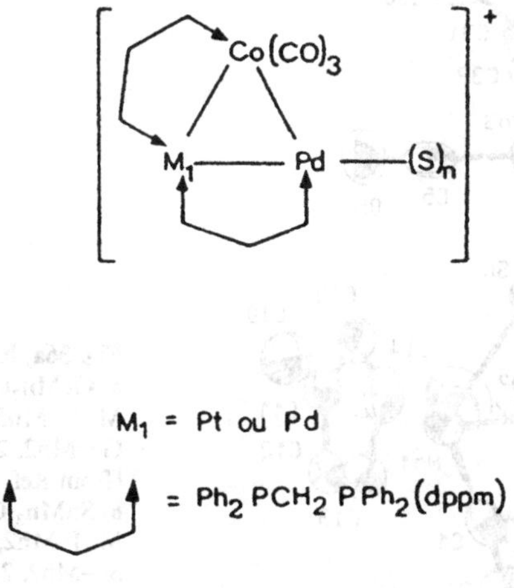

M$_1$ = Pt ou Pd

= Ph$_2$PCH$_2$PPh$_2$ (dppm)

Scheme 7

reduction ($E^{o\prime} = -1.13$ V). However, as previously pointed out for the exocyclic bond Pd–X (X = I, CO), also the Pd–Co(CO)$_4$ bond is notably labile in polar solvents, so that the responses shown in Fig. 35 are really attributable to the solvated trinuclear cations depicted in Scheme 7, due to the release of the anion [Co(CO)$_4$]$^-$ (clearly evident by the oxidation peak at Ep = + 0.2 V) [69, 70].

Two other metalloligated clusters of known electrochemistry are shown in Fig. 36, namely GeMn$_3$(C$_5$H$_4$Me)$_3$(CO)$_6$ [86] and SnMn$_3$(C$_5$H$_4$Me)$_3$(CO)$_6$ [87].

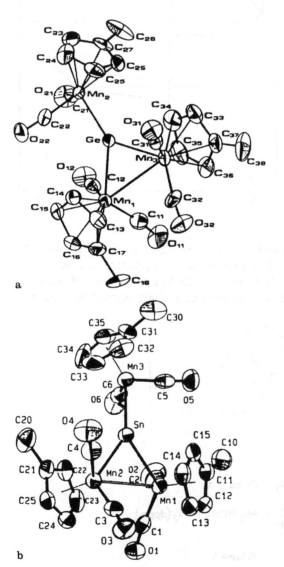

Fig. 36a, b. Molecular structure of:
a) GeMn$_3$(C$_5$H4Me)$_3$(CO)$_6$.
Mn1– Mn3, 2.98 Å; Ge–Mn1, 2.36 Å;
Ge–Mn2, 2.26 Å; Ge–Mn3, 2.38 Å
[from Ref. 86];
b) SnMn$_3$(C$_5$H$_4$Me)$_3$(CO)$_6$.
Mn1–Mn2, 3.06 Å; Sn–Mn1,
Sn–Mn2, 2.55 Å; Sn–Mn3, 2.44 Å
[from Ref. 87]

The isosceles MMn$_2$ triangles (M = Ge, Sn) are bound to an outer Mn(CO)$_2$(C$_5$H$_4$Me) fragment through a formal double-bond M=Mn [2].

Both the complexes, in Tetrahydrofuran solution, undergo declustering reduction (M = Ge, E$^{o\prime}$ = − 1.24 V; M = Sn, E$^{o\prime}$ = − 1.13 V) and oxidation (M = Ge, Ep = + 0.21 V; M = Sn, Ep = +0.22 V) processes [30], so indicating they are the only stable members of their potential redox families.

3.3 Planar Frames

The tetrametallic planar assembly will be subdivided into square-planar and triangulated-rhomboidal geometries.

3.3.1 Square Planar Cores

Fig. 37a shows the structure of RhFe$_3$(μ_4–PPh)$_2$(C$_5$Me$_5$)(CO)$_8$. It consists of an almost square planar RhFe$_3$ core (Rh–Fe1, Rh–Fe3, 2.77 Å; Fe1–Fe2, 2.51 Å; Fe2–Fe3, 2.51 Å), quadruply bridged, above and below, by two phosphinidene groups, giving rise to an octahedral M$_4$P$_2$ assembly [88]. By reacting with an equivalent amount of CO, this cluster converts, reversibly, to RhFe$_3$(μ_4–PPh)$_2$(C$_5$Me$_5$)(CO)$_9$, the structure of which is illustrated in Fig. 37b [88].

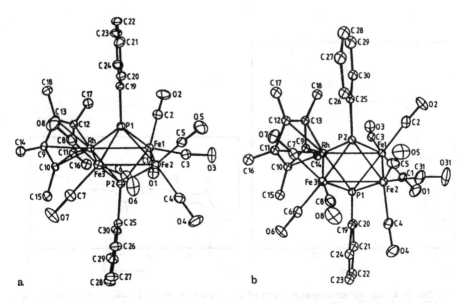

Fig. 37a, b. Molecular structure of: a) RhFe$_3$(PPh)$_2$(C$_5$Me$_5$)(CO)$_8$; b) RhFe$_3$(PPh)$_2$(C$_5$Me$_5$)(CO)$_9$ [from Ref. 88]

This addition causes some rearrangement within the M_4P_2 core. In fact, the Rh–Fe3 bond length shortens by about 0.07 Å (2.70 Å), while the Fe1–Fe2 and Fe2–Fe3 distances increase by more than 0.1 Å (mean distance, 2.66 Å). As a consequence, even if the M–P$_{(capping)}$ distances do not vary (average = 2.30 Å), a compression of the octahedral core occurs, in that the distance from the two capping phosphorus atoms decreases from 2.67 Å to 2.58 Å.

As far as the bonding mode of the carbonyl groups is concerned, it is likely that the unsaturation of the Fe2 atom in $RhFe_3(PPh)_2(C_5Me_5)(CO)_8$ is compensated by two semibridging bonds with C3 and C6.

The reversible addition/removal of the two-electron donor CO leads to a clean redox ability of these species. Accordingly, their redox propensity is shown in Fig. 38 [88]. $RhFe_3(PPh)_2(C_5Me_5)(CO)_8$, undergoes, in tetrahydrofuran solution, apart from a sequence of irreversible electron removals, two quasireversible one-electron reduction steps ($E^{\circ\prime}_{(A)} = -0.46$ V, $\Delta E_{p(0.01\ Vs^{-1})} = 80$ mV; $E^{\circ\prime}_{(B)} = -1.34$ V, $\Delta E_{p(0.01\ Vs^{-1})} = 80$ mV)), which generate the stable (at least in the cyclic voltammetric period) mono- and dianion congeners, respectively, Fig. 38a.

On the other hand, $RhFe_3(PPh)_2(C_5Me_5)(CO)_9$ undergoes two quasireversible one-electron removals ($E^{\circ\prime}_{(X)} = +0.42$ V, $\Delta E_{p(0.1\ Vs^{-1})} = 80$ mV; $E^{\circ\prime}_{(Y)} = +0.97$ V, $\Delta E_{p(0.1\ Vs^{-1})} = 70$ mV), which generate the stable (at least in the cyclic voltammetric timescale) mono- and dication congeners. In addition, as shown in Fig. 38b, the carbonylated species displays a series of reduction steps which must be interpreted as follows. The one-electron reduction at peak D (Ep = − 1.1 V) primarily generates the monoanion $[RhFe_3(PPh)_2(C_5Me_5)(CO)_9]^-$, which

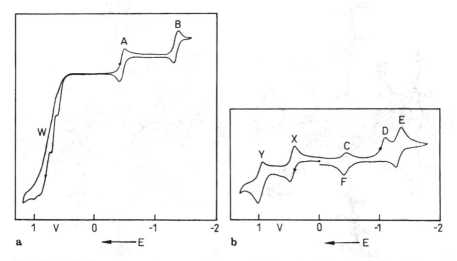

Fig. 38a, b. Cyclic voltammograms recorded at a platinum electrode with THF solutions of: a) $RhFe_3(PPh)_2(C_5Me_5)(CO)_8$, scan rate 0.01 Vs^{-1}; b) $RhFe_3(PPh)_2(C_5Me_5)(CO)_9$, scan rate 0.1 Vs^{-1} [from Ref. 88]

is unstable and loses CO, affording the unsaturated monoanion [RhFe$_3$(PPh)$_2$(C$_5$Me$_5$)(CO)$_8$]$^-$. In accordance with the previously discussed behavior of RhFe$_3$(PPh)$_2$(C$_5$Me$_5$)(CO)$_8$, such a monoanion is able to be reduced to the corresponding dianion (at the peak-system, E, which coincides with the peak-system B in Fig. 38a) or reoxidized to the neutral parent (at the peak pair F/C, which coincides with the peak-system A of Fig. 38a).

Without any doubt, the ability of unsaturated species to add reversibly two electrons, and the ability of saturated species to lose reversibly two electrons account for their chemical propensity to add/remove the two-electron donor CO ligand.

The maintenance of an octahedral geometry within the RhFe$_3$(PPh)$_2$(C$_5$Me$_5$)(CO)$_8$/RhFe$_3$(PPh)$_2$(C$_5$Me$_5$)(CO)$_9$ interconversion, formally caused by a two-electron addition/removal, can be understood if one considers that, in these complexes, the electrons enter or leave nonbonding frontier orbitals centred on the square-planar tetrametal fragment stabilized by the interaction with the apical phosphorus atoms [89].

Isoelectronic (and substantially isostructural) with the unsaturated RhFe$_3$(PPh)$_2$(C$_5$Me$_5$)(CO)$_8$ are two other phosphinidene clusters: RuFe$_3$(μ_4-PPh)$_2$(CO)$_{11}$ and Ru$_2$Fe$_2$(μ_4-PPh)$_2$(CO)$_{11}$ [90], which have likely the structure of Fe$_4$(μ_4-PR)$_2$(CO)$_{11}$ [91]. As such, they are expected to be able to add reversibly two electrons. Even if their redox ability is not completely known (their electrochemistry has not been explored), they afford, by chemical reduction with cobaltocene, the corresponding monoanions [90].

3.3.2 Triangulated Rhomboidal Cores

A wide series of planar triangulated tetrametallic complexes of general formula M$_2$M'$_2$(μ_3-CO)$_2$(μ-CO)$_4$(C$_5$H$_5$)$_2$(PR$_3$)$_2$(M=Pd, Pt; M'=Cr, Mo, W) have been characterized. Their typical structures are illustrated in Fig. 39 [92,93]. The tetrametallic core forms a triangulated rhombus (selected bond lengths are reported in Table 15), in which each MPd$_2$ or MPt$_2$ isosceles triangle is asymmetrically triply-bridged by a carbonyl group, as well as each MPd or MPt vector is asymmetrically bridged by a carbonyl group. The metal-carbonyl bonding systems in the two symmetric triangulated units are placed above and below the tetrametallic plane. The complexes are considered to contain Pd(I)–Pd(I) and Pt(I)–Pt(I) centres, respectively.

The electrochemistry of these clusters has been accurately studied [94, 95]. They undergo a single-stepped two-electron reduction, irreversible in character because of the fast declusterification reaction following the Pd$_2^I$/Pd$_2^0$ or Pt$_2^I$/Pt$_2^0$ redox change. One of the products of this chemical complex is certainly the anion [M(C$_5$H$_5$)(CO)$_3$]$^-$. More interesting is the oxidation behaviour of these molecules. As representatively shown in Fig. 40, they undergo, in 1,2-dichloroethane solution, two distinct one-electron removals, the first of which is always chemically reversible, whereas the second one is generally complicated

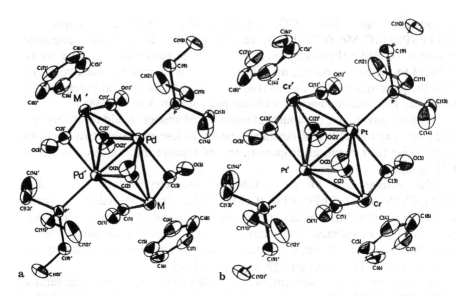

Fig. 39a, b. Perspective view of the geometrical assembly in: **a)** Pd$_2$M$_2$(CO)$_6$(C$_5$H$_5$)$_2$(PR$_3$)$_2$ (M = Cr, Mo, W; R = Et) [from Ref. 92]; **b)** Pt$_2$M$_2$(CO)$_6$(C$_5$H$_5$)$_2$(PR$_3$)$_2$ (M = Cr, Mo, W; R = Et) [from Ref. 93].

Table 15. Selected bond lengths (in Å) in Pd$_2$M$_2$(CO)$_6$(C$_5$H$_5$)$_2$(PEt$_3$)$_2$ (M = Cr, Mo, W) and Pt$_2$M$_2$(CO)$_6$(C$_5$H$_5$)$_2$(PEt$_3$)$_2$ (M = Cr, Mo, W) [92, 93]

Bond distances in the palladium complexes	M = Cr	M = Mo	M = W
Pd–Pd'	2.58	2.58	2.57
Pd–M	2.78	2.86	2.87
Pd'–M	2.74	2.83	2.83
Pd–C2	2.34	2.38	2.41
Pd'–C2	2.26	2.31	2.35
M–C2	1.92	2.04	2.05
Pd–C3	2.34	2.41	2.44
M–C3	1.84	1.97	1.98
Pd–P	2.34	2.34	2.33
Bond distances in the platinum complexes			
Pt–Pt'	2.61	2.68[a]	2.66[a]
Pt–M	2.75	2.78[a]	2.77[a]
Pt'–M	2.71	2.83[a]	2.84[a]
Pt–C2	2.42	2.38[a]	2.38[a]
Pt'–C2	2.27	2.74[a]	2.75[a]
M–C2	1.93	2.04[a]	2.06[a]
Pt–C3	2.26	2.30[a]	2.33[a]
M–C3	1.86	1.97[a]	1.97[a]
Pt–P	2.29	2.28[a]	2.28[a]

[a] Value referred to one of the two independent molecules present in the unit cell

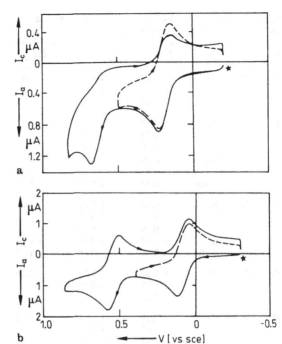

Fig. 40a, b. Anodic portions of the cyclic voltammetric responses exhibited, in 1,2-Dichloroethane solution, by: a) Pd$_2$W$_2$(CO)$_6$(C$_5$H$_5$)$_2$(PPh$_3$)$_2$; b) Pt$_2$W$_2$(CO)$_6$(C$_5$H$_5$)$_2$(PPh$_3$)$_2$. Platinum working electrode. Scan rate 0.2 Vs^{-1} [from Ref. 95]

by following reactions (which, for instance, are fast in the Pd$_2$W$_2$ complex, but slow in the Pt$_2$W$_2$ complex).

The substantial electrochemical reversibility of the neutral/monoation step would indicate that no gross structural change accompanies the first electron removal. In addition, the ESR absorption pattern exhibited by the electrogenerated monocations suggests that the HOMO of these species is localized on the tetrametal plane.

It is noteworthy that the lifetime of the cations is strongly dependent on solvent effects. In dimethyl sulfoxide and in dichloromethane solution, the monocations are only transient.

The electrode potentials of the discussed redox changes are summarized in Table 16. The effects of the phosphine substituent R on the redox potentials are also shown. They follow the trend expected on the basis of their electron donating ability.

A somewhat structurally related complex is the planar isomer Mo$_2$Fe$_2$(μ_3–S)$_2$(μ–CO)$_2$(C$_5$H$_4$Me)$_2$(CO)$_6$, whose structure, with some selected bond-lengths, is shown in Fig. 41 [96]. The isomeric butterfly analogue will be discussed in the next Section.

Each triangulated Mo$_2$Fe subunit of the tetrametallic rhombus is triply bridged, on opposite sides, by a Sulfur atom. One Mo–Fe edge of each subunit is asymmetrically μ–CO bonded.

In Acetonitrile solution, this cluster undergoes a first irreversible one-electron reduction (Ep = − 0.68 V), followed by a further, chemically reversible

Table 16. Electrode potentials (V) for the redox changes exhibited by $M_2M'_2(CO)_6(C_5H_5)_2(PR_3)_2$ [94, 95]

M	M'	R	$E^{o'}_{2+/+}$ [a]	$E^{o'}_{+/0}$	$Ep_{0/2-}$	Solvent
Pd	Cr	Ph	+1.08	+0.31[a]	−0.85	DMF
			+1.04	+0.26[a]	—	CH_2Cl_2
			+1.00	+0.25	—	$C_2H_4Cl_2$
Pd	Mo	Ph	+0.77	+0.32[a]	−0.99	DMF
			+0.73	+0.29[a]	—	CH_2Cl_2
			+0.70	+0.27	—	$C_2H_4Cl_2$
		Me	+0.39	+0.24[a]	−1.17	DMF
		Et	+0.53	+0.25[a]	−1.20	DMF
		n–Bu	+0.56	+0.26[a]	−1.23	DMF
Pd	W	Ph	+0.56	+0.25[a]	−1.07	DMF
			+0.66	+0.18[a]	—	CH_2Cl_2
			+0.66	+0.19	—	$C_2H_4Cl_2$
		Et	+0.44	+0.16[a]	−1.27	DMF
		n–Bu	+0.47	+0.18[a]	−1.32	DMF
Pt	Cr	Ph	+0.86	+0.23[a]	−1.16	DMF
			+0.95	+0.21[a]	—	CH_2Cl_2
			+0.88	+0.19	—	$C_2H_4Cl_2$
Pt	Mo	Ph	+0.54	+0.24[a]	−1.32	DMF
			+0.61	+0.23[a]	—	CH_2Cl_2
			+0.69	+0.20	—	$C_2H_4Cl_2$
Pt	W	Ph	+0.45	+0.14[a]	−1.43	DMF
			+0.44	+0.08[a]	—	CH_2Cl_2
			+0.54	+0.08	—	$C_2H_4Cl_2$
		Et	+0.39	+0.12[a]	−1.57	DMF

[a] Complicated by decomposition reaction

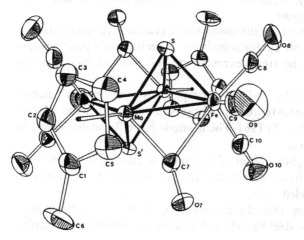

Fig. 41. Molecular structure of the planar isomer $Mo_2Fe_2(S)_2(CO)_8(C_5H_4Me)_2$. Mo–Mo', 2.82 Å; Mo–Fe, 2.78 Å; Mo–Fe', 2.80; Fe–S, 2.21; Mo–S (average), 2.36 Å; Mo–C7, 2.02 Å; Fe–C7, 2.27 Å [from Ref. 96]

one-electron reduction ($E^{o\prime} = -1.14$ V). This pattern is thought to be due to an overall reduction scheme of the type [96]:

$$Mo_2Fe_2 \xrightarrow{+e} [Mo_2Fe_2]^- \xrightarrow{fast} [Mo_2Fe_2]^{*-} \underset{-e}{\overset{+e}{\rightleftarrows}} [Mo_2Fe_2]^{*2-}$$

It indicates that the primarily electrogenerated monoanion $[Mo_2Fe_2(S)_2(CO)_8(C_5H_4Me)_2]^-$ undergoes a fast geometrical reorganization to a structurally unknown monoanion $[Mo_2Fe_2(S)_2(CO)_8(C_5H_4Me)_2]^{*-}$ (the fast reorganization seems however to prevent more rearrangements [96]), which, in turn, is able to support, without framework destruction, a further one-electron addition.

As far as the ability of $Mo_2Fe_2(S)_2(C_5H_4Me)_2(CO)_8$ to undergo electron removals is concerned, it only exhibits irreversible anodic steps (Ep = +0.99 V, and +1.39 V, respectively).

This last cluster offers the opportunity to include in the present Section the complex $Mo_2Fe_2(\mu_3-S)_4(C_5H_4Me)_2(CO)_6$, which, as shown in Fig. 42, really displays an open rhomboidal tetrametallic core [97, 98].

Each one of the two open-triangulated Mo_2Fe subunits is capped, at opposite sides, by triply-bridging sulfur atoms. With respect to the preceding compound, the absence of the edge-bridging carbonyls induces opening of the metallic triangles.

In Acetonitrile solution, the present cluster has been reported to display first, a substantially reversible, one-electron reduction ($E^{o\prime} = -1.06$ V; ΔEp (0.1 Vs^{-1}) = 70 mV), followed by second a cathodic step ($E^{o\prime} = -1.47$ V), complicated by decomposition of the electrogenerated dianion. A transient monocation can be also generated at a first anodic process ($E^{o\prime} = +0.5$ V) [97]. Really, further examination [96, 98] proved that also the first reduction step involves a relatively slow reorganization of the initially electrogenerated monoanion to a new species, which it has been proposed to be a cubane isomer [96].

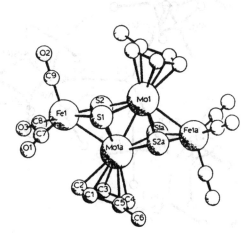

Fig. 42. Perspective view of $Mo_2Fe_2(S)_4(C_5H_4Me)_2(CO)_6$. Mo–Mo, 2.62 Å; Mo–Fe, 2.85 Å; Mo---Fe, 3.61 Å; Mo1–S1, 2.46 Å; Mo–S$_{(capping)}$ average, 2.46 Å; Fe–S$_{(capping)}$ average, 2.23 Å [from Ref. 97]

3.3.3. Trigonal Planar Cores

As illustrated in Fig. 43, the family $M[M'(NR_2)_2]_3$ (M = Pt, Pd; M' = Ge, Sn; R = SiMe$_3$) possesses a trigonal planar geometry [99, 100].

These complexes show no tendency to support redox changes, in that they undergo only irreversible reduction steps [101].

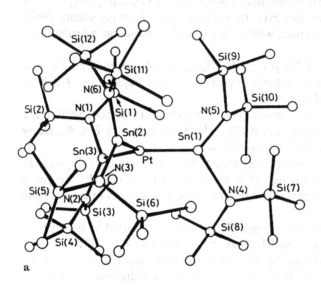

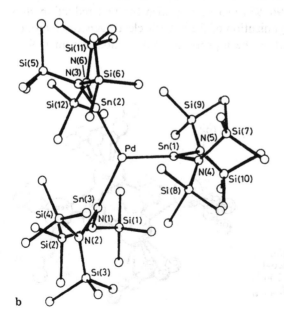

Fig. 43a, b. Molecular structures of: **a)** Pt{Sn[N(SiMe$_3$)$_2$]$_2$}$_3$. Pt–Sn mean bond length, 2.49 Å [from Ref. 99]; **b)** Pd{Sn[N(SiMe$_3$)$_2$]$_2$}$_3$. Pd–Sn mean bond length, 2.53 Å [from Ref. 100]

3.4 Butterfly Frames

Figure 44, which refers to $Pt_2Co_2(\mu-CO)_3(CO)_5(PPh_3)_2$ [102, 103], illustrates the typical tetrametallic assembly in the so-called "open butterfly" geometry [4]. All the edges of the $Co_2Pt(1)$ subunit are bridged by carbonyl groups. The remaining carbonyls are terminally bonded to the hinge cobalt atoms and to one wingtip platinum atom, respectively.

It has been briefly reported that the present cluster undergoes reduction, in Dichloromethane solution, at $Ep = -1.25$ V, as well as oxidation, in propylene carbonate, at $Ep = +1.75$ V [49]. The lifetime of the corresponding monoanion and monocation are unknown. The most interesting electrochemical aspect associated with this species is that it can be electrogenerated as a consequence of the chemical complications following the one-electron reduction of the previously discussed trinuclear complex $PtCo_2(CO)_8(PPh_3)$ (Sect. 2.4.1). Such a redox pathway can be represented by the following equations [49]:

$$PtCo_2(CO)_8(PPh_3) \xrightarrow{+e} [PtCo_2(CO)_8(PPh_3)]^-$$

$$[PtCo(CO)_4(PPh_3)]^{\cdot} + [Co(CO)_4]^-$$

$$\downarrow \text{dimerization}$$

$$1/2 \; Pt_2Co_2(CO)_8(PPh_3)_2$$

Two other electrochemically studied clusters have open butterfly structures: $Fe_2M_2(\mu_3-S)_2(\mu_3-CO)_2(C_5H_4Me)_2(CO)_6$ (M = Mo, W) [96]. The Fe_2Mo_2 complex is just the butterfly isomer of the planar analogue discussed in the preceding

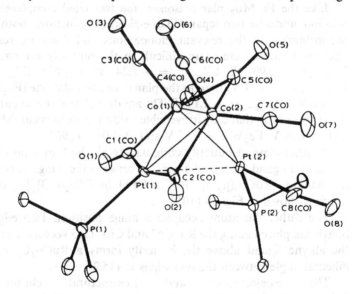

Fig. 44. Perspective view of $Pt_2Co_2(CO)_8(PPh_3)_2$. Co1–Co2, 2.50 Å; Co–Pt1 (average), 2.56 Å; Co–Pt2 (average), 2.54 Å; Pt1---Pt2, 2.99 Å [from Ref. 102]

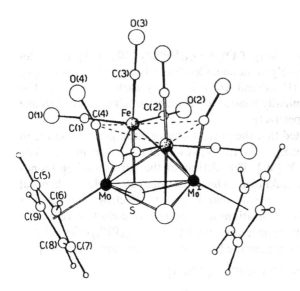

Fig. 45. Perspective view of Fe$_2$Mo$_2$(S)$_2$(CO)$_8$(C$_5$H$_5$)$_2$. Mo–Mo, 2.85 Å; Mo–Fe (average), 2.81 Å; Mo–S (average), 2.33 Å; Fe–S, 2.16 Å; Fe---Fe, 3.83 Å [from Ref. 104]

Sect. 3.3.2. On the basis of spectroscopic measurements, they are thought to be isostructural with Fe$_2$Mo$_2$(μ_3-S)$_2$(μ_3-CO)$_2$(C$_5$H$_5$)$_2$(CO)$_6$, the structure of which is shown in Fig. 45 [104].

Each one of the two Mo$_2$Fe triangles is capped by one triply-bridging sulfur atom. In addition, the two open triangles Fe$_2$Mo are triply-bridged by one carbonyl group. The dihedral angle between the two Mo$_2$Fe wings is 104.1°.

Like the Fe$_2$Mo$_2$ planar isomer, the two cited complexes in Acetonitrile solution, undergo, two separated one-electron reductions, both complicated by the instability of the relevant monoanions and dianions, respectively. Such electron additions are more difficult in the butterfly arrangement (Fe$_2$Mo$_2$: $E°'_{(0/-)} = -0.90$ V, $E°'_{(-/2-)} = -1.24$ V; Fe$_2$W$_2$: $E°'_{(0/-)} = -0.95$ V, $E°'_{(-/2-)} = -1.22$ V) than in the planar rhomboidal one (Fe$_2$Mo$_2$: $E°'_{(0/-)} = -0.68$ V; $E°'_{(-/2-)} = -1.14$ V). The anodic behaviour also parallels that of the planar isomer, in that only irreversible oxidation steps occur (Mo$_2$Fe$_2$: + 1.01 V and + 1.38 V; Fe$_2$W$_2$: + 0.67 V and + 1.08 V) [96].

Another class of butterfly core, having all the four atoms bridged by two atoms of a ligand (or ligands) lying in between the wings, is constituted by the monoanion [RuCo$_3$(μ_3-η^2-PhC≡CPh) (μ-CO)$_2$(CO)$_8$]$^-$, the structure of which is shown in Fig. 46 [105].

The ruthenium atom occupies a hinge position. Two edge-bridging carbonyls are placed along the Ru–Co2 and Co1–Co3 vectors. The arrangement of the alkyne ligand above the butterfly forms a RuCo$_3$C$_2$ octahedron. The dihedral angle between the two wings is 115.2°.

The isoelectronic and isostructural cluster compounds [MCo$_3$(C$_2$Ph$_2$)(CO)$_{10}$]$^-$ (M = Fe, Ru) display multiple redox changes [105, 106]. Typically, Fig. 47 shows that [RuCo$_3$(C$_2$Ph$_2$)(CO)$_{10}$]$^-$ undergoes firstly a

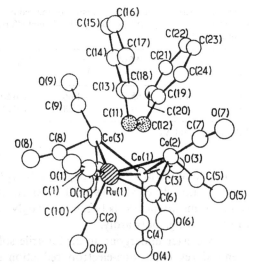

Fig. 46. Molecular structure of [RuCo₃(C₂Ph₂)(CO)₁₀]⁻. Ru–Co1, 2.72 Å; Ru–Co2, 2.52 Å; Ru–Co3, 2.49 Å; Co1–Co2, 2.48 Å; Co1–Co3, 2.52 Å; Ru–C11, 2.13 Å; Co1–C12, 2.14 Å; Co2–C11, 2.07 Å; Co2–C12, 2.14 Å; Co3–C11, 2.12 Å; Co3–C12, 2.06 Å; C11–C12, 1.34 Å; Co2----Co3, 3.55 Å [from Ref. 105]

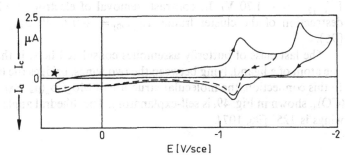

Fig. 47. Cyclic voltammetric response exhibited at a gold electrode by [RuCo₃(C₂Ph₂)(CO)₁₀]⁻ in DMF solution. Scan rate 0.1 Vs⁻¹ [from Ref. 106]

one-electron reduction, chemically reversible in the period of cyclic voltammetry. The initially electrogenerated dianion [RuCo₃(C₂Ph₂)(CO)₁₀]²⁻ displays a second one-electron cathodic step, complicated by fast decarbonylation of the instantaneously generated [RuCo₃(C₂Ph₂)(CO)₁₀]³⁻.

With a longer period of macroelectrolysis, the dianion [RuCo₃(C₂Ph₂)(CO)₁₀]²⁻ undergoes decomposition reactions. By performing such exhaustive one-electron reduction in the presence of PPh₃, the stable dianion [RuCo₃(C₂Ph₂)(CO)₉(PPh₃)]²⁻ may be obtained. The cluster also undergoes an irreversible four-electron oxidation step.

The redox potentials for the Fe and Ru clusters are summarized in Table 17.

Based on the substantial reversibility of the −/2− electron transfer, it is likely that the dianion [MCo₃(C₂Ph₂)(CO)₁₀]²⁻ (M = Fe, Ru), even if short-lived ($t_{1/2}$ ≈ some seconds), maintains a structure not significantly different from that illustrated for the monoanion.

Table 17. Electrochemical characteristics of the redox changes exhibited by [MCo$_3$(C$_2$Ph$_2$)(CO)$_{10}$]$^-$ in Dimethylformamide solution [106]

M	$E^{o'}_{(-/2-)}$ (V)	ΔEp^a (mV)	$Ep_{(2-/3-)}$ (V)	$Ep_{(3+/-)}$ (V)
Fe	−1.06	65	−1.66	+0.50
Ru	−1.14	70	−1.68	+0.55

^{a}Measured at 0.1 Vs^{-1}

The complex Mo$_2$Ni$_2$(μ$_3$–S)$_4$(C$_5$H$_4$Me)$_2$ (CO)$_2$ also belongs to this second class of butterfly clusters [96, 107]. As shown in Fig. 48, two triply-bridging sulfur atoms are placed between the Mo$_2$Ni wings. The two other sulfur atoms cap these wings outwardly.

This cluster undergoes, in acetonitrile solution, two chemical and electrochemical reversible one-electron reduction steps to the corresponding (and likely isostructural) monoanion and dianion, respectively (E$^{o'}_{(0/-)}$ = −1.17 V; E$^{o'}_{(-/2-)}$ = −1.70 V). In contrast, removal of electrons leads to irreversible destruction of the cluster frame (Ep$_{(+/0)}$ = +0.30 V; Ep$_{(2+/+)}$ = +0.50 V) [96].

The last class of butterfly assemblies considered here, is that in which only one atom of a ligand, lying between the two wings, bridges the four metal atoms. In this connection, the molecular structure of Mo$_2$Co$_2$(μ$_4$–S)(μ$_3$–S)$_2$(C$_5$H$_4$Me)$_2$ (CO)$_4$, shown in Fig. 49, is self-explanatory. The dihedral angle between the two wings is 125° [96, 107].

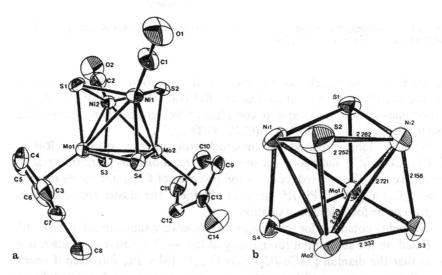

Fig. 48a, b. Perspective view **a)** and selected core bond lengths **b)** of Mo$_2$Ni$_2$(S)$_4$(C$_5$H$_4$Me)$_2$(CO)$_2$. Ni----Ni, 2.96 Å [from Refs. 96, 107]

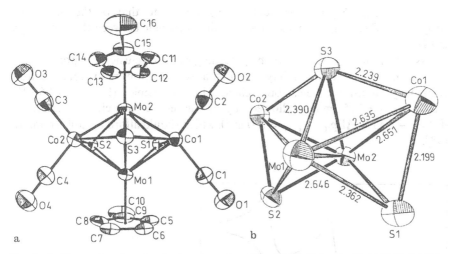

Fig. 49a, b. Perspective view a) and selected core bond lengths b) of $Mo_2Co_2(S)_3(C_5H_4Me)_2(CO)_4$ [from Refs. 96, 107]

In acetonitrile solution, the cluster undergoes a chemical and electrochemical reversible two-electron reduction to $[Mo_2Co_2(\mu_4-S)(\mu_3-S)_2(C_5H_4Me)_2(CO)_4]^{2-}$ ($E^{o'} = -0.99$ V), and a further one-electron addition to the corresponding trianion which likely undergoes significant geometrical reorganization ($E^{o'} = -1.33$ V). Irreversible oxidations occurs at Ep = +0.38 V and +0.68 V, respectively [96].

Strictly related to this species is $Mo_2Fe_2(\mu_4-S)(\mu_3-S)_2(C_5H_4Me)_2(CO)_6$, the structure of which is shown in Fig. 50 [98]. The geometrical assembly of the

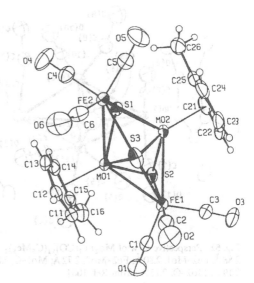

Fig. 50. Perspective view of $Mo_2Fe_2(S)_3(C_5H_4Me)_2(CO)_6$. Mo–Mo, 2.67 Å; Fe1–Mo (average), 2.80 Å; Fe2–Mo (average), 2.80 Å; Fe–S3 (average), 2.55 Å; Mo–S3 (average), 2.31 Å; Mo–S1, Mo–S2 (average), 2.37 Å; Fe1–S1, Fe2–S1, 2.23 Å; Fe----Fe, 4.64 Å [from Ref. 98]

Mo$_2$Fe$_2$(μ_4-S)(μ_3-S)$_2$ core is qualitatively similar to that of the preceding Mo$_2$Co$_2$(μ_4-S)(μ_3-S)$_2$ complex (now, the dihedral angle between the two wings is 141°).

As shown in Fig. 51, just like the preceding Mo$_2$Co$_2$ cluster, the present one in Acetonitrile solution, undergoes, first a two-electron reduction, electrochemically quasireversible, but chemically reversible (E$^{o\prime}$ = − 0.96 V, ΔE$p_{(0.1\ Vs^{-1})}$ = 215 mV), followed by a successive one-elecron addition (E$^{o\prime}$ = − 1.55 V, ΔE$p_{(0.1\ Vs^{-1})}$ = 100 mV), which should afford a stable trianion [98].

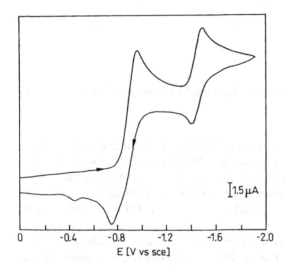

Fig. 51. Cyclic voltammogram recorded at a platinum electrode on a MeCN solution of Mo$_2$Fe$_2$(S)$_3$(C$_5$H$_4$Me)$_2$(CO)$_6$. Scan rate 0.1 Vs^{-1} [from Ref. 98]

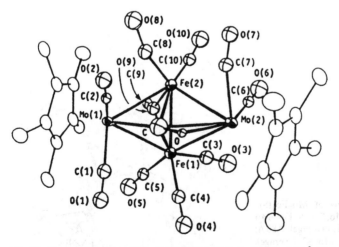

Fig. 52. Perspective view of Mo$_2$Fe$_2$(CO)$_{11}$(C$_5$Me$_5$)$_2$. Fe–Fe, 2.55 Å; Fe1–Mo1, 2.86 Å; Fe1–Mo2, 2.88 Å; Fe2–Mo1, 2.80 Å; Fe2–Mo2, 2.72 Å; Mo1–C, 2.06 Å; Mo2–C, 2.54 Å; Fe1–C, 2.25 Å; Fe2–C, 2.19 Å; Mo2–O, 2.15 Å [from Ref. 108]

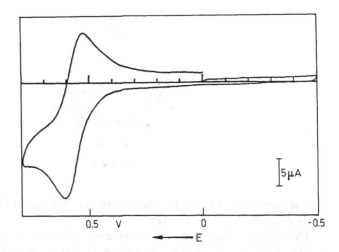

Fig. 53. Cyclic voltammetric response exhibited by $Mo_2Fe_2(CO)_{11}(C_5Me_5)_2$ in CH_2Cl_2 solution. Platinum working electrode. Scan rate 0.1 Vs^{-1} [from Ref. 108 (Supplementary Material)]

Also in this case, electron removal steps are framework destroying (Ep = + 0.57 V, and + 0.99 V, respectively).

Finally, the cluster $Mo_2Fe_2(\mu_4-CO)(\mu-CO)(C_5Me_5)_2(CO)_9$ also belongs to this third class of butterfly geometries. In fact, as shown in Fig. 52, the Mo_2Fe_2 butterfly is quadruply-bridged by a Carbon-carbonyl atom; the Oxygen atom of such capping carbonyl group binds one wingtip molybdenum atom [108]. Structural asymmetry in the pseudo bipyramidal trigonal Mo_2Fe_2C core is induced by the presence of three terminal carbonyls coordinated to Fe1, but only two to Fe2. One edge-bridging carbonyl is present along the Fe-Fe hinge.

As shown in Fig. 53, this cluster in dichloromethane solution, undergoes, a one-electron removal, chemically reversible in the short period of cyclic voltammetry ($E^{\circ\prime}$ = + 0.57 V; $\Delta Ep_{(0.1\ Vs^{-1})}$ = 81 mV). Nevertheless, chemical oxidation by $AgPF_6$ shows that the corresponding monocation rapidly decomposes in solution [108].

3.5 Tetrahedral Frames

A number of "pure carbonyl" terahedral clusters have been characterized both from the structural and electrochemical viewpoint.

$Co_2Rh_2(CO)_9(\mu-CO)_3$ has assigned the structure shown in Scheme 8 [106–109], with the basal Rh_2Co face defined by the three coplanar, edge-bridging carbonyl groups.

This cluster is unable to undergo non-destructive redox changes. In fact, in 1,2-Dichloroethane solution, it exhibits two irreversible one-electron reductions ($Ep_{(0/-)}$ = -0.35 V; $Ep_{(-/2-)}$ = -1.20 V), as well as an irreversible four-electron oxidation (Ep = + 1.34 V) [113].

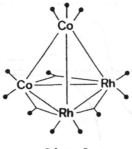

Scheme 8

The geometry of the isostructural anions $[MCo_3(\mu-CO)_3(CO)_9]^-$ (M = Fe, Ru) is illustrated in Fig. 54 [114].

The $RuCo_3$ tetrahedron (Ru–Co (average), 2.63 Å; Co–Co (average), 2.53 Å) has the apical site occupied by the Ruthenium atom, which bears three terminal carbonyls. Each one of the three basal cobalt atoms has two terminal carbonyls. In addition, each Co–Co edge is μ-bridged by a carbonyl group lying in the basal plane.

Fig. 55, which refers to $[FeCo_3(CO)_{12}]^-$, shows the typical electrochemical response exhibited by these anions in Dimethylformamide solution [106].

Two closely-spaced one-electron reductions, with features of chemical reversibility, are displayed, suggesting the existence of $[FeCo_3(CO)_{12}]^{2-}$ and $[FeCo_3(CO)_{12}]^{3-}$, at least as transients. Nevertheless, some doubt exists that this may be the true response of $[MCo_3(CO)_{12}]^-$. In fact, by changing the solvent (acetonitrile, dichloroethane, propylene carbonate), the reduction pattern changes in a single-stepped two-electron process, irreversible in character. Since the addition in such solutions of PPh_3 (which affords

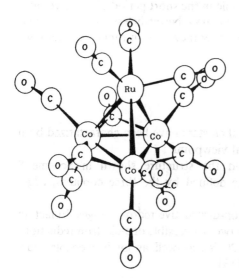

Fig. 54. Schematic representation of the structure of the monoanion $[RuCo_3(CO)_{12}]^-$

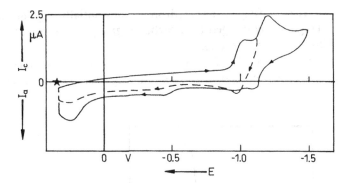

Fig. 55. Cyclic voltammogram exhibited by [FeCO$_3$(CO)$_{12}$](NEt$_4$) in DMF solution. Gold working electrode. Scan rate 0.1 Vs^{-1} [from Ref. 106]

[MCo$_3$(CO)$_{11}$(PPh$_3$)]$^-$) again causes a splitting in two distinct one-electron transfers, it is presumed that the response in Fig. 55 may be attributed to the preliminary formation of [FeCo$_3$(CO)$_{11}$(DMF)]$^-$ [106]. Finally, these species undergo an irreversible four-electron oxidation step. The redox potentials of these electron transfers are summarized in Table 18.

Figure 56 shows the molecular structure of [BiFe$_3$(μ_3–CO)(CO)$_9$]$^-$ [115].

Table 18. Electrochemical characteristics of the redox changes exhibited by [MCo$_3$(CO)$_{12}$]$^-$ [106]

M	$E^{o'}_{(-/2-)}$ (V)	$\Delta E p^a$ (mV)	$E^{o'}_{(2-/3-)}$ (V)	$\Delta E p^a$ (mV)	$E p_{(-/3-)}$ (V)	$E p_{(3+/-)}$ (V)	Solvent
Fe	−0.99	80	−1.24	110	—	+0.75	DMF
	—	—	—	—	−1.22	—	C$_2$H$_4$Cl$_2$
	—	—	—	—	−1.14	—	MeCN
	—	—	—	—	−1.08	—	PC^b
Ru	−1.12	95	−1.32^c	—	—	+0.72	DMF

^{a}Measured at 0.1 Vs^{-1}; ^{b}PC = Propylene carbonate; ^{c}peak potential value for an irreversible step

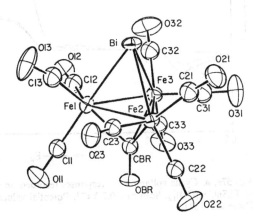

Fig. 56. Perspective view of [BiFe$_3$(CO)$_{10}$]$^-$. Bi–Fe (average), 2.65 Å; Fe–Fe (average), 2.64 Å; Fe–C$_{(capping)}$ (average), 2.07 Å [from Ref. 115]

The Fe₃Bi₍apical₎ tetrahedron is substantially regular, with the basal Fe₃ triangle triply-bridged by a carbonyl group. Each iron atom is coordinated to three terminal carbonyls.

It has been briefly reported that this cluster is able in dichloromethane solution, to undergo, only a declustering oxidation process at Ep = 0.6 V [116].

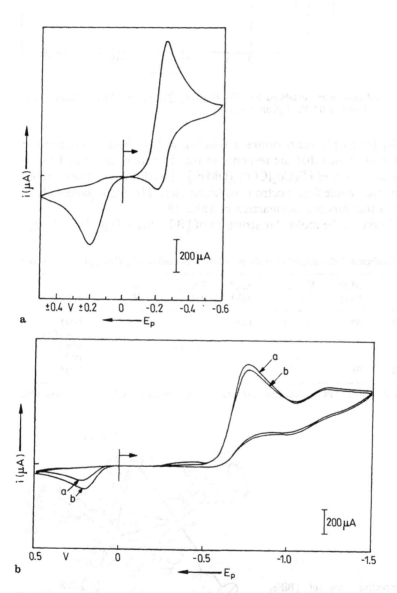

Fig. 57a, b. Cyclic voltammetric responses recorded on CH_2Cl_2 solutions of: **a)** $PhSiCo_3(CO)_{11}$; **b)** $PhSnCo_3(CO)_{12}$. Scan rate 0.2 Vs^{-1}. Potential values refer to a pseudoreference Ag electrode [from Ref. 122]

A few hybrid (transition metal-main group element) clusters, namely PhSiCo$_3$(μ-CO)(CO)$_{10}$, PhSnCo$_3$(CO)$_{12}$, and PhSnMn$_3$(CO)$_{15}$, have been assigned [117–121] the "open" tetrahedral assemblies illustrated in Scheme 9.

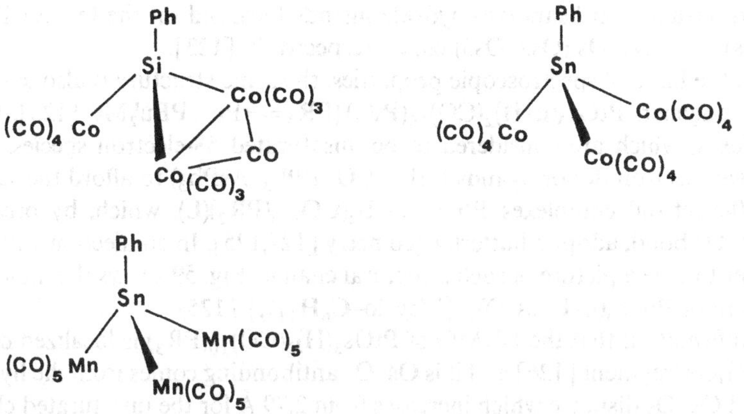

Scheme 9

Figure 57 shows the most significant redox change exhibited by the two cobalt derivatives, in Dichloromethane solution [122].

Both the complexes undergo a one-electron reduction step (SiCo$_3$: $E_{p(0/-)} = -0.3$ V; SnCo$_3$: $E_{p(0/-)} = -0.8$ V), which exhibits some degree of chemical reversibility for PhSiCo$_3$(CO)$_{11}$, but it is not at all reversible for PhSnCo$_3$(CO)$_{12}$. In both cases, the presence of the reoxidation peak at $+0.2$ V is indicative that the decomposition of the corresponding monoanions generates, among a few indefinable products, [Co(CO)$_4$]$^-$.

It has been suggested that this redox pattern arises because the presence of one Co–Co bond in the SiCo$_3$ cluster seems to increase the stability towards electron addition, with respect to the totally open SnCo$_3$ cluster. This would suggest that the propensity of clusters to undergo multiple redox changes is related to their metal-metal bonding network. In the present case, this (perhaps too naive) speculation seems reinforced by the fact that the "closed" tetrahedral PhCCo$_3$(CO)$_9$, with three Co-Co bonds, is able to afford a particularly stable monoanion [13].

PhSnMn$_3$(CO)$_{15}$ in acetonitrile solution, undergoes, two separate two-electron removals, irreversible in character (Ep = $+1.40$ V; $+1.86$ V), which involve the sequential cleavage of the tin-manganese bonds, according to [35]:

$$\text{PhSn[Mn(CO)}_5\text{]}_3 \xrightarrow[\text{MeCN}]{-2e} [\{\text{Mn(CO)}_5\}_2\text{SnPh(MeCN)}_2]^+ + [\text{Mn(CO)}_5(\text{MeCN})]^+$$

$$\downarrow -2e \quad \text{MeCN}$$

$$[\text{Mn(CO)}_5\text{SnPh(MeCN)}_3]^{2+} + [\text{Mn(CO)}_5(\text{MeCN})]^+$$

It also undergoes a totally declustering four electron reduction process (Ep = −1.69 V), which mainly affords [Mn(CO)$_5$]$^-$ [35].

Figure 58 shows the structure of PtOs$_3$(μ-H)$_2$(CO)$_{10}$ {P(cyclo-C$_6$H$_{11}$)$_3$}, which is the first of a series of "closo" tetrahedral, mixed-ligand clusters. The Pt-capped tetrahedron bears two hydride ligands localized on the longest Pt–Os (Pt–Os1) and Os–Os (Os2–Os3) edges, respectively [123].

On the basis of spectroscopic properties, the same structure is also assigned to the congeners PtOs$_3$(μ-H)$_2$(CO)$_{10}$(PR$_3$) (PR$_3$ = PPh$_3$, PBu$^t_2$Me) [123]. These complexes, which are considered to be unsaturated 58-electron species, react with two-electron donor ligands L (L = CO, PPh$_3$, AsPh$_3$) to afford the saturated (60-electron) complexes PtOs$_3$(μ-H)$_2$(CO)$_{10}$(PR$_3$)(L), which, by breaking one Pt–Os bond, adopt a butterfly geometry [124,125]. In connection with this, in order to give a picture of such structural change, Fig. 59 shows the molecular structure of PtOs$_3$(μ-H)$_2$(CO)$_{11}$ {P(cyclo-C$_6$H$_{11}$)$_3$} [125].

Confirmation that the LUMO of PtOs$_3$(H)$_2$(CO)$_{10}$(PR$_3$) is localized on the Os(μ-H)Os fragment [126] and it is Os–Os antibonding comes from the hydride bridged Os–Os distance which increases from 2.79 Å for the unsaturated cluster to 2.87 Å and 3.04 Å for the saturated clusters with L = CO and PPh$_3$, respectively.

Paralleling the chemical pathway, PtOs$_3$(H)$_2$(CO)$_{10}$(PPh$_3$) displays, in Dichloromethane solution, a chemically reversible, two-electron reduction (E$^{o\prime}$ = −0.81 V), which is probably accompanied by the concomitant tetrahedral/butterfly reorganization and is quasireversible in character. The peak-to-

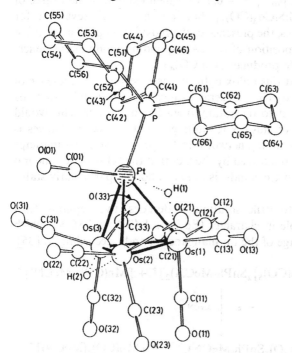

Fig. 58. Molecular structure of PtOs$_3$(H)$_2$(CO)$_{10}$ {P(C$_6$H$_{11}$)$_3$}. Pt–Os1, 2.86 Å; Pt–Os2, 2.79 Å; Pt–Os3, 2.83 Å; Os1–Os2, 2.78 Å; Os2–Os3, 2.79 Å; Os1–Os3, 2.74 Å [from Ref. 123]

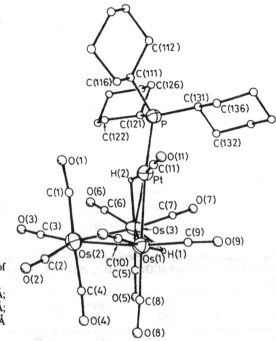

Fig. 59. Molecular structure of PtOs$_3$(H)$_2$(CO)$_{11}$\{P(C$_6$H$_{11}$)$_3$\}. Os1–Os3 hinge distance, 2.87 Å; Pt–Os1, 2.73 Å; Pt–Os3, 2.91 Å; O2–Os1, 2.88 Å; Os2–Os3, 2.88 Å [from Ref. 125]

peak separation (at 0.1 Vs^{-1}) is 180 mV, markedly higher than the value of 29 mV expected for a non-stereodynamic two-electron transfer [124].

The redox propensity of a series of germanium-capped tetrahedral clusters of general formula Co$_2$M(μ_3–GeR)(CO)$_{9-n}$(C$_5$H$_5$)$_n$ (R = Me, Ph; M = Co, n = 0; M = Mo, n = 1) has been investigated [51]. They are structurally similar to the heterotrinuclear complexes capped by alkylidyne and phosphinidene units described in Sects. 2.4.2, 2.4.3. This can be confirmed by the structure of the complex with R = Co(CO)$_4$ [127].

At variance with the C$_{(capped)}$ and P$_{(capped)}$ species, which generally gave unstable anions upon reduction, the germylidyne complexes undergo an initial one-electron reduction step, which affords the quite stable monoanions [Co$_2$M(Ge–R)(CO)$_{9-n}$(C$_5$H$_5$)$_n$]$^-$. The successive reduction to the dianion is complicated by following decomposition. As an example, Fig. 60 illustrates the cathodic behaviour of Co$_2$Mo(Ge–Me)(CO)$_8$(C$_5$H$_5$) [51].

Table 19 summarizes the relevant redox potentials.

Since it is commonly assumed that the LUMO of the CM$_3$, PM$_3$, GeM$_3$ complexes is centred on the basal metallic triangle, one would expect no influence of the capping units on their redox behaviour. Nevertheless, one must also consider that the different capping units interact to a different extent with the M$_3$ fragment, in accordance with their ability to push or to pull the electron density. This means that the M–M bonding interactions can be indirectly perturbed by the capping unit, and this is reflected in the thermodynamic and kinetic aspects of their redox properties [51].

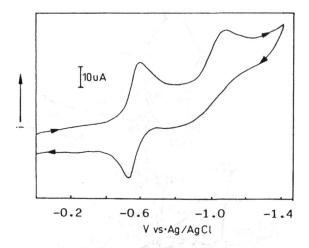

Fig. 60. Cyclic voltammogram exhibited by $Co_2Mo(Ge-Me)(CO)_8(C_5H_5)$ in CH_2Cl_2 solution. Scan rate $0.1\ Vs^{-1}$ [from Ref. 51]

Table 19. Electrochemical characteristics for the reduction steps exhibited by $Co_2M(Ge-R)(CO)_{9-n}(C_5H_5)_n$ in CH_2Cl_2 solution [51]

M	n	R	$E^{o'}_{(0/-)}$ (V)	ΔE_p^a (mV)	$Ep_{(-/2-)}^a$ (V)
Co	0	Me	−0.36	80	−1.22
		Ph	−0.35	85	−1.31
Mo	1	Me	−0.60	100	−1.16
		Ph	−0.65	140	—

[a] Measured at $0.2\ Vs^{-1}$

Scheme 10 depicts the structure assigned to $Co_2Rh_2(CO)_6(\mu-CO)_3\{HC(PPh_2)_3\}$ [128]. The tripodal polyphosphine $HC(PPh_2)_3$ binds to the basal Rh_2Co face of the metallic tetrahedron.

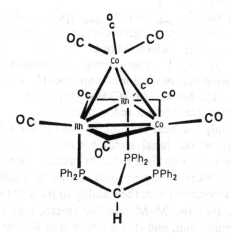

Scheme 10

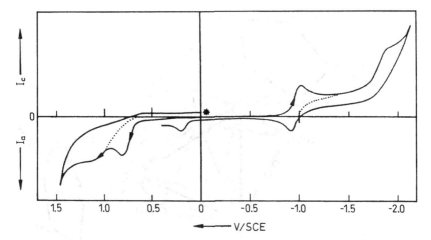

Fig. 61. Cyclic voltammogram recorded at a platinum electrode on a $C_2H_4Cl_2$ solution of $Co_2Rh_2(CO)_9\{HC(PPh_2)_3\}$. Scan rate 0.1 Vs^{-1} [from Ref. 113]

As illustrated in Fig. 61, such a cluster in 1,2-dichloroethane solution, undergoes, an initial (one-electron) reduction, with features of chemical reversibility ($E°'_{(0/-)} = -0.94$ V, $\Delta Ep_{(0.1\ Vs^{-1})} = 100$ mV), followed by a second irreversible step ($Ep_{(-/2-)} = -1.9$ V). In addition, an irreversible one-electron oxidation occurs ($Ep_{(+/0)} = +0.76$ V) [113].

Comparison with the response exhibited by the precursor $Co_2Rh_2(CO)_{12}$ immediately suggests that the presence of the basal-clasping tripodal ligand significantly enhances the stability of $[Co_2Rh_2(CO)_9\{HC(PPh_2)_3\}]^-$ with respect to that of $[Co_2Rh_2(CO)_{12}]^-$. Even if the monoanion is not indefinitely stable, its lifetime is sufficiently long to allow recording of its EPR spectrum [113].

The addition of one electron to the metal-metal antibonding LUMO of the tetrahedral cluster must induce important geometrical reorganizations, but there are no structural determinations to confirm it, unless one takes into account the quasireversibility of the relevant electrochemical response.

Fig. 62 shows the molecular structure of $Cr_3Co(\mu_3-S)_4(CO)(C_5H_4Me)_3$ [129].

As in the case of the unmethylated-cyclopentadienyl analogue $Cr_3Co(\mu_3-S)_4(CO)(C_5H_5)_3$ [130], each face of the metallic Cr_3Co tetrahedron is capped by a triply-bridging sulfur atom according to a pseudo-cubane assembly. The Cr–Co bond lengths are significantly shorter (0.13 Å) than the Cr–Cr ones.

This compound displays a remarkable ability to lose electrons, in that in Dimethyl formaide solution, it undergoes three reversible oxidation steps ($E°'_{(0/+)} = -1.07$ V; $E°'_{(+/2+)} = -0.23$ V; $E°'_{(2+/3+)} = +0.23$ V). A further single-stepped two-electron removal (Ep = +0.80 V) results declustering [129].

Analogously, the cubane trianion $[Fe_3Mo(S)_4(SEt)_3(CO)_3]^{3-}$, the structure of which is illustrated in Fig. 63, also exhibits a remarkable redox propensity. In fact, in dichloromethane solution, it undergoes both a reversible one-electron

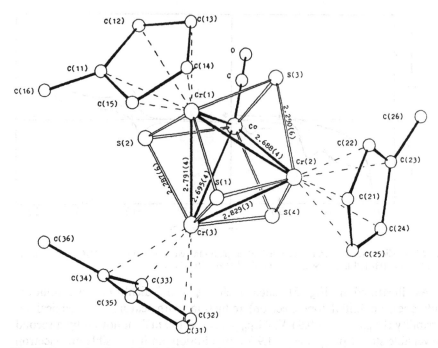

Fig. 62. Perspective view of Cr$_3$Co(S)$_4$(CO)(C$_5$H$_4$Me)$_3$, with selected bond lengths [from Ref. 129]

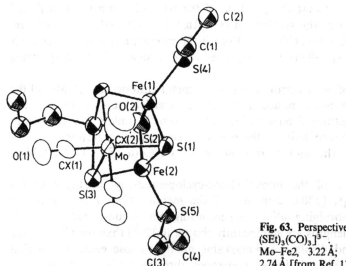

Fig. 63. Perspective view of [Fe$_3$Mo(S)$_4$(SEt)$_3$(CO)$_3$]$^{3-}$. Mo–Fe1, 3.27 Å; Mo–Fe2, 3.22 Å; Fe1–Fe2, Fe2–Fe2′, 2.74 Å [from Ref. 131]

oxidation (E$^{o'}_{(2-/3-)}$ = −0.36 V) and a reversible one-electron reduction (E$^{o'}_{(3-/4-)}$ = −0.89 V) [131].

Finally, Fig. 64 illustrates the molecular structure of Mo$_2$Co$_2$(µ–CO)(CO)$_6$(C$_5$H$_5$){1,2-(P–Bu^t)$_2$C$_6$H$_4$} [132].

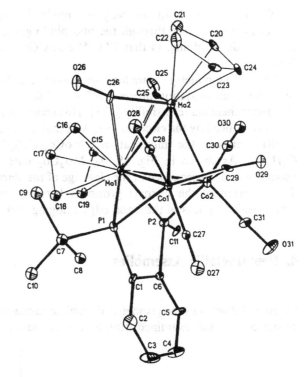

Fig. 64. Perspective view of $Mo_2Co_2(CO)_7(C_5H_5)_2\{1,2-(P-Bu^t)_2C_6H_4\}$. Mean bond lengths in the metal core: Mo1–Co, 2.89 Å; Mo2–Co, 2.79 Å; Co–Co, 2.49 Å; Mo–Mo, 2.87 Å [from Ref. 132]

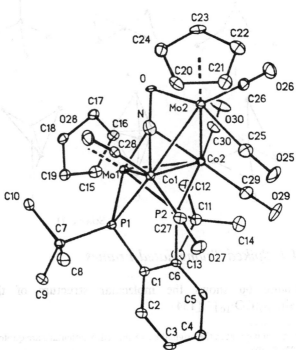

Fig. 65. Molecular structure of $[Mo_2Co_2(CO)_6(C_5H_5)_2\{1,2-(P-Bu^t)_2C_6H_4\}(NO)]^+$. Mean bond lengths in the metal core: Mo1–Co, 2.68 Å; Mo2–Co, 2.78 Å; Co–Co, 2.54 Å [from Ref. 133]

The o-Phenylenebis(μ–tert-butylphosphido) ligand coordinates to the basal face of the tetrahedron through the phosphido groups. Only the basal CO(29) group is edge-bridging, in that CO(25) and CO(26) are semibridging towards Mo1.

Cyclic voltammetry shows that this complex, in Acetonitrile solution, is able to undergo an oxidation process ($E^{o\prime} = +1.03$ V*) markedly departing from the electrochemical reversibility [133]. This means that a significant geometrical strain is induced by the electron loss. As a matter of fact, chemical oxidation by NOBF$_4$ affords the monocation [MoCo$_2$(CO)$_6$(C$_5$H$_5$)$_2$(NO){1,2-(P–Bu^t)$_2$C$_6$H$_4$}]$^+$. As shown in Fig. 65, the Mo$_2$Co$_2$ core now assumes a butterfly geometry as a consequence of the cleavage of the starting Mo–Mo bond [133]. The insertion of the nitrosyl group in between the wingtip Mo atoms is accompanied by loss of the basal edge-bridging carbonyl group.

4. Pentametallic Assemblies

Scheme 11 illustrates the geometrical architectures assumed by the pentanuclear cluster compounds examined from an electrochemical viewpoint.

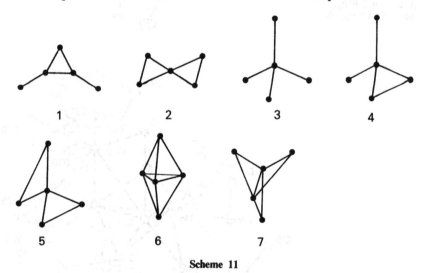

Scheme 11

4.1 Spiked Triangulated Frames

Figure 66 shows the molecular structure of the monoanion [HRu$_3$(SiEt$_3$)$_2$(CO)$_{10}$]$^-$ [134].

---

* Assuming that in the original paper the redox potentials are quoted versus the ferrocenium/ferrocene couple

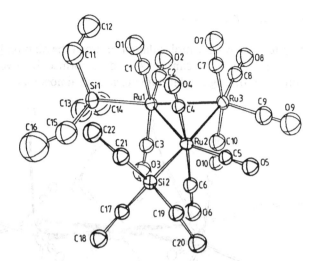

Fig. 66. Perspective view of $[HRu_3(SiEt_3)_2(CO)_{10}]^-$.
Mean bond lengths:
Ru3–Ru1 ≈ Ru3–Ru2, 2.90 Å; Ru1–Ru2, 3.00 Å;
Ru1–Si1 ≈ Ru2–Si2, 2.45 Å
[from Ref. 134]

Two SiEt$_3$ fragments coordinate to two vertices of the isosceles Ru$_3$ triangle.

Figure 67 shows that the cluster in Dichloromethane solution, undergoes, a one-electron oxidation ($E^{\circ\prime}$ = +0.05 V), which, as suggested by the lack of a peak on the reverse scan at low scan rate, generates the transient neutral congener HRu$_3$(SiEt$_3$)$_2$(CO)$_{10}$ (0.2 s ⩽ $t_{1/2}$ ⩽ 2 s) [134].

It is useful to note that the precursor $[HRu_3(CO)_{11}]^-$ undergoes a one-electron oxidation just at the same potential value [13], indicating that the electron density of the metallic triangle, from which the electron is removed, is not significantly altered by substituting one edge-bridging CO group for two SiEt$_3$ groups.

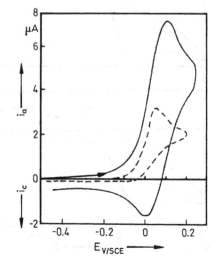

Fig. 67. Cyclic voltammograms recorded at a platinum electrode on a CH$_2$Cl$_2$ solution of $[HRu_3(SiEt_3)_2(CO)_{10}]^-$. Scan rate: (---) 0.1 Vs^{-1}; (——) 1 Vs^{-1} [from Ref. 134]

4.2 Planar Frames

Figure 68 shows the molecular structure of the dianion $[Fe_4Pt(CO)_{16}]^{2-}$ [135].

The framework can be viewed as formed by two $Fe_2(CO)_8$ units interconnected through a platinum atom in an almost planar fashion (the two Fe_2–Pt

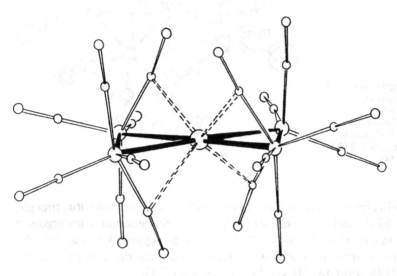

Fig. 68. Perspective view of $[Fe_4Pt(CO)_{16}]^{2-}$. Fe–Fe, 2.71 Å; Fe–Pt, 2.60 Å [from Ref. 135]

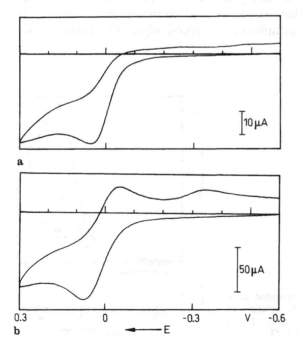

Fig. 69a, b. Cyclic voltammograms recorded at a platinum electrode on a MeCN solution of $[Fe_4Pt(CO)_{16}]^{2-}$. Scan rate: a) 0.2 Vs^{-1}; b) 2.0 Vs^{-1}

triangles form a dihedral angle of about 7°; in the homologue $[Fe_4Pd(CO)_{16}]^{2-}$ the geometry is quite planar). Each iron atom bears three terminal carbonyls, while the fourth carbonyl is semibridging the central platinum atom.

The most interesting redox change exhibited by the present cluster is its one-electron oxidation process, which is illustrated in Fig. 69. As indicated by the lack of the relevant backward response at slow scan rates, the primarily electrogenerated monoanion $[Fe_4Pt(CO)_{16}]^-$ is unstable, but the increase in the scan rate is sufficient to compete with the rate of such a decomposition [136].

An $E^{o\prime}$ value of + 0.01 V is attributable to the $[Fe_4Pt(CO)_{16}]^{-/2-}$ electron transfer, and the monoanion is assigned a lifetime of 1 sec $\leqslant t_{1/2} \leqslant 0.1$ s.

The dianion also undergoes, in Acetonitrile solution, an irreversible one-electron reduction (Ep = − 1.57 V), likely causing a deep structural modification to a species, able to undergo almost reversibly, a further one-electron reduction ($E^{o\prime}$ = − 1.68 V).

4.3 Tetrahedral-like Frames

Figure 70 shows the molecular structure of the dianion $[PbFe_4(CO)_{16}]^{2-}$ [134]. It can be viewed as a distorted tetrahedral complex of Pb (IV), bearing as ligands, two mononuclear $[Fe(CO)_4]^{2-}$ units and one dinuclear $[Fe_2(CO)_8]^{2-}$ fragment. Only this last moiety possesses two edge-bridging carbonyl groups.

Chemical oxidation by $[Cu(NCMe)_4](BF_4)$ affords the neutral $PbFe_4(CO)_{16}$ [138], the structure of which is shown in Fig. 71.

The main change induced by the two-electron removal is the fusion of the two mononuclear iron ligands in a further dinuclear fragment. In addition, all the carbonyls become terminal.

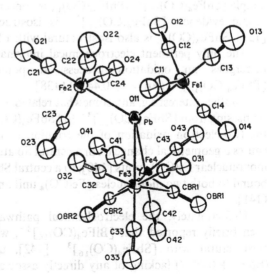

Fig. 70. Perspective view of $[PbFe_4(CO)_{16}]^{2-}$. Mean bond lengths: Pb–Fe1 ≈ Pb–Fe2, 2.66 Å; Pb–Fe3 ≈ Pb–Fe4, 2.83 Å; Fe3–Fe4, 2.62 Å [from Ref. 137]

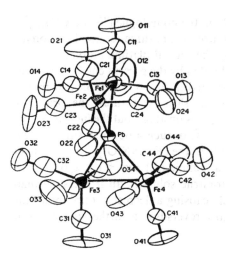

Fig. 71. Molecular structure of PbFe$_4$(CO)$_{16}$. Pb–Fe mean distance, 2.62 Å; Fe1–Fe2 ≈ Fe3–Fe4, average, 2.90 Å [from Ref. 138]

The subsequent chemical reduction by cobaltocene restores the starting dianion [PbFe$_4$(CO)$_{16}$]$^{2-}$ [138].

In principle, the chemical reversibility of the 2 − /0 redox change cannot be paralleled by the electrochemical reversibility, because of the underlying large structural reorganization. As a matter of fact, PbFe$_4$(CO)$_{16}$ in Dichloromethane solution, undergoes, a single two-electron reduction (Ep = − 0.64 V), which displays an associated response at very positive potentials (Ep = + 0.26 V). This last value coincides with the peak potential value for the oxidation of [PbFe$_4$(CO)$_{16}$]$^{2-}$ [138]. The peak-to-peak separation of 0.9 V can be qualitatively assumed to be an index of the high reorganizational energy accompanying the described Fe–Fe bond breaking/making process.

A similar stereochemical-redox relationship probably holds as far as the couple [SnFe$_4$(CO)$_{16}$]$^{2-}$/SnFe$_4$(CO)$_{16}$ is concerned. In fact, based on spectroscopic evidence, [SnFe$_4$(CO)$_{16}$]$^{2-}$ is isostructural with [PbFe$_4$(CO)$_{16}$]$^{2-}$ [138]. SnFe$_4$(CO)$_{16}$ is also isostructural with PbFe$_4$(CO)$_{16}$ [139].

The only pertinent electrochemical information is that [SnFe$_4$(CO)$_{16}$]$^{2-}$ undergoes an oxidation process at potentials slightly higher than [PbFe$_4$(CO)$_{16}$]$^{2-}$ (Ep = + 0.44 V) [138].

Another interesting, and somewhat related, structural reorganization occurs on passing from [SbFe$_4$(CO)$_{16}$]$^{3-}$ to [SbFe$_4$(CO)$_{16}$]$^{-}$. As illustrated in Fig. 72, the two-electron oxidation of the trianion to the corresponding monoanion causes a geometrical change from a central Sb atom tetrahedrally bound to four mononuclear Fe(CO)$_4$ units [140], to a central Sb atom (distorted) tetrahedrally bound to both two mononuclear Fe(CO)$_4$ unit and one dinuclear Fe$_2$(CO)$_8$ unit [141].

Unfortunately, the electrochemical pathway is still ill defined. It has been briefly reported that [BiFe$_4$(CO)$_{16}$]$^{3-}$, which, as shown in Fig. 73, is isostructural with [SbFe$_4$(CO)$_{16}$]$^{3-}$ [142], undergoes an oxidation step (Ep = + 0.44 V) lacking of any directly associated reduction response in the

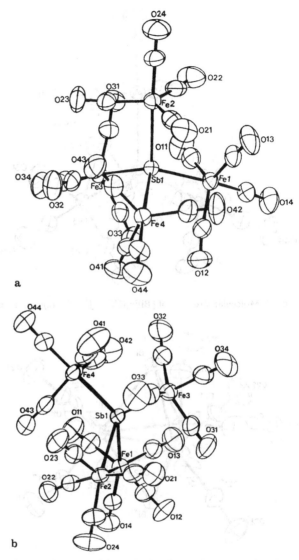

Fig. 72a, b. Perspective view of: a) $[SbFe_4(CO)_{16}]^{3-}$. Average Sb–Fe distance, 2.67 Å [from Ref. 140]; b) $[SbFe_4(CO)_{16}]^-$. Sb–Fe3 ≈ Sb–Fe4, 2.57 Å; Sb–Fe1 ≈ Sb–Fe2, 2.64 Å; Fe1–Fe2, 2.72 Å [from Ref. 141]

reverse scan [116]. This is to be expected, if we assume that the oxidation process may generate $[BiFe_4(CO)_{16}]^-$, isostructural with $[SbFe_4(CO)_{16}]^-$.

4.4 Trigonal Bipyramidal Frames

Figure 74 shows the molecular structure of $Bi_2Fe_3(CO)_9$ [143].

In the trigonal bipyramid, the Fe-Fe distances are significantly longer than the Bi-Fe distances ($\geqslant 0.1$ Å).

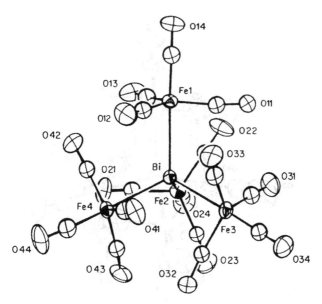

Fig. 73. Molecular structure of $[BiFe_4(CO)_{16}]^{3-}$. Average Bi–Fe distance, 2.75 Å [from Ref. 142]

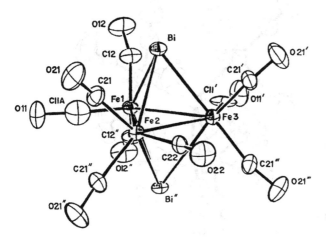

Fig. 74. Perspective view of $Bi_2Fe_3(CO)_9$. Mean bond lengths: Bi–Fe, 2.63 Å; Fe–Fe, 2.74 Å [from Ref. 143]

In Dichloromethane solution, $Bi_2Fe_3(CO)_9$ undergoes, in a stepwise fashion, an initial one-electron reduction, chemically and electrochemically reversible ($E^{\circ\prime} = -0.44$ V), and a second irreversible step (Ep = -1.15 V) [116]. Based on spectroscopic data, it has been proposed that both the generated $[Bi_2Fe_3(CO)_9]^-$ and $[Bi_2Fe_3(CO)_9]^{2-}$ anions assume a square-pyramidal geometry [116]. Indeed, the electrochemical reversibility of the 0/− redox change induce us to foresee that the monoanion probably maintains the trigonal bipyramidal geometry of the neutral parent.

4.5 Star-like Frames

It has been recently shown that the cluster $(\mu-H)_2Os_3(CO)_{10}$, which is one of the most classical electron-deficient molecules, is able to add, stepwisely, two electrons. The first addition leads to the corresponding, di-µ-hydrido-bridged, monoanion, which is moderately stable. The second electron addition generates the relevant dianion, which immediately decomposes [13, 144]. Because of the isolobal analogy between an hydrido ligand and the $Au(PR_3)$ unit, it was expected that the species $(\mu-AuPR_3)_2Os_3(CO)_{10}$, the structure of which is shown in Fig. 75 ($R = PEt_3$) [145], might electrochemically behave like the dihydrido precursor.

As a matter of fact, the complex $(AuPPh_3)_2Os_3(CO)_{10}$ undergoes, in acetonitrile solution (which really does not seem the best suited solvent, because of the

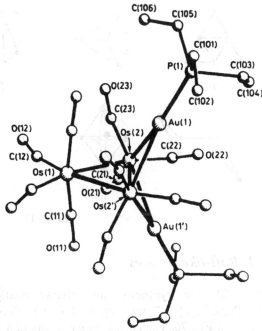

Fig. 75. Molecular structure of $Os_3(AuPEt_3)_2(CO)_{10}$. Os1–Os2, 2.83 Å; Os2–Os2′, 2.68 Å; average Os–Au distance, 2.76 Å [from Ref. 145]

Table 20. Comparison between the redox potentials for the reduction steps exhibited by $H_2Os_3(CO)_{10}$ and $(AuPPh_3)_2Os_3(CO)_{10}$

Redox Change	$E^{o\prime}$ (V)	Solvent	Reference
$[H_2Os_3(CO)_{10}]^{0/-}$	− 0.65	Me_2CO	144
$[(AuPPh_3)_2Os_3(CO)_{10}]^{0/-}$	− 1.35	MeCN	12
$[H_2Os_3(CO)_{10}]^{-/2-}$	− 1.12[a]	Me_2CO	144
$[(AuPPh_3)_2Os_3(CO)_{10}]^{-/2-}$	− 1.62[a]	MeCN	12

[a] Peak potential value for irreversible processes

possibility of nucleophilic attack), an initial one-electron addition, reversible at
− 20 °C, followed by a second one, irreversible in character [12].

As shown in Table 20, a remarkable difference in redox potentials exists
between the dihydrido and the gold-phosphine homologues. The electron
donating power of the phosphine groups, which disfavour the electron addition,
must contribute to this difference.

5 Hexametallic Assemblies

Scheme 12 shows the molecular geometries of the hexanuclear clusters discussed
in this review.

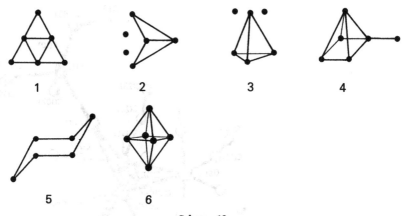

Scheme 12

5.1 Raft-like Frames

In 1980, the synthesis and characterization of the two congeners
$[Fe_3Pt_3(CO)_{15}]^{2-,-}$ was reported [146]. The paramagnetic monoanion is
shown in Fig. 76. The cluster core consists of a planar "raft" metallic array, in
which the platinum atoms form an equilateral triangle, each edge of which is
bridged by one Fe(CO)$_4$ unit.

With respect to the isostructural dianion, removal of one electron causes a
shortening of the Pt-Pt distance by 0.09 Å, whereas the Fe-Pt distance remains
almost unaltered (see Table 21).

This data, together with the results of an ESR investigation on the monoanion, suggested that the unpaired electron resides in an antibonding orbital
centred on the Pt$_3$ triangle [146, 147].

These suggestions have been fully confirmed by a recent theoretical analysis,
which showed that the frontier orbital (the HOMO) of the dianion has a

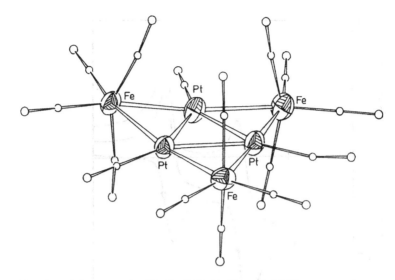

Fig. 76. Perspective view of the monoanion [Fe$_3$Pt$_3$(CO)$_{15}$]$^-$ [from Ref. 146]

predominant antibonding character among the three platinum atoms. In addition it is slightly antibonding (or even nonbonding) with respect to the Fe–Pt interactions [148]. Electrochemical indications suggested the existence of the neutral congener. In fact, as shown in Fig. 77a, the dianion in Dichloromethane solution, undergoes, two successive, chemically reversible, one-electron oxidation steps ($E^{\circ\prime}_{2-/-} = -0.40$ V; $E^{\circ\prime}_{-/0} = +0.19$ V), quasireversible in character [148].

At the same time, the neutral species Fe$_3$Pt$_3$(CO)$_{15}$ has been obtained by reacting Pt(1,5-cyclooctadiene)$_2$ with Fe(CO)$_5$ [149]. As shown in Fig. 77b, its redox fingerprint, which consists of two subsequent, chemically reversible (one-electron) reductions, is fully complementary to that of the dianion [150]. In addition, its X-ray characterization quantified the shortening of the Pt–Pt distances. As illustrated in Table 21, the second one-electron removal causes a further shortening of the Pt–Pt bond length by 0.07 Å, leaving, once again, the Fe–Pt distance almost unchanged.

It is noteworthy that, even for redox changes which do not alter the skeletal architecture, a simple, but marked variation in the bond lengths induces the relevant electron transfers to depart from pure electrochemical reversibility.

Table 21. Metal-metal bond lengths (in Å) in the series [Fe$_3$Pt$_3$(CO)$_{15}$]^n

n	Pt–Pt	Pt–Fe	Reference
2 –	2.75	2.60	146
1 –	2.66	2.59	146
0	2.59	2.58	150

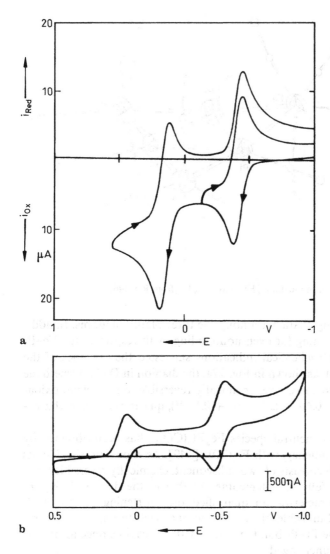

Fig. 77a, b. Cyclic voltammetric responses recorded at a platinum electrode on a CH_2Cl_2 solution of: a) $[Fe_3Pt_3(CO)_{15}]^{2-}$ [from Ref. 148]; b) $Fe_3Pt_3(CO)_{15}$ [from Ref. 150]

Structurally related to the preceding Fe_3Pt_3 compounds are $M_3Pt_3(CO)_3[NR_2]_6$ and $M_3Pd_3(CO)_3[NR_2]_6$ (M = Ge, Sn; R = SiMe$_3$). Their molecular assembly is illustrated in Fig. 78 [101].

All the M_3Pt_3 and M_3Pd_3 derivatives in tetrahydrofuran solution, undergo, an initial reversible one-electron reduction, followed by two other irreversible cathodic steps [101]. The redox potentials for the neutral/monoanion process are summarized in Table 22.

In a longer period of macroelectrolysis, only the paramagnetic monoanions $[M_3Pt_3(CO)_3\{N(SiMe_3)_2\}_6]^-$ are completely stable, whereas the M_3Pd_3 analogues undergo slow decomposition.

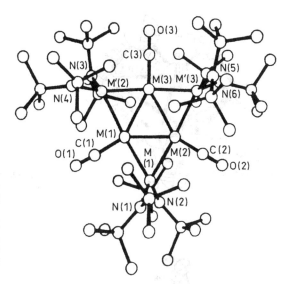

Fig. 78. Molecular structure of $M_3M'_3(CO)_3\{N(SiMe_3)_2\}_6$ (M = Pt, Pd; M' = Sn). Average M–M distances: M = Pt, 2.76 Å; M = Pd, 2.81 Å. Average M–M' distances: M = Pt, 2.62 Å; M = Pd, 2.62 Å; M = Pd, 2.63 Å [from Ref. 101]

Table 22. Redox potentials (V) for the reduction step $[M_3M'_3(CO)_3\{N(SiMe_3)_2\}_6]^{0/-}$, in Tetrahydrofuran solution [101]

M	M'	$E^{\circ}_{0/-}$
Pt	Ge	− 1.13
	Sn	− 1.04
Pd	Ge	− 1.36
	Sn	− 1.16

5.2 Butterfly-Based Frames

Despite the isolobal analogy between the two fragments H^+ and $AuPR_3^+$, replacement of two protons in $Fe_4(CO)_{12}BH_3$, for two $AuPPh_3$ units gives rise to a formal migration of the hydride group bridging the Fe-Fe hinge position towards a $Fe_{(wing)}$ − B position. This is illustrated in Figs. 79 and 80, which give the molecular structures of $Fe_4(CO)_{12}BH_3$ [151] and $Fe_4(CO)_{12}\{Au(PPh_3)\}_2BH$ [152], respectively.

The gold(I)-phosphine fragment, nominally substituting the $Fe_{(hinge)}$-$Fe_{(hinge)}$ bridging hydride of the tetrairon butterfly, has migrated towards the Boride atom.

It must however be noted that such a migration depends upon the steric requirements of the phosphine group, in that $AuPR_3$ fragments with less bulky alkyl substituents maintain the geometry of the parent $Fe_4(CO)_{12}BH_3$. This is the case of $HFe_4(CO)_{12}\{Au(PEt_3)\}_2B$ [153].

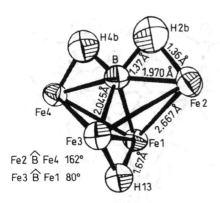

Fig. 79. Perspective view of the core of Fe$_4$(CO)$_{12}$BH$_3$ [from Ref. 151]

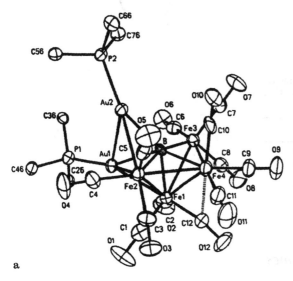

a

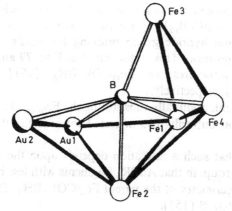

b

Fig. 80a. Perspective view and **b)** cluster core of Fe$_4$(CO)$_{12}$ {Au(PPh$_3$)}$_2$BH. Mean bond lengths: Fe2–Fe$_{(hinge)}$, 2.71 Å; Fe3–Fe$_{(hinge)}$, 2.66 Å. Fe$_{(hinge)}$–Fe$_{(hinge)}$, 2.58 Å; Au1–Fe1, 2.63 Å; Au1–Fe2, 2.85 Å; Au2–Fe2, 2.61 Å; Au1–Au2, 2.94 Å; Fe$_{(wing)}$–B, 2.10 Å; Au–B ≈ 2.35 Å [from Ref. 152]

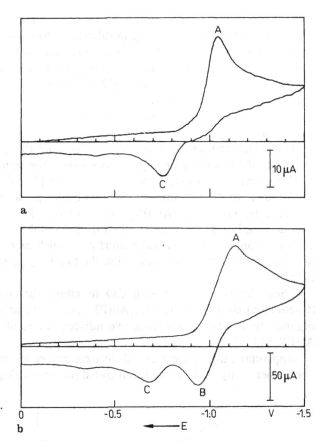

Fig. 81a,b. Cyclic voltammograms recorded at a platinum electrode on a CH$_2$Cl$_2$ solution of Fe$_4$(CO)$_{12}$\{Au(AsPh$_3$)\}$_2$BH. Scan rate: **a)** 0.2 Vs^{-1}; **b)** 5.12 Vs^{-1} [from Ref. 154]

The redox propensity of the two isostructural borido clusters Fe$_4$(CO)$_{12}$\{Au(PPh$_3$)\}$_2$BH and Fe$_4$(CO)$_{12}$\{Au(AsPh$_3$)\}$_2$BH has been studied [154]. Fig. 81, which refers to the latter, typifies the ability of these clusters to add electrons. They undergo, in Dichloromethane solution, a one-electron reduction, which generates the corresponding, short-lived, monoanion [Fe$_4$(CO)$_{12}$\{Au(XPh$_3$)\}$_2$BH]$^-$ (X = As, P). The relevant redox potentials are reported in Table 23 together with the monoanion lifetimes.

Table 23. Redox characteristics for Fe$_4$(CO)$_{12}$\{Au(XPh$_3$)\}$_2$BH in Dichloromethane solution [154]

	Redox changes					
	(0/−)		(+/0)		(2+/+)	
X	E$^{o\prime}$ (V)	t$_{1/2}$ (monoanion) (s)	E$^{o\prime}$ (V)	t$_{1/2}$ (monocation) (s)	E$^{o\prime}$ (V)	t$_{1/2}$ (dication) (s)
As	−1.04	0.1	+0.69	3	—	—
P	−1.10	0.01	+0.39	0.05	+0.65	1

It is interesting to note that exhaustive reoxidation of the decomposition products regenerates the starting auraferraborido clusters.

Their ability to lose electrons highlights a difference between the two species. In fact, as illustrated in Fig. 82, Fe$_4$(CO)$_{12}${Au(AsPh$_3$)}$_2$BH undergoes a single one-electron oxidation, whereas Fe$_4$(CO)$_{12}${Au(PPh$_3$)}$_2$BH undergoes two subsequent one-electron removals.

Also in this case, the electrogenerated cations tend to decompose (see Table 23).

It is evident that the different electron donating ability of the phosphine with respect to the arsine ligand significantly lowers the spin pairing energy of the HOMO level of the former, and permits its complete depopulation.

Another series of redox active gold(I)-phosphine substituted clusters is provided by Os$_4$(CO)$_{12}${Au(PR$_3$)}$_2$(PR$_3$ = PEt$_3$, PPh$_3$, PMePh$_2$) [155]. As illustrated in Fig. 83, the framework of these, formally unsaturated (58-electron) species is dominated by an Os$_4$ butterfly, in which each one of the two AuPR$_3$ units triply bridges, on opposite sides, the two Os$_{(wing)}$ and one Os$_{(hinge)}$ atoms [155].

These derivatives react with CO to afford the corresponding saturated (60-electron) clusters Os$_4$(CO)$_{13}${Au(PR$_3$)}$_2$, which, as shown in Fig. 84, reorganize to an Os$_4$ tetrahedron, two adjacent edges of which are bridged by Au(PR$_3$) [155].

Apparently, this 58-electron/60-electron conversion may also be performed electrochemically. In fact, the unsaturated molecules Os$_4$(CO)$_{12}${Au(PR$_3$)}$_2$ in

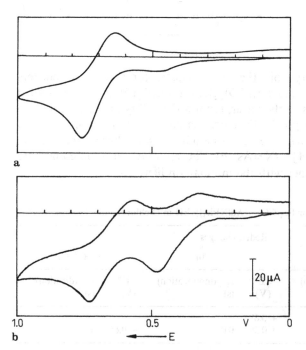

Fig. 82a, b. Cyclic voltammograms recorded at a platinum electrode on a CH$_2$Cl$_2$ solution of: **a)** Fe$_4$(CO)$_{12}${Au(AsPh$_3$)}$_2$BH; **b)** Fe$_4$(CO)$_{12}${Au(PPh$_3$)}$_2$BH. Scan rate, 1.00 Vs^{-1} [from Ref. 154]

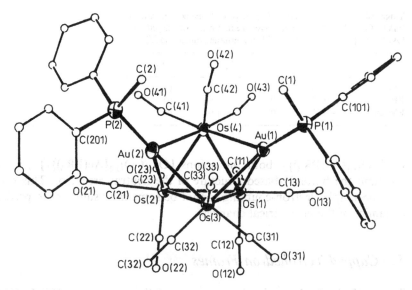

Fig. 83. Molecular structure of $Os_4(CO)_{12}\{Au(PMePh_2)\}_2$. Os1–Os2, 2.92 Å; Os3–$Os_{(hinge)} \approx$ Os4–$Os_{(hinge)}$, average, 2.76 Å; mean Au–Os bond length, 2.81 Å; Au...Au, 4.05 Å [from Ref. 155]

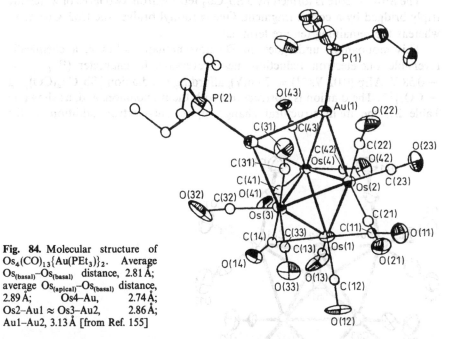

Fig. 84. Molecular structure of $Os_4(CO)_{13}\{Au(PEt_3)\}_2$. Average $Os_{(basal)}$–$Os_{(basal)}$ distance, 2.81 Å; average $Os_{(apical)}$–$Os_{(basal)}$ distance, 2.89 Å; Os4–Au, 2.74 Å; Os2–Au1 ≈ Os3–Au2, 2.86 Å; Au1–Au2, 3.13 Å [from Ref. 155]

Dichloromethane solution, undergo, two separated, chemically reversible, one-electron additions [155]. However, as suggested by the relevant peak-to-peak separations reported in Table 24, these electron transfers appear substantially reversible. This would mean that, at variance with the molecular assembly of

Table 24. Redox potentials (V) and peak-to-peak separations (mV) for the two successive reductions displayed by Os$_4$(CO)$_{12}${Au(PR$_3$)}$_2$, in Dichloromethane solution (155)

PR$_3$	$E^{o'}_{(0/-)}$	ΔEp	$E^{o'}_{(-/2-)}$	ΔEp
PEt$_3$	−0.85	65	−1.13	60
PPh$_3$	−0.82	70	−1.05	65
PMePh$_2$	−0.81	60	−1.03	65

Os$_4$(CO)$_{13}${Au(PR$_3$)}$_2$, both the anions [Os$_4$(CO)$_{12}${Au(PR$_3$)}$_2$]$^-$, $^{2-}$ likely maintain the butterfly-based structure of the neutral precursor. This suggests that the simple electron-counting cannot be sometimes sufficient to predict the dynamics of the geometrical changes.

5.3 Capped-Tetrahedron Frames

Figure 85 shows the molecular geometry of the monoanion [Sb$_2$Co$_4$(CO)$_{10}$)(μ-CO)]$^-$ [156, 157].

The Sb$_2$Co$_4$ core is formed by a Sb$_2$Co$_2$ tetrahedron, two faces of which are triply bridged by a cobalt fragment. One carbonyl bridges the Co2–Co3 edge, whereas the remaining ten are terminal.

This monoanion undergoes, in Dichloromethane solution, a chemically reversible one-electron reduction, near reversible in character ($E^{o'}_{(-/2-)}$ = −0.58 V; ΔEp (0.02 Vs^{-1}) = 70 mV), affording the dianion [Sb$_2$Co$_4$(CO)$_{10}$(μ−CO)]$^{2-}$. The dianion is isostructural with the monoanion, and, as shown in Table 25, the main structural change, upon one-electron addition, is the

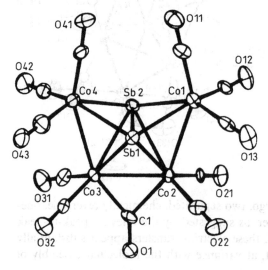

Fig. 85. Perspective view of [Sb$_2$Co$_4$(CO)$_{11}$]$^-$ [from Ref. 156]

Table 25. Selected bond distances (in Å) for the redox congeners $[Sb_2Co_4(CO)_{11}]^{n-}$ [156, 157]

n	Sb1–Sb2	Sb–Co (average)	Co2–Co3	Co1–Co2	Co3–Co4
1	2.91	2.64	2.51	2.72	2.69
2	2.88	2.64	2.65	2.72	2.72

elongation of 0.14 Å of the carbonyl bridged Co2–Co3 edge [156, 157]. The immediate indication that the LUMO of the monoanion (alternatively, the SOMO of the dianion) is substantially based on the Co2–Co3 antibonding interaction has been theoretically confirmed [157].

A quite similar behaviour is displayed by the isoelectronic and isostructural monoanion $[Bi_2Co_4(CO)_{10}(\mu - CO)]^-$, the structure of which is shown in Fig. 86 [158].

In fact, it also exhibits, in dichloromethane solution, the quasi-reversible reduction to the corresponding dianion ($E^{o'}_{(-/2-)} = -0.65$ V, $\Delta E_{p(0.02\ Vs^{-1})} = 70$ mV) [157]. It is likely that in this case the carbonyl-bridged Co2–Co3 edge significantly elongates upon one-electron addition.

Figure 87 shows the structure of $MoFe_5(CO)_6(S)_6(PEt_3)_3$ [159].

It can be viewed as a $MoFe_3$ tetrahedron capped by a $Fe_2(S)_2(CO)_6$ unit. Each face of the tetrahedron is triply bridged by a Sulfur atom.

In tetrahydrofuran solution, this cluster exhibits two sequential one-electron oxidations, quasireversible in character ($E^{o'}_{(0/+)} = -1.09$ V; $E^{o'}_{(+/2+)} = +0.13$ V) [159].

An analog of the one-electron oxidized form has been isolated as $[MoFe_5(CO)_6(S)_6(I)_3]^{2-}$, which, based on spectroscopic evidence, has been

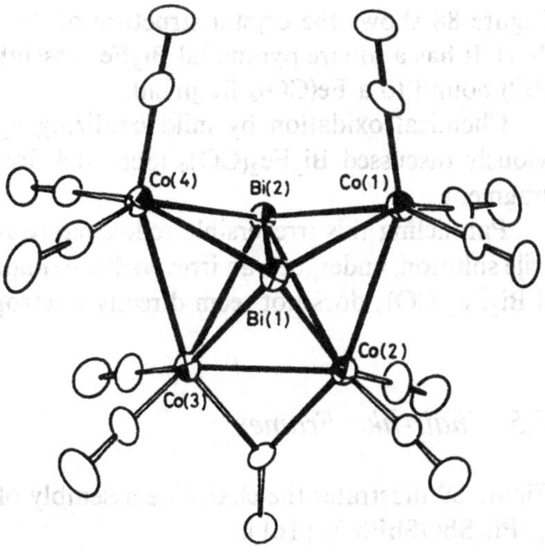

Fig. 86. Perspective view of $[Bi_2Co_4(CO)_{11}]^-$; $[NMe_4]^+$ counteranion). Bi–Bi, 3.09 Å; Bi–Co (average), 2.74 Å; Co2–Co3, 2.54 Å; Co1–Co2 ≈ Co3–Co4, 2.74 Å [from Ref. 158]

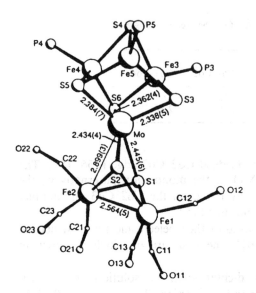

Fig. 87. Molecular structure (with selected bond lengths) of MoFe$_5$(CO)$_6$(S)$_6$(PEt$_3$)$_3$ [from Ref. 159]

assigned a structure similar to the phosphine derivative [159]. In further confirmation of the reversibility of the core redox change [MoFe$_5$(CO)$_6$(S)$_6$]$^{0/+}$, [MoFe$_5$(CO)$_6$(S)$_6$(I)$_3$]$^{2-}$ in Acetonitrile solution, undergoes, a reversible one-electron reduction ($E°'_{(2-/3-)} = -0.77$ V) [159].

It is thought that the frontier orbitals responsible for the one-electron addition/removal processes are localized on the cubane MoFe$_3$(S)$_4$ unit, which is able to exhibit the reversible couple [MoFe$_3$(S)$_4$]$^{3+/2+}$.

5.4 Metallo-Ligated Square Pyramidal Frames

Figure 88 shows the crystal structure of the dianion [Bi$_2$Fe$_4$(CO)$_{13}$]$^{2-}$ [116, 160]. It has a square pyramidal Bi$_2$Fe$_3$ assembly, with one basal Bismuth atom (Bi1) bound to a Fe(CO)$_4$ fragment.

Chemical oxidation by mild oxidizing agents (e.g. Cu$^+$) affords the previously discussed Bi$_2$Fe$_3$(CO)$_9$ (Sect. 4.4) by expulsion of the outer Fe(CO)$_4$ fragment.

Paralleling this irreversible redox pathway, [Bi$_2$Fe$_4$(CO)$_{13}$]$^{2-}$ in Acetonitrile solution, undergoes, an irreversible oxidation process (Ep = +0.29 V), even if Bi$_2$Fe$_3$(CO)$_9$ does not seem directly electrogenerated [116].

5.5 Chair-like Frames

Figure 89 illustrates the chair-like assembly of the Ni$_2$Sb$_4$O$_2$ core of Ni$_2$(CO)$_4$ (μ-Ph$_2$SbOSbPh$_2$)$_2$ [161].

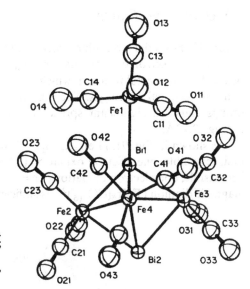

Fig. 88. Perspective view of [Bi$_2$Fe$_4$(CO)$_{13}$]$^{2-}$. Bi1–Fe2 ≈ Bi1–Fe3, 2.59 Å; Bi1–Fe4, 2.64 Å; Bi2–Fe2 ≈ Bi2–Fe3 ≈ Bi2–Fe4, 2.68 Å; Fe2–Fe4 ≈ Fe3–Fe4, 2.80 Å; Bi1–Fe1, 2.67 Å [from. Ref. 160]

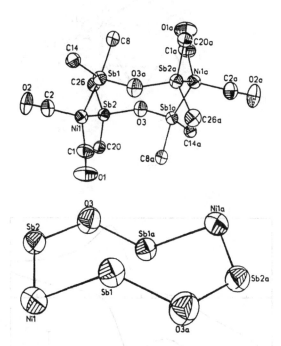

Fig. 89. Molecular structure of Ni$_2$(CO)$_4$(Ph$_2$SbOSbPh$_2$)$_2$. Ni–Sb, 2.45 Å; Sb–O, 1.94 Å [from Ref. 161]

This molecule exhibits no redox flexibility. It has been briefly reported that in Tetrahydrofuran solution, undergoes, only irreversible oxidation processes [161].

5.6 Octahedral Frames

Figure 90 shows the molecular structure of the trianion $[FeIr_5(CO)_{15}]^{3-}$ [162]. Like the isoelectronic homonuclear $[Ir_6(CO)_{15}]^{2-}$ [163], it is formed by an octahedral hexametal frame, in which each metal atom binds one edge-bridging and two terminal carbonyls. The non-uniform bonding mode of the carbonyl groups causes a significant spread in the metal-metal bond lengths, which average 2.77 Å.

As shown in Fig. 91, $[FeIr_5(CO)_{15}]^{3-}$ in acetonitrile solution, undergoes, a first, reversible, one-electron oxidation ($E^{\circ\prime} = -0.57$ V), followed by a second, irreversible, one-electron oxidation (Ep = -0.23 V). The electrogenerable paramagnetic dianion $[FeIr_5(CO)_{15}]^{2-}$ is sufficiently stable to allow recording of its

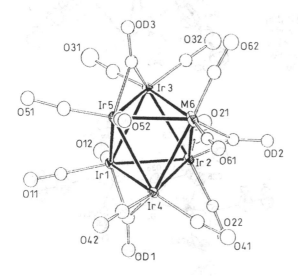

Fig. 90. Perspective view of the trianion $[FeIr_5(CO)_{15}]^{3-}$ [from Ref. 162]

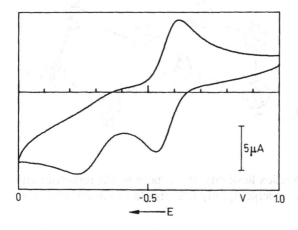

Fig. 91. Cyclic voltammogram recorded at a platinum electrode on a deaerated MeCN solution of $[FeIr_5(CO)_{15}]^{3-}$. Scan rate 0.2 Vs^{-1} [from Ref. 162].

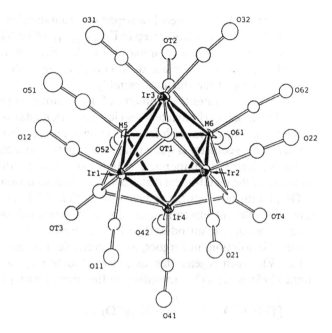

Fig. 92. Molecular structure of the monoanion [FeIr$_5$(CO)$_{16}$]$^-$ [from Ref. 162]

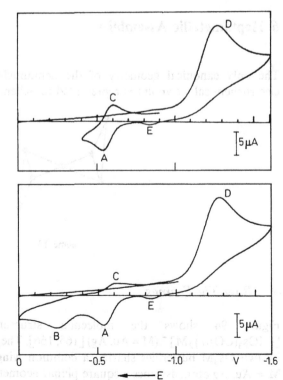

Fig. 93. Cyclic voltammograms exhibited at a platinum electrode by a deaerated MeCN solution of [FeIr$_5$(CO)$_{16}$]$^-$. Scan rate 0.2 Vs^{-1} [from Ref. 162]

EPR spectrum. The relatively low departure from the electrochemical reversibility of the 3 − /2 − oxidation step ($\Delta E p_{(0.2\ vs-1)} = 86$ mV) suggests that, even if some geometrical strain is induced by the electron removal, the starting octahedral frame is still retained. In contrast, the further one-electron release causes breaking of the starting assembly.

Figure 92 illustrates the geometry of the monoanion $[FeIr_5(CO)_{16}]^-$ [162].

In this more carbonylated species the hexametal assembly assumes an octahedral architecture, with four carbonyl groups μ_3-face bridging and twelve terminal. This geometry is quite reminiscent of that exhibited by the red isomer of the isoelectronic homonuclear $Ir_6(CO)_{16}$ [164]. Although the expected spreading of the metal-metal distances, their averaged value still remains 2.77 Å.

Despite the fact that the conversion of $[FeIr_5(CO)_{15}]^{3-}$ to $[FeIr_5(CO)_{16}]^-$ leaves the relevant electron count formally unaltered, deep modifications in redox propensity are introduced. In fact, as shown in Fig. 93, $[FeIr_5(CO)_{16}]^-$ in acetonitrile solution, undergoes, an irreversible, two-electron reduction (Ep = − 1.27 V), which regenerates, as put in evidence by the reverse profiles, the parent $[FeIr_5(CO)_{15}]^{3-}$, according to the overall path [162].

$$[FeIr_5(CO)_{16}]^- \xrightarrow[-CO]{+2e} [FeIr_5(CO)_{15}]^{3-}$$

6 Heptametallic Assemblies

The only canonical geometry of the heptanuclear clusters studied by the electrochemical viewpoint is represented in Scheme 13.

Scheme 13

6.1 Bow-Tie Frames

Figure 94 shows the molecular structure of the monoanions $[\{HOs_3(CO)_{10}\}_2M]^-$ (M = Au, Ag) [165, 166]. The metallic frame is constituted by two Os_3M butterflies sharing a common wing-tip vertex M. The central M = Au, Ag atom assumes a square planar geometry. The relative shortness of

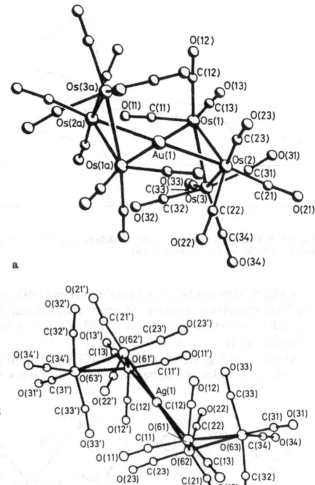

Fig. 94a, b. Perspective view of: a) [{HOs₃(CO)₁₀}₂Au]⁻.
Au–Os1 ≈ Au–Os2, 2.80 Å; Os1–Os2, 2.70 Å; Os3–Os1 ≈ Os3–Os2, 2.83 Å [from Ref. 165]; b) [{HOs₃(CO)₁₀}₂Ag]⁻. Ag–Os1 ≈ Ag–Os2, 2.86 Å; Os1–Os2, 2.68 Å; Os3–Os1 ≈ Os3–Os2, 2.83 Å [from Ref. 166]

the Os–Os hinge edges (which are likely hydride bridged) induces one to think, that here resides the electronic unsaturation.

As shown in Fig. 95a, [{HOs₃(CO)₁₀}₂Ag]⁻ undergoes a reduction process represented by an irreversible, single-stepped, two-electron transfer (Ep = − 1.35 V), at ambient temperature. Nevertheless, Fig. 95b shows that the decrease of temperature changes the reduction path in two separated one-electron steps, with features of chemical reversibility ($E°'_{(-/2-)}$ = − 1.29 V, ΔEp = 90 mV; $E°'_{(2-/3-)}$ = − 1.38 V; ΔEp = 80 mV) [166].

[{HOs₃(CO)₁₀}₂Au]⁻ displays a rather similar redox propensity, in that it is able to add, stepwisely, two electrons, in an almost reversible manner even at ambient temperature ($E°'_{(-/2-)}$ ≈ − 1.15 V; $E°'_{(2-/3-)}$ ≈ − 1.37 V) [12].

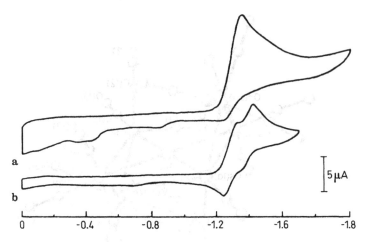

Fig. 95a, b. Cyclic voltammetric responses exhibited by [{HOs₃(CO)₁₀Ag]⁻ in Acetonitrile solution. **a)** 20 °C; **b)** − 25 °C [from Ref. 166]

A slightly different behaviour is exhibited by [{HOs₃(CO)₁₀}₂Tl]⁻, in that, in Dicholoromethane solution, at room temperature, it undergoes both an irreversible reduction step (Ep = − 1.07 V), and a significantly reversible oxidation step ($E^{\circ\prime}_{(0/-)} \approx$ + 0.35 V) [12].

As shown in Fig. 96, {(C₂ − Buᵗ)Ru₃(CO)₉}₂Hg also possesses a molecular architecture in which two butterflies share a common wing-tip vertex [167].

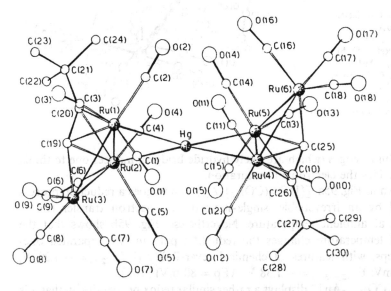

Fig. 96. Perspective view of {Ru₃(CO)₉(C₂-Buᵗ)}₂Hg. Average Hg–Ru distance, 2.82 Å; Ru1–Ru2 ≈ Ru4–Ru5, 2.85 Å; Ru3–Ru1 ≈ Ru3–Ru2 ≈ Ru6–Ru4 ≈ Ru6–Ru5, 2.80 Å [from Ref. 167]

Even if the electrochemical behaviour of this cluster is presently unknown, it has been reported that it undergoes, chemically, a first one-electron reduction to the corresponding monoanion $[\{(C_2 - Bu^t)Ru_3(CO)_9\}_2Hg]^-$, and a second one-electron addition, which causes declustering of the central Hg metal (with formation of triruthenium anions) [168].

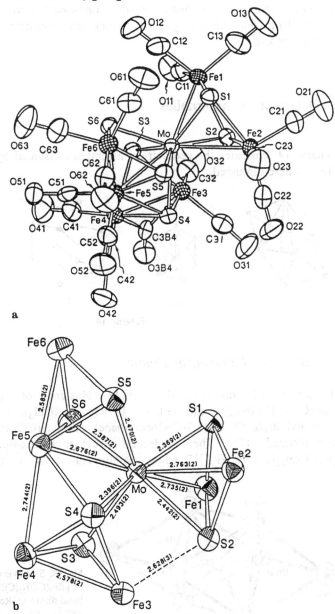

Fig. 97a. Perspective view of $[MoFe_6(CO)_{16}(S)_6]^{2-}$. **b)** Stereoview of its $MoFe_6S_6$ core [from Ref. 170]

6.2 Uncommon Frames

A quite unusual molecular assembly is displayed by the dianion [MoFe$_6$(CO)$_{16}$(S)$_6$]$^{2-}$, Fig. 97 [169, 170].

It has been briefly reported that this low-symmetry cluster in Acetonitrile solution, undergoes, a quasireversible one-electron reduction (E$^{o\prime}_{(0/-)}$ = −1.00 V) [169], which suggests that some reorganization accompanies such electron transfer.

7 Octametallic Assemblies

Scheme 14 illustrates the geometrical assemblies assumed by the octametallic clusters here discussed.

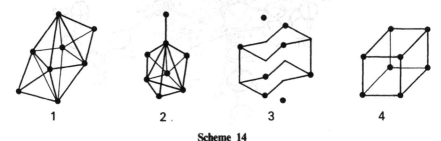

Scheme 14

7.1 Bicapped Octahedral Frames

A series of mixed-metal carbide clusters of general formula [Re$_7$(CO)$_{21}$(C)M(L)]$^{2-}$ [M = Pt, Pd, Ir, Rh; L = (CH$_3$)$_3$, C$_3$H$_5$ (allyl); C$_4$H$_7$ (2-methyl-allyl), COD (1,5-Cyclooctadiene), (CO)$_2$, (CO)(PPh$_3$)] have been characterized [171]. Their molecular structure is exemplified by that of [Re$_7$(CO)$_{21}$(C)Pt(C$_4$H$_7$)]$^{2-}$, whose metallic core is shown in Fig. 98 [171].

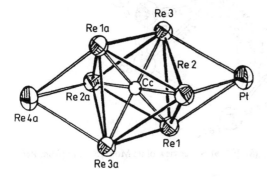

Fig. 98. Stereoview of the Re$_7$Pt(C) core of [Re$_7$(CO)$_{21}$(C)Pt(C$_4$H$_7$)]$^{2-}$. Average bond distances: Re–Re$_{(within\ the\ octahedron)}$, 2.99 Å; Re$_{(octahedron)}$–C$_{(carbide)}$, 2.12 Å; Re–Pt ≈ Re–Re4a, 2.90 Å [from. Ref. 171]

It consists of a Re₇ octahedron, encapsulating a carbide carbon atom, two opposite faces of which are capped by a Pt(L) fragment and a Re(CO)₃ group, respectively.

As shown in Fig. 99, which again refers to $[Re_7(CO)_{21}(C)Pt(C_4H_7)]^{2-}$, these clusters in dichloromethane solution, undergo, two subsequent one-electron removals, the first of which only displays marked features of chemical reversibility [171]. There is evidence that the corresponding monoanions $[Re_7(CO)_{21}(C)Pt(L)]^-$ are relatively long-lived [172].

Strictly related to these clusters is the series $[Re_7(CO)_{21}(C)Hg(L)]^{2-}$ (L=CN, Cl, Ph, Bu^t) [173]. On the basis of spectroscopic evidence, these derivatives have also been assigned a bicapped octahedral geometry similar to that illustrated in Fig. 98, with the Pt(L) fragment now substituted by an Hg(L) fragment.

As shown in Fig. 100, which refers to $[Re_7(CO)_{21}(C)Hg(CN)]^{2-}$, these species also undergo two subsequent one-electron removals, only the first one being chemically reversible [173].

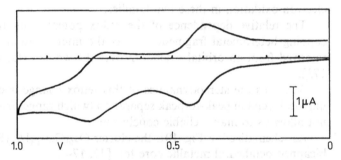

Fig. 99. Cyclic voltammogram recorded at a platinum electrode on a CH₂Cl₂ solution of $[Re_7(CO)_{21}(C)Pt(C_4H_7)]^{2-}$. Scan rate 0.1 Vs^{-1}. Potential values are against an Ag/AgCl reference couple [from Ref. 171]

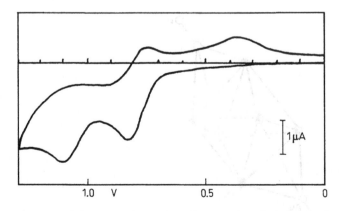

Fig. 100. Cyclic voltammogram recorded at a platinum electrode on a CH₂Cl₂ solution of $[Re_7(CO)_{21}(C)Hg(CN)]^{2-}$. Scan rate 0.1 Vs^{-1}. Potential values vs. Ag/AgCl [from Ref. 173]

Table 26. Redox potentials (V) and peak-to-peak separation (mV) for the electron transfer $[Re_7(CO)_{21}(C)M(L)]^{2-/-}$, in Dichloromethane solution

M(L)	$E^{o\prime}$	ΔE_p[a]	Reference
Rh(CO)$_2$	+0.58	125	171
Rh(CO)(PPh$_3$)	+0.38	93	171
Rh(COD)	+0.44	87	171
Ir(COD)	+0.44	92	171
Pd(C$_3$H$_5$)	+0.43	72	171
Pt(CH$_3$)$_3$	+0.60	64	171
Pt(C$_4$H$_7$)	+0.37	65	171
Hg(CN)	+0.74	95	173
Hg(Cl)	+0.69	136	173
Hg(Ph)	+0.60	74	173
Hg(Bu^t)	+0.64	104	173

[a]Measured at 0.1 Vs^{-1}

Table 26 summarizes the electrochemical characteristics of the first one-electron oxidation in these two families.

The relative dependence of the redox potential upon the nature of the capping heterometal fragment favours the interpretation that the electron is removed from an orbital extending over the whole octametallic moiety [171, 173].

As far as the stereodynamics of this redox change is concerned, the nonunivocal trend in peak-to-peak separation (which ranges from 60–140 mV) does not allow us to make reliable conclusions.

As schematized in Fig. 101, the cluster $Os_6(CO)_{18}\{Au(PMe_3)\}_2$ possesses a bicapped octahedral metallic core too [12, 174].

It has been briefly reported that, in dichloromethane solution, this species undergoes an irreversibe (hence, declustering) one-electron oxidation (Ep = +0.61 V) (12).

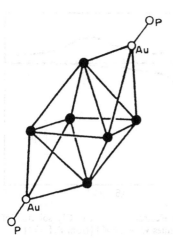

Fig. 101. Schematic representation of the molecular structure of $Os_6(CO)_{18}\{Au(PMe_3)\}_2$ [from Ref. 12]

7.2 Metallo–Ligated Tricapped Tetrahedral Frames

Figure 102 shows the molecular structure of the dianion $[Bi_4Fe_4(CO)_{13}]^{2-}$ [175,176].

It is formed by a tetrahedron of bismuth atoms, three faces of which are μ_3-bridged by a $Fe(CO)_3$ fragment. A fourth $Fe(CO)_4$ unit is externally linked to the apical bismuth atom.

In acetonitrile solution, this complex undergoes a series of irreversible electron removals (Ep = + 0·13 V, + 0·43 V, + 0·60 V), which represent degradation processes towards more or less known Bi/Fe carbonyl clusters [116]. Really, this easy decomposition does not seem to conform to the theoretical picture which foresees that the HOMO of the core cluster is a $(\mu_3-Fe)_3$-centred nonbonding orbital [176].

7.3 Rhombohedral Frames

A rich family of mixed-metal octanuclear clusters of general formula $[Fe_6(S)_6(X)_6\{M(CO)_3\}_2]^{n-}$ (M = Mo, W; X = Cl, Br, I, O–C$_6$H$_4$-p-R, ·S–C$_6$H$_5$; n = 3, 4) have been characterized [177–182]. Figure 103 shows the two possible views of the structure of the couple $[Fe_6(S)_6(Cl)_6\{Mo(CO)_3\}]^{3-,\,4-}$ [181]. They can be described as "bicapped prismanes". The term stems from the hexagonal prismatic geometry of the central $Fe_6(\mu-S)_6$ moiety, which is constituted by two identical cyclohexane-chair-like Fe_3S_3 units. Alternatively, they

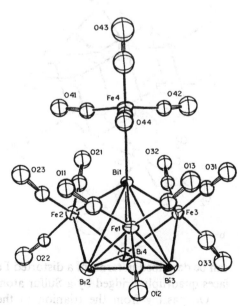

Fig. 102. Perspective view of the dianion $[Bi_4Fe_4(CO)_{13}]^{2-}$. Average M–M distances: Bi1–Bi$_{(basal)}$, 3.47 Å; Bi$_{(basal)}$–Bi$_{(basal)}$, 3.16 Å; Bi–(μ_3)Fe, 2.72 Å. Bi1–Fe4, 2.75 Å [from Ref. 176]

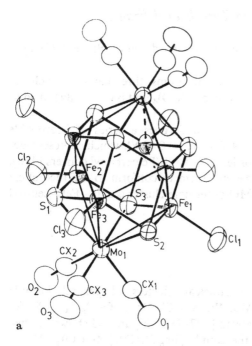

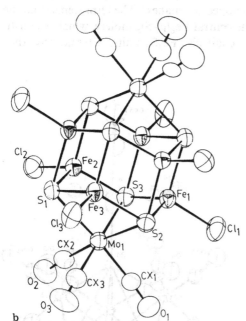

Fig. 103. Perspective views of the structure of the anions $[Fe_6(S)_6(Cl)_6\{Mo(CO)_3\}_2]^{3-,4-}$ [from Ref. 181]

can be described in terms of a distorted Fe_6M_2 cube (M = Mo, W), with the six faces quadruply bridged by a Sulfur atom.

On passing from the trianion to the corresponding tetranion, the most significant structural change is the elongation of the Mo–Fe and Mo–S

Table 27. Electrochemical characteristics for the two one-electron transfers exhibited by [Fe$_6$(S)$_6$(X)$_6${M(CO)$_3$}$_2$]$^{3-}$

M	X	$E^{o'}_{(3-/4-)}$ (V)	ΔEp[a] (mV)	$E^{o'}_{(4-/5-)}$ (V)	ΔEp[a] (mV)	Solvent	Reference
Mo	Cl	+0.05	92	−0.54	86	CH$_2$Cl$_2$	181
		−0.05	–	−0.55	–	MeCN	179
	Br	+0.08	90	−0.50	90	CH$_2$Cl$_2$	181
	I	+0.08	121	−0.49	120	CH$_2$Cl$_2$	181
	O–C$_6$H$_4$–p–Me	−0.29	110	−0.87	103	CH$_2$Cl$_2$	181, 182
		−0.39	–	–	–	MeCN	179
	O–C$_6$H$_4$–p–OMe	−0.31	130	−0.87	130	CH$_2$Cl$_2$	182
	O–C$_6$H$_4$–p–NMe$_2$	−0.40	110	−0.91	110	CH$_2$Cl$_2$	182
	O–C$_6$H$_4$–p–COMe	−0.16	100	−0.67	130	CH$_2$Cl$_2$	182
	S–C$_6$H$_5$	−0.30	110	−0.75	108	CH$_2$Cl$_2$	181
W	Cl	+0.04	84	−0.55	103	CH$_2$Cl$_2$	181
	Br	+0.06	96	−0.51	103	CH$_2$Cl$_2$	181
	O–C$_6$H$_4$–p–Me	−0.38	141	−0.87	130	CH$_2$Cl$_2$	181, 182
	S–C$_6$H$_5$	−0.31	114	−0.73	114	CH$_2$Cl$_2$	181

[a] Measured at 0.2 Vs^{-1}

distances (from 2.93 Å to 3.00 Å, and from 2.58 Å to 2.62 Å, respectively), leaving substantially intact the central prismane core [181]. This suggests that the frontier orbitals of the redox couple are mainly centred on the heterometal capping fragments.

The occurrence of such redox congeners encouraged the search for the relevant thermodynamic redox potential. As a matter of fact, electrochemistry showed that the Fe$_6$M$_2$ assembly is able to support, without molecular destruction, two subsequent one-electron transfers, namely [Fe$_6$(S)$_6$(X)$_6${M(CO)$_3$}$_2$]$^{3-/4-/5-}$. The relevant redox potentials are summarized in Table 27.

As suggested by the relevant ΔEp values, these electron transfers are quasireversible in character. This, once again, proves that the simple, but significant, variation of the bonding distances within a same molecular architecture is sufficient to cause departure from pure electrochemical reversibility, i.e., to slow down the electron-transfer rate.

Although the corresponding pentaanions have not been isolated, it is conceivable that a further elongation of the Mo–Fe and Mo–S bonding distances takes place.

Figure 104 shows the molecular structure of the dicarbide cluster dianion [Co$_6$Ni$_2$(CO)$_{16}$(C)$_2$]$^{2-}$ [183].

The rhombohedral Co$_6$Ni$_2$ core can be thought to be formed by the fusion, along the Co$_4$ square face, of two trigonal-prismatic Co$_5$Ni fragments, each one encapsulating one carbide carbon atom. The edges of the pseudocubane assembly have a mean value of 2.57 Å, whereas the Co2–Co4 diagonal is 2.77 Å. Within the trigonal-prismatic halves, the M–C$_{(carbide)}$ distance averages 1.99 Å, the C$_{(carbide)}$–C$_{(carbide)}$ distance is 1.49 Å.

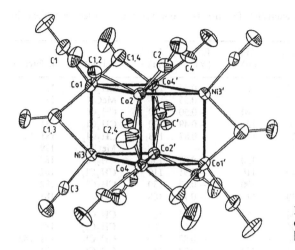

Fig. 104. Pespective view of the dianion $[Co_6Ni_2(CO)_{16}(C)_2]^{2-}$ [from. Ref. 183]

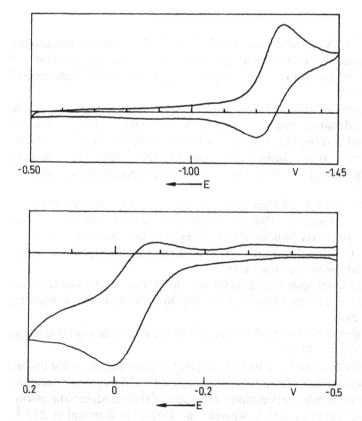

Fig. 105. Cyclic voltammetric responses recorded at a platinum electrode on a THF solution of $[Co_6Ni_2(CO)_{16}(C)_2]^{2-}$. Scan rate 0.02 Vs^{-1}

As shown in Fig. 105, the dianion in Tetrahydrofuran solution, undergoes, both an oxidation process ($E^{\circ\prime} = -0.04$ V) and a reduction step ($E^{\circ\prime} = -1.24$ V) [184]. The monoanion generated in correspondence to the anodic process is transient ($t_{1/2} \approx 20$), whereas the corresponding trianion $[Co_6Ni_2(CO)_{16}(C)_2]^{3-}$ results decidedly longer lived.

8 Decametallic Assemblies

Scheme 15 shows the geometry of the only decanuclear cluster, which has been electrochemically investigated.

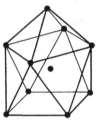

Scheme 15

8.1 Capped Square-Antiprismatic Frames

The molecular structure of the dianion $[SiCo_9(CO)_{21}]^{2-}$ is illustrated in Fig 106 [185]. It can be described as an octacobalt tetragonal antiprism, encapsulating a silicon atom, and capped by a ninth cobalt atom. The carbonyl groups are divided in three classes: (i) four bridge the Co-Co capping fragment; (ii) four bridge, alternately, the Co-Co bonds linking the basal-to-top Co_4 rhombs; (iii) thirteen are terminal.

As shown in Fig. 107, in agreement with its electron deficiency (129 valence electrons instead of the 130 expected for a capped square antiprism [185]), $[SiCo_9(CO)_{21}]^{2-}$, in dichloromethane solution, is able to undergo, without immediate decomposition, two successive reduction processes. The first one, occurring at peak A ($E^{\circ\prime} = -0.47$ V), generates the corresponding trianion $[SiCo_9(CO)_{21}]^{3-}$, which is stable and can be obtained also by chemical reduction. The second one, occurring at peak B ($E^{\circ\prime} = -1.62$ V), generates the corresponding pentaanion $[SiCo_9(CO)_{21}]^{5-}$, which is transient and probably undergoes deep structural modifications. Finally, the irreversibility of the oxidation steps occuring at peaks E and F suggests that electron removal causes fast breakage of the cluster framework [186].

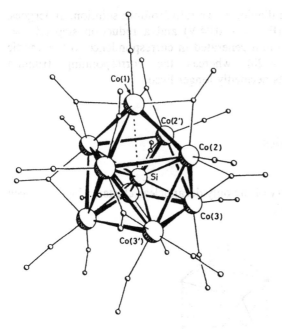

Fig. 106. Perspective view of $[SiCo_9(CO)_{21}]^{2-}$. Co1–Co, 2.61 Å; Co2–Co2', 2.94 Å; Co2–Co3, 2.69 Å; Co2–Co3', 2.59 Å; Co3–Co3', 2.81 Å; average Si–Co, 3.37 Å, [from Ref. 185]

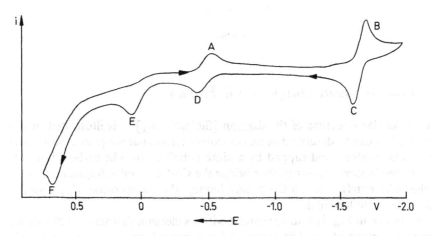

Fig. 107. Cyclic voltammogram exhibited by a CH_2Cl_2 solution of $[SiCo_9(CO)_{21}]^{2-}$. Platinum working electrode [from Ref. 186]

9 Undecametallic Assemblies

The metallic core of the only undecanuclear cluster electrochemically studied is represented in Scheme 16.

Redox Behaviour of Heterometal Carbonyl Clusters

Scheme 16

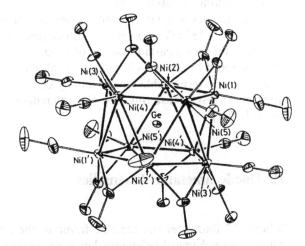

Fig. 108. Perspective view of $[Ni_{10}Ge(CO)_{20}]^{2-}$. Average bond lengths: intralayer Ni–Ni, 2.55 Å; interlayer Ni–Ni, 2.72 Å; Ni–Ge, 2.47 Å [from Ref. 187]

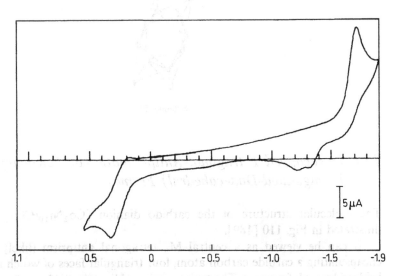

Fig. 109. Cyclic voltammogram displayed by $[Ni_{10}Ge(CO)_{20}]^{2-}$ in THF solution. Platinum working electrode. Scan rate 0.2 Vs^{-1}

9.1 Pentagonal-Antiprismatic Frames

Figure 108 shows the molecular structure of the dianion $[Ni_{10}(\mu_{10}\text{-}Ge)(CO)_{20}]^{2-}$ [187].

It consists of a pentagonal antiprism of nickel atoms encapsulating a germanium atom. Each nickel atom possesses one terminal and one edge-bridging (along the pentagons) carbonyl.

Figure 109 shows the redox changes exhibited by the present dianion, in Tetrahydrofuran solution [188].

It undergoes a quasireversible one-electron oxidation ($E^{\circ\prime} = +0.26$ V; $\Delta E_{p(0.2\, Vs^{-1})} = 160$ mV), coupled to subsequent chemical complications, which impart to the monoanion a lifetime of about 3 s. Speculatively, the geometrical strain induced by the one-electron removal (see the large peak-to-peak separation) might be responsible for the lability of this monoanion. The dianion also undergoes a fast declustering, two-electron reduction at low potential values (Ep = -1.70 V).

10 Duodecametallic Assemblies

Scheme 17 illustrates the metallic frame of the only duodecanuclear cluster, whose electrochemical behaviour has been investigated.

Scheme 17

10.1 Tetracapped Tetragonal-Antiprismatic (or Tetracapped Triangulated-Dodecahedral) Frames

The molecular structure of the carbido dianion $[Co_2Ni_{10}(CO)_{20}(C)]^{2-}$ is illustrated in Fig. 110 [189].

It can be viewed as a central M_8 tetragonal antiprism (likely Co_2Ni_6), encapsulating a carbide carbon atom, four triangular faces of which are triply-bridged by a Ni fragment. The twenty carbonyl ligands divide as: four terminal (bonded to the $Ni_{(capping)}$ atoms) and sixteen edge-bridging groups. Because of

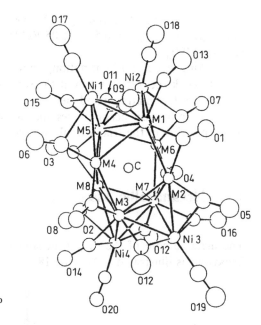

Fig. 110. Perspective view of [Co$_2$Ni$_{10}$(CO)$_{20}$(C)]$^{2-}$ [from Ref. 189]

the structural complexity, there is a significant spreading in the relevant metal-metal bond lengths, which vary from an averaged value of 2.41 Å for the (four) CO-bridged M–M distances to 2.82 Å for the (two) unbridged M–M bonds through a value of 2.58 Å for the eight M–Ni$_{(capping)}$ distances and of 2.64 Å for the (two) M–M distances spanned by two vicinal, capping Ni atoms.

Figure 111 shows that, in Tetrahydrofuran solution, the present dianion undergoes a one-electron reduction to the corresponding trianion

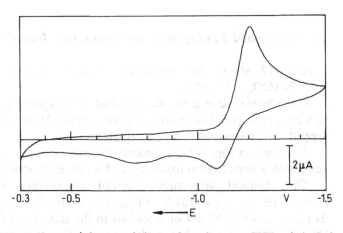

Fig. 111. Cyclic voltammogram recorded at a platinum electrode on a THF solution of [Co$_2$Ni$_{10}$(CO)$_{20}$(C)] (PPh$_4$)$_2$. Scan rate 0.05 Vs^{-1}

($E°' = -1.14$ V, $\Delta E_{p(0.5 Vs^{-1})} = 130$ mV), which, even if not long-lived ($i_{p(backward)}/i_{p(foreward)} \approx 0.5$, at 0.05 Vs^{-1}), has a lifetime of some tenths of seconds [190].

Lacking structural information, we foresee, on the basis of the rather large peak-to-peak separation (even at the slowest scan rate of 0.02 Vs^{-1}, it equals 110 mV, both at platinum and mercury electrode), that some significant geometrical reorganization accompanies the one-electron addition, before the slow decomposition of the trianion takes place.

11 Tridecametallic Assemblies

The metallic core of the only tridecanuclear cluster with documented redox behaviour is illustrated in Scheme 18.

Scheme 18

11.1 Bicapped Pentagonal-Antiprismatic (Icosahedral) Frames

Figure 112 shows the molecular structure of the trianion [Ni$_{10}$Bi$_2$ (μ_{12}- Ni)(CO)$_{18}$]$^{3-}$ [191].

It can be viewed as a Ni$_{10}$Bi$_2$ icosahedron encapsulating a Ni atom, as well as a Ni$_{10}$ pentagonal antiprism, with an eleventh Ni atom placed in its cavity, capped, on opposite sides, by two Bi atoms.

As shown in Fig. 113, the present trianion in acetonitrile solution, undergoes, both a one-electron oxidation and a one-electron reduction [191].

The chemical and electrochemical reversibility of the two redox changes ($E°'_{(2-/3-)} = 0.67$ V, $\Delta E_{p(0.2 Vs^{-1})} = 70$ mV; $E°'_{(3-/4-)} = -1.42$ V, $\Delta E_{p(0.2 Vs^{-1})} = 68$ mV) not only testifies to the stability of the redox congeners [Ni$_{11}$Bi$_2$(CO)$_{18}$]$^{2-,\ 4-}$, but it also suggests that they maintain a molecular assembly quite similar to that of the parent trianion.

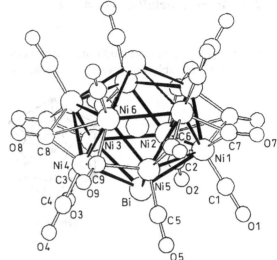

Fig. 112. Perspective view of the trianion $[Ni_{11}Bi_2(CO)_{18}]^{3-}$. Average bonding lengths: intralayer Ni–Ni, 2.81 Å; interlayer Ni–Ni, 2.51 Å; Ni–Ni$_{(interstitial)}$, 2.60 Å; Bi–Ni$_{(pentagon)}$, 2.82 Å; Bi–Ni$_{(interstitial)}$, 2.51 Å [from Ref. 191]

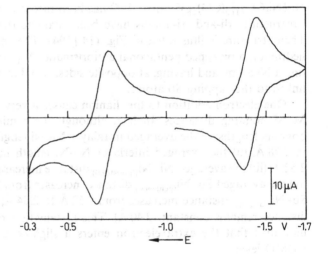

Fig. 113. Cyclic voltammogram recorded at a platinum electrode on a MeCN solution of $[Ni_{11}Bi_2(CO)_{18}]^{3-}$. Scan rate 0.2 Vs^{-1} [from Ref. 191]

This behaviour parallels that exhibited by the core-isostructural $[Ni_{10}Sb_2(\mu_{12}-Ni)(E)_2(CO)_{18}]^{3-}$ (E = Ni(CO)$_3$), which will be discussed in the next Section. In contrast, the isostructural $[Ni_{12}(\mu_{12}-Sn)(CO)_{22}]^{2-}$ [187] does not possess redox flexibility (it, in fact, undergoes, in Tetrahydrofuran solution, a fast declustering oxidation (Ep = +0.4 V) and reduction (Ep = −1.45 V) processes [192]).

12 Pentadecametallic Assemblies

The only electrochemical study on a pentadecametallic derivative concerns the geometrical assembly illustrated in Scheme 19.

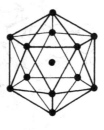

Scheme 19

12.1 Metallo-Ligated Icosahedral Frames

A family of redox congeners of formula $[Ni_{10}Sb_2(\mu_{12}\text{-}Ni)\{Ni(CO)_3\}_2(CO)_{18}]^{n-}$ (n = 2–4) has been recently reported [193, 194]. The isostructural di- and tri-anions have been characterized by X-ray techniques. Their structure is illustrated in Fig. 114 [194]. The core consists of an icosahedral (or, a bicapped pentagonal antiprismatic) $Ni_{10}Sb_2$ assembly, encapsulating a Ni atom, and having, at opposite sides, two dangling $Ni(CO)_3$ fragments linked to the capping Sb atoms.

One-electron addition to the dianion causes a very slight elongation of the M–M bonding distances. In fact, although the significant spreading of the bonding lengths: (i) the averaged intralayer Ni–Ni length increases from 2.77 Å to 2.78 Å; (ii) the averaged interlayer Ni–Ni length increases from 2.51 Å to 2.52 Å; (iii) the averaged Ni–Ni$_{(interstitial)}$ distance increases from 2.57 Å to 2.58 Å; (iv) the averaged Sb–Ni$_{(pentagon)}$ distance increases from 2.73 Å to 2.74 Å; (v) the Sb–Ni$_{(dangling)}$ distance increases from 2.52 Å to 2.54 Å; (vi) the Sb–Ni$_{(interstitial)}$ distance remains constant (2.40 Å). This systematic elongation is conceivably indicative that the extra electron enters a slightly antibonding metal-metal LUMO level.

In accordance with the substantial lack of structural reorganization accompanying the $[Ni_{13}Sb_2(CO)_{24}]^{2-/3-}$ redox change, the electrochemical response illustrated in Fig. 115 shows that, in Acetonitrile solution, the relevant electron transfer is almost reversible (E°′ = − 0.56 V; $\Delta Ep_{(0.2\ Vs-1)}$ = 68 mV). Also the successive reduction $[Ni_{13}Sb_2(CO)_{24}]^{3-/4-}$ is electrochemically reversible (E°′ = − 1.28 V; $\Delta Ep_{(0.2\ Vs-1)}$ = 66 mV) [194]. This means that, even if not X-ray characterized, the tetraanion $[Ni_{13}Sb_2(CO)_{24}]^{4-}$ possesses the geometry of the di- and tri-anions precursors, likely with a further, very slight elongation of its metal-metal bond lengths.

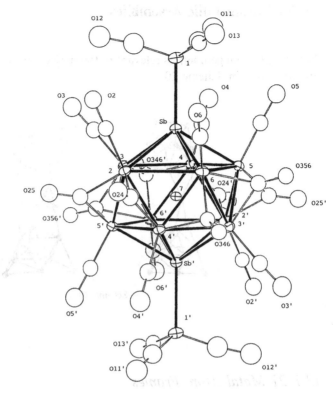

Fig. 114. Perspective view of the trianion $[Ni_{13}Sb_2(CO)_{24}]^{3-}$ [from Ref. 194]

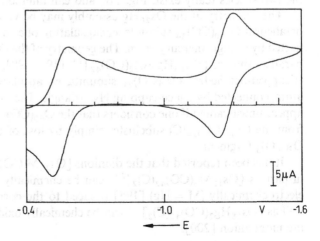

Fig. 115. Cyclic voltammogram recorded at a platinum electrode by a MeCN solution of $[Ni_{13}Sb_2(CO)_{24}]^{2-}$. Scan rate 0.2 Vs^{-1} [from Ref. 194]

13 Eneicosametallic Assemblies

The two 21-atom geometries relevant to the subject matter of the present review are illustrated in Scheme 20.

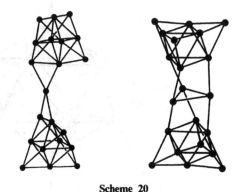

Scheme 20

13.1 21-Metal Atom Frames

Some uncertainty still exists on the exact formulation and structure of a series of redox-active giant heteronuclear clusters formulable as $Os_{20}M$ and/or $Os_{18}M_3$ (M = Hg, Au, Ag). Accordingly, the previously formulated cluster dianion $[Os_{20}Hg(C)_2(CO)_{48}]^{2-}$ [195] has been now reassigned as $[Os_{18}Hg_3(C)_2(CO)_{42}]^{2-}$ [196]. The uncertainty arises from the fact that both the two species really exist, Fig. 116, and can interconvert [196].

The geometry of the $Os_{20}Hg$ assembly may be viewed as two tetracapped-octahedral $Os_{10}(CO)_{24}$ subunits, encapsulating one carbide carbon atom, connected by a single mercury atom. The geometry of the $Os_{18}Hg_3$ assembly, which parallels that of $[Ru_{18}Hg_3(C)_2(CO)_{42}]^{2-}$ [197, 198], may be viewed as two tricapped-octahedral $Os_9(CO)_{21}$ subunits, encapsulating one carbide carbon atom, connected by a face-capping Hg_3 triangle. The interconversion might not appear unwarranted if one considers that the $Os_9(CO)_{21}(C)$ subcluster can form from the $Os_{10}(CO)_{24}(C)$ subcluster simply by loss of one octahedron-capping $Os(CO)_3$ fragment.

It has been reported that the dianions $[Os_{20}M(CO)_{48}(C)_2]^{2-}$, reformulated in part as $[Os_{18}M_3(CO)_{42}(C)_2]^{2-}$, can be chemically (M = Au, Ag) [199] or electrochemically (M = Hg) [195] reduced to the corresponding trianions, as well as $[Os_{18}Hg_3(CO)_{42}(C)_2]^{2-}$ can be chemically oxidized to the corresponding monoanion [200].

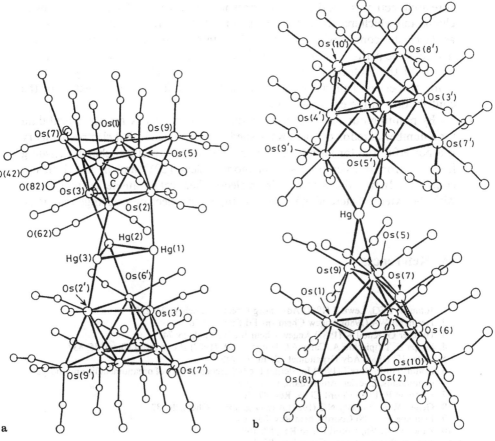

Fig. 116a, b. Perspective view of: a)$[Os_{18}Hg_3(C)_2(CO)_{42}]^{2-}$; b) $[Os_{20}Hg(C)_2(CO)_{48}]^{2-}$ [from Ref. 196]

14 Summary

Metal cluster compounds are commonly thought to be able to act as electron reservoirs. A detailed examination of the electrochemical studies performed on a wide range of either homometallic [13] or heterometallic carbonyl cluster assemblies, ranging from trinuclear to giant molecules, seems to discredit in part such a belief. In fact, metal-carbonyl clusters which are able to undergo multiple redox changes without framework destruction are rather rare; the majority of them can tolerate a very limited number of electron additions/removals. On the other hand, for instance, metal-sulfur clusters seem generally more equipped to support multisequential redox processes [10, 11], thus indicating that the metal core architecture is not the only factor which makes accessible electron transfer chains and which may relieve the concomitant structural reorganizations. As far as the cluster connectivity is concerned, an important role is also played by the nature of the peripheral ligands. This is not unexpected, if one considers that the nature of the HOMO/LUMO frontier orbitals, which ultimately govern the redox properties of all the molecules, is commonly affected to some extent by the contribution of the external ligands.

By pointing out the distinctive features of solid-state X-ray structural data for a number of redox couples, we wished to give evidence that electrochemistry is a powerful tool for monitoring the occurrence of structural changes ensuing from electron transfer processes in mononuclear [201] and in metal cluster complexes. In fact, the inferences from electrochemical parameters usually agree with the extent of stereodynamism accompanying electron exchange reactions.

15. References

1. Johnson BFG, Lewis J (1981) Adv Inorg Chem Radiochem 24: 225
2. Hermann WA (1986) Angew Chem Int Ed Engl, 25: 56
3. Huttner G, Knoll K (1987) Angew Chem Int Ed Engl, 26: 743
4. Sappa E, Tiripicchio A, Carty AJ, Toogood GE (1987) Progr Inorg Chem, 35: 437
5. Salter ID (1989) Adv Organomet Chem, 29: 249
6. Braunstein P, Rose J (1989) In: Bernal I (ed) Stereochemistry of organometallic and inorganic compounds, Elsevier, Amsterdam, vol 3; p 2
7. Lemoine P (1982) Coord Chem Rev, 47: 56
8. Geiger WE, Connelly NG (1987) Adv Organomet Chem, 24: 87
9. Lemoine P (1988) Coord Chem Rev, 83: 169
10. Zanello P (1988) Coord Chem Rev, 83: 199
11. Zanello P (1988) Coord Chem Rev, 87: 1
12. Drake SR (1990) Polyhedron, 9: 455
13. Zanello P, In: Bernal I (ed) Stereochemistry of organometallic and inorganic compounds, Elsevier, Amsterdam; vol 5, in the press
14. Geiger WE (1985) Progr Inorg Chem, 33: 275
15. Brown ER, Sandifer JR (1986) In: Rossiter BW, Hamilton JF (eds) Physical methods of chemistry. Electrochemical Methods, Wiley J, New York
16. Sheldrick GM, Simpson RNF (1968) J Chem Soc (A), 1005
17. Katcher ML, Simon GL (1972) Inorg Chem, 11: 1651
18. Sosinsky BA, Shong RG, Fitzgerald BJ, Norem N, O'Rourke C (1983) Inorg Chem, 22: 3124
19. Alvarez S, Ferrer M, Reina R, Rossell O, Seco M, Solans X (1989) J Organomet Chem, 377: 291
20. Stephens FS (1972) J Chem Soc, Dalton Trans, 2257
21. Dessy RE, Weissman PM, Pohl RL (1966) J Am Chem Soc, 88: 5117
22. Murgia SM, Paliani G, Cardaci G (1972) Z Naturforsch, B27: 134
23. de Montauzon D, Poilblanc R (1976) J Organomet Chem, 104: 99
24. Lemoine P, Giraudeau A, Gross M, Braunstein P (1980) J Chem Soc, Chem Commun, 77
25. Giraudeau A, Lemoine P, Gross M, Braunstein P (1980) J Organomet Chem, 202: 447
26. Uson R, Laguna A, Laguna M, Jones PG, Sheldrick G (1981) J Chem Soc, Dalton Trans, 306
27. Moras D, Dehand J, Weiss R, (1968) CR Acad Sc Paris, 267: 1471
28. Bars O, Braunstein P, Jud J-M (1984) Nouv J Chem, 8: 771
29. Urbancic MA, Wilson SR, Shapley JR (1984) Inorg Chem, 23: 2954
30. Hermann WA, Kneuper H-J, Herdtweck E (1989) Chem Ber, 122: 445
31. Strube A, Huttner G, Zsolnai L (1990) J Organomet Chem, 399: 267
32. Strube A, Huttner G, Zsolnai L (1988) Angew Chem Int Ed Engl, 27: 1529

33. Strube A, Huttner G, Zsolnai L (1989) Z Anorg Allg Chem, 577: 263
34. Hermann WA, Hecht C, Ziegler ML, Balbach B (1984) J Chem Soc, Chem Commun, 686
35. Bullock JP, Palazzotto MC, Mann KR (1990) Inorg Chem, 29: 4413
36. Kilbourn BT, Powell HM, (1964) Chem Ind (London), 1578
37. Struchkov Yu T, Anisimov KN, Osipova OP, Kolobova NE, Nesmeyanov AN (1967) Dokl Akad Nauk SSSR, 172: 107
38. Bir'Yukov BP, Struchkov Yu T, Anisimov KN, Kolobova NE, Skripkin VV (1968) Chem Commun, 159
39. Preut H, Wolfes W, Haupt H-J (1975) Z Anorg Allg Chem, 412: 121
40. Zubieta JA, Zuckerman JJ, (1978) Progr Inorg Chem, 24: 251
41. Holt MS, Wilson WL, Nelson JH (1989) Chem Rev, 89: 11
42. Young DA (1981) Inorg Chem, 20: 2049
43. Bender R, Braunstein P, Metz B, Lemoine P (1984) Organometallics, 3: 381
44. Cotton FA, Troup JM (1976) J Am Chem Soc, 96: 4155
45. Churchill MR, Hollander FJ, Hutchinson JP (1977) Inorg Chem, 16: 2655
46. Dawson PA, Peake BM, Robinson BH, Simpson J (1980) Inorg Chem, 19: 465
47. Bender R, Braunstein P, Fisher J, Ricard L, Mitschler A (1981) Nouv J Chim, 5: 81
48. Adams RD, Chen G, Wu W, Yin J (1990) Inorg Chem, 29: 4208
49. Lemoine P, Giraudeau A, Gross M, Bender R, Braunstein P (1981) J Chem Soc, Dalton Trans, 2059
50. Honrath U, Vahrenkamp H (1984) Z Naturforsch, B39: 545
51. Lindsay PN, Peake BM, Robinson BH, Simpson J, Honrath U, Vahrenkamp H (1984) Organometallics, 3: 413
52. Beurich H, Vahrenkamp H (1982) Chem Ber, 115: 2385
53. Chetcuti MJ, Chetcuti PAM, Jeffery JC, Mills RM, Mitrprachachon P, Pickering SJ, Stone FGA, Woodward P (1982) J Chem Soc, Dalton Trans, 699
54. Beurich H, Richter F, Vahrenkamp H (1982) Acta Cryst, B38: 3012
55. Muller M, Vahrenkamp H (1983) Chem Ber, 116: 2748
56. Muller M, Vahrenkamp H (1983) Chem Ber, 116: 2765
57. Kubat-Martin KA, Spencer B, Dahl LF (1987) Organometallics, 6: 2580
58. Stevenson DL, Wei CH, Dahl LF (1971) J Am Chem Soc, 93: 6027
59. Peake BM, Rieger PH, Robinson BH, Simpson J (1981) Inorg Chem, 20: 2540
60. Williams PD, Curtis MD (1988) J Organomet Chem, 352: 169
61. Richter F, Vahrenkamp H (1982) Chem Ber, 115: 3243
62. Strouse CE, Dahl LF (1971) J Am Chem Soc, 93: 6032
63. Einstein FWB, Tyers KG, Tracey AS, Sutton D (1986) Inorg Chem, 25: 1631
64. Osella D, Raithby PR (1989)In: Bernal I (ed) Stereochemistry of organometallic and inorganic compounds, Elsevier, Amsterdam, Vol 3, p 301
65. Aime S, Milone L, Osella D, Tiripicchio A, Manotti Lanfredi AM (1982) Inorg Chem, 21: 501
66. Osella D, Gobetto R, Montangero P, Zanello P, Cinquantini A (1986) Organometallics, 5: 1247
67. Green M, Mills RM, Pain GN, Stone FGA, Woodward P (1982) J Chem Soc, Dalton Trans, 1309
68. Braunstein P, de Meric de Bellefon C, Ries M, Fischer J, Bouaoud S-E, Grandjean D (1988) Inorg Chem, 27: 1327
69. Nemra G, Lemoine P, Braunstein P, de Meric de Bellefon C, Ries M (1986) J Organomet Chem, 304: 245
70. Nemra G, Lemoine P, Gross M, Braunstein P, de Meric de Bellefon C, Ries M (1986) Electrochimica Acta, 31: 1205
71. Byers LR, Uchtman VA, Dahl LF (1981) J Am Chem Soc, 103: 1942
72. Maj JJ, Rae AD, Dahl LF (1982) J Am Chem Soc, 104: 3054
73. Rosenhein LD, Newton WE, McDonald JW (1987) Inorg Chem, 26: 1695
74. Rosenhein LD, McDonald JW (1987) Inorg Chem, 26: 3414
75. Haines RJ, Steen NDCT, English RB (1981) J Chem Soc, Chem Commun, 587
76. Marchino MLN, Sappa E, Manotti Lanfredi AM, Tiripicchio A (1984) J Chem Soc, Dalton Trans, 1541
77. Bender R, Braunstein P, de Meric de Bellefon C (1988) Polyhedron, 7: 2271
78. Richter F, Vahrenkamp H (1978) Angew Chem Int Ed Engl, 17: 444
79. Colbran S, Robinson BH, Simpson J (1982) J Chem Soc, Chem Commun, 1361
80. Colbran SB, Hanton LR, Robinson BH, Robinson WT, Simpson J (1987) J Organomet Chem, 330: 415

81. Leung P, Coppens P, McMullan RK, Koetzle TF (1981) Acta Cryst, B37: 1347
82. Colbran SB, Robinson BH, Simpson J (1983) Organometallics, 2: 943
83. Colbran SB, Robinson BH, Simpson J (1983) Organometallics, 2: 952
84. Colbran SB, Robinson BH, Simpson J (1984) J Organomet Chem, 265: 199
85. Braunstein P, Jud J-M, Dusausoy Y, Fischer J (1983) Organometallics, 2: 180
86. Gade W, Weiss E (1981) J Organomet Chem, 213: 451
87. Hermann WA, Kneuper H-J, Herdtweck E (1989) Chem Ber, 122: 437
88. Ohst HH, Kochi JK (1986) Organometallics, 5: 1359
89. Halet J-F, Hoffmann R, Saillard J-Y (1985) Inorg Chem, 24: 1695
90. Jaeger JT, Field JS, Collison D, Speck GP, Peake BM, Hanle J, Vahrenkamp H (1988) Organometallics, 7: 1753
91. Vahrenkamp H, Wucherer EJ, Wolters D (1983) Chem Ber, 116: 1219
92. Bender R, Braunstein P, Jud J-M, Dusausoy Y (1983) Inorg Chem, 22: 3394
93. Bender R, Braunstein P, Jud J-M, Dusausoy Y (1984) Inorg Chem, 23: 4489
94. Jund R, Lemoine P, Gross M, Bender R, Braunstein P (1983) J Chem Soc, Chem Commun, 86
95. Jund R, Lemoine P, Gross M, Bender R, Braunstein P (1985) J Chem Soc, Dalton Trans, 711
96. Curtis MD, Williams PD, Butler WM (1988) Inorg Chem, 27: 2853
97. Cowans B, Noordik J, Rakowski DuBois M (1983) Organometallics, 2: 931
98. Cowans BA, Haltiwanger RC, Rakowski DuBois M (1987) Organometallics, 6: 995
99. Al-Aliaf TAK, Eaborn C, Hitchcock PB, Lappert MF, Pidcock A (1985) J Chem Soc, Chem Commun, 548
100. Hitchcock PB, Lappert MF, Misra MC (1985) J Chem Soc, Chem Commun, 863
101. Campbell GK, Hitchcock PB, Lappert MF, Misra MC (1985) J Organomet Chem, 289: C1
102. Fischer J, Mitscheler A, Weiss R, Dehand J, Nennig JF (1975) J Organomet Chem, 91: C37
103. Braunstein P, Dehand J, Nennig JF (1975) J Organomet Chem, 92: 117
104. Braunstein P, Jud J-M, Tiripicchio A, Tiripicchio-Camellini M, Sappa E (1982) Angew Chem, Int Ed Engl, 21: 307
105. Braunstein P, Rose J, Bars O (1983) J Organomet Chem, 252: C101
106. Jund R, Rimmelin J, Gross M (1990) J Organomet Chem, 381: 239
107. Curtis MD, Williams PD (1983) Inorg Chem, 22: 2661
108. Gibson CP, Dahl LF (1988) Organometallics, 7: 535
109. Martinengo S, Chini P, Cariati F, Salvatori T (1973) J Organomet Chem, 59: 379
110. Albano V, Ciani G, Martinengo S (1974) J Organomet Chem, 78: 265
111. Horvath IT (1986) Organometallics, 5: 2333
112. Bojczuk M, Heaton BT, Johnson S, Ghilardi CA, Orlandini A (1988) J Organomet Chem, 341: 473
113. Rimmelin J, Lemoine P, Gross M, Bahsoun AA, Osborn JA (1985) Nouv J Chim, 9: 181
114. Hidai M, Orisaku M, Ue M, Koyasu Y, Kodama T, Uchida Y (1983) Organometallics 2: 292
115. Whitmire KH, Lagrone CB, Churchill MR, Fettinger JC, Biondi LV (1984) Inorg Chem, 23: 4227
116. Whitmire KH, Shieh M, Lagrone CB, Robinson BH, Churchill MR, Fettinger JC, See RF (1987) Inorg Chem, 26: 2798
117. Ball R, Bennet MJ, Brooks EH, Graham WAG, Hoyano J, Illingworth SM (1970) Chem Commun, 592
118. Patmore DJ, Graham WAG (1966) Inorg Chem, 5: 2222
119. Bir' Yukov BP, Kukhtenkova EA, Struchkov Yu T, Anisimov KN, Kolobova NE, Khandozhko VI (1971) J Organomet Chem, 27: 337
120. Ball RD, Hall D (1973) J Organomet Chem, 52: 293
121. Tsai JH, Flynn JJ, Boer FP (1967) Chem Commun, 702
122. Bonny AM, Crane TJ, Kane-Maguire NAP (1985) J Organomet Chem, 289: 157
123. Farrugia LJ, Howard JAK, Mitrprachachon P, Stone FGA, Woodward P (1981) J Chem Soc, Dalton Trans, 155
124. Farrugia LJ, Howard JAK, Mitrprachachon P, Stone FGA, Woodward P (1981) J Chem Soc, Dalton Trans, 162
125. Farrugia LJ, Green M, Hankey DR, Murray M, Orpen AG, Stone FGA (1985) J Chem Soc, Dalton Trans, 177
126. Ewing P, Farrugia L (1988) New J Chem, 12: 409
127. Schmid G, Etzrodt G, (1977) J Organomet Chem, 137: 367
128. Bahsoun AA, Osborn JA, Voelker C, Bonnet JJ, Lavigne G (1982) Organometallics, 1: 1114
129. Pasynskii AA, Eremenko IL, Katugin AS, Gasanov G Sh, Turchanova EA, Ellert OG,

Struchkov Yu T, Shklover VE, Berberova NT, Sogomonova AG, Okhlobystin O Yu (1988) J Organomet Chem, 344: 195
130. Pasynskii AA, Eremenko IL, Orazsakhatov B, Kalinnikov VT, Aleksandrov GG, Struchkov Yu T (1981) J Organomet Chem, 214: 367
131. Coucouvanis D, Al-Ahmad S, Salifoglou A, Dunham WR, Sands RH (1988) Angew Chem Int Ed Engl, 27: 1353
132. Kyba EP, Kerby MC, Kashyap RP, Mountzouris JA, Davis RE (1989) Organometallics, 8: 852
133. Kyba EP, Kerby MC, Kashyap RP, Mountzouris JA, Davis RE (1990) J Am Chem Soc, 112: 905
134. Klein H-P, Thewalt U, Herrmann G, Suss-Fink G, Moinet C (1985) J Organomet Chem, 286: 225
135. Longoni G, Manassero M, Sansoni M (1980) J Am Chem Soc, 102: 3242
136. Zanello P, Garlaschelli L, unpublished results
137. Lagrone CB, Whitmire KH, Churchill MR, Fettinger J (1986) Inorg Chem, 25: 2080
138. Whitmire KH, Lagrone CB, Churchill MR, Fettinger JC (1986) Inorg Chem, 26: 3491
139. Lindley PF, Woodward P (1967) J Chem Soc (A) 382
140. Luo S, Whitmire KH (1989) Inorg Chem, 28: 1424
141. Luo S, Whitmire KH (1989) J Organomet Chem, 376: 297
142. Churchill MR, Fettinger JC, Whitmire KH, Lagrone CB (1986) J Organomet Chem, 303: 99
143. Churchill MR, Fettinger JC, Whitmire KH (1985) J Organomet Chem, 284: 13
144. Osella D, Stein E, Nervi C, Zanello P, Laschi F, Cinquantini A, Rosenberg E, Fiedler J (1991) Organometallics, 10: 1929
145. Burgess K, Johnson BFG, Kaner DA, Lewis J, Raithby PR, Syed-Mustaffa SNAB (1983) J Chem Soc, Chem Commun, 455
146. Longoni G, Manassero M, Sansoni M (1980) J Am Chem Soc, 102: 7973
147. Longoni G, Morazzoni F (1981) J Chem Soc, Dalton Trans, 1735
148. Della Pergola R, Garlaschelli L, Mealli C, Proserpio DM, Zanello P (1990) J Cluster Sci, 1: 93
149. Adams RD, Chen G, Wang J-G (1989) Polyhedron, 8: 2521
150. Adams RD, Arafa I, Chen G, Lii J-C, Wang J-G (1990) Organometallics, 9: 2350
151. Fehlner TP, Housecroft CE, Scheidt WR, Wong KS (1983) Organometallics, 2: 825
152. Housecroft CE, Rheingold AL, (1987) Organometallics, 6: 1332
153. Housecroft CE, Shongwe MS, Rheingold AL (1989) Organometallics, 8: 2651
154. Housecroft CE, Shongwe MS, Rheingold AL, Zanello P (1991) J Organomet Chem, 408: 7
155. Hay CM, Johnson BFG, Lewis J, McQueen RCS, Raithby PR, Sorrell RM, Taylor MJ (1985) Organometallics, 4: 202
156. Leigh JS, Whitmire KH, Yee KA, Albright TA (1989) J Am Chem Soc, 111: 2726
157. Albright TA, Yee KA, Saillard J-Y, Kahlal S, Halet J-F, Leigh JS, Whitmire KH (1991) Inorg Chem, 30: 1179
158. Martinengo S, Ciani G (1987) J Chem Soc, Chem Commun, 1589
159. Bose KS, Chmielewski SA, Eldredge PA, Sinn E, Averill BA (1989) J Am Chem Soc, 111: 8953
160. Whitmire KH, Raghuveer KS, Churchill MR, Fettinger JC, See RF (1986) J Am Chem Soc, 108: 2778
161. DesEnfants RE, Gavney JA, Hayashi RK, Rae AD, Dahl LF, Bjarnason A (1990) J Organomet Chem, 383: 543
162. Ceriotti A, Della Pergola R, Garlaschelli L, Laschi F, Manassero M, Masciocchi N, Sansoni M, Zanello P, (1991) Inorg Chem, 30: 3349
163. Demartin F, Manassero M, Sansoni M, Garlaschelli L, Martinengo S, Canziani F (1980) J Chem Soc, Chem Commun, 903
164. Garlaschelli L, Martinengo S, Bellon PL, Demartin F, Manassero M, Chiang MY, Wei C-Y, Bau R (1984) J Am Chem Soc, 106: 6664
165. Johnson BFG, Kaner DA, Lewis J, Raithby PR (1981) J Chem Soc, Chem Commun, 753
166. Fajardo M, Gomez-Sal MP, Holden HD, Johnson BFG, Lewis J, McQueen RCS, Raithby PR, (1984) J Organomet Chem, 267: C25
167. Ermer S, King K, Hardcastle KI, Rosenberg E, Lanfredi AMM, Tiripicchio A, Camellini MT (1983) Inorg Chem, 22: 1339
168. Hajela S, Novak BM, Rosenberg E (1989) Organometallics, 8: 468
169. Eldredge PA, Bryan RF, Sinn E, Averill BA (1988) J Am Chem Soc, 110: 5573
170. Eldredge PA, Bose KS, Barber DE, Bryan RF, Sinn E, Rheingold A, Averill BA (1991) Inorg Chem, 30: 2365
171. Henly TJ, Shapley JR, Rheingold AL, Gelb S (1988) Organometallics, 7: 441

172. Simerly SW, Shapley JR (1990) Inorg Chem, 29: 3634
173. Henly TJ, Shapley JR (1989) Organometallics, 8: 2729
174. Diebold MP, Johnson BFG, Lewis J, McPartlin M, Powell HR, (1990) Polyhedron, 9: 75
175. Whitmire KH, Churchill MR, Fettinger JC (1985) J Am Chem Soc, 107: 1056
176. Whitmire KH, Albright TA, Kang S-K, Churchill MR, Fettinger JC (1986) Inorg Chem, 25: 2799
177. Coucouvanis D, Kanatzidis MG (1985) J Am Chem Soc, 107: 5005
178. Salifoglou A, Kanatzidis MG, Coucouvanis D, (1986) J Chem Soc, Chem Commun, 559
179. Kanatzidis MG, Coucouvanis D (1986) J Am Chem Soc, 108: 337
180. Coucouvanis D, Salifoglou A, Kanatzidis MG, Simopoulos A, Kostikas A (1987) J Am Chem Soc, 109: 3807
181. Coucouvanis D, Salifoglou A, Kanatzidis MG, Dunham WR, Simopoulos A, Kostikas A (1988) Inorg Chem, 27: 4066
182. Al-Ahmad SA, Salifoglou A, Kanatzidis MG, Dunham WR, Coucouvanis D (1990) Inorg Chem, 29: 927
183. Arrigoni A, Ceriotti A, Della Pergola R, Longoni G, Manassero M, Masciocchi N, Sansoni M (1984) Angew Chem Int Ed Engl, 23: 322
184. Della Pergola R, Garlaschelli L, Longoni G, Zanello P work in progress
185. Mackay KM, Nicholson BK, Robinson WT, Sims AW (1984) J Chem Soc, Chem Commun, 1276
186. Barris GC, (1990) D Phil Thesis, University of Waikato (New Zealand)
187. Ceriotti A, Demartin F, Heaton BT, Ingallina P, Longoni G, Manassero M, Marchionna M, Masciocchi N, (1989) J Chem Soc, Chem Commun, 786
188. Longoni G, Zanello P unpublished results
189. Ceriotti A, Della Pergola R, Longoni G, Manassero M, Masciocchi N, Sansoni M (1987) J Organomet Chem, 330: 237
190. Garlaschelli L, Zanello P, unpublished results
191. Albano VG, Demartin F, Iapalucci MC, Longoni G, Monari M, Zanello P, J Chem Soc, Dalton Trans, in press
192. Longoni G, Zanello P, unpublished results
193. Albano VG, Demartin F, Iapalucci MC, Longoni G, Sironi A, Zanotti V (1990) J Chem Soc, Chem Commun, 547
194. Albano VG, Demartin F, Iapalucci MC, Laschi F, Longoni G, Sironi A, Zanello P (1991) J Chem Soc, Dalton Trans, 739
195. Drake SR, Henrick K, Johnson BFG, Lewis J, McPartlin M, Morris J (1986) J Chem Soc, Chem Commun, 928
196. Gade LH, Johnson BFG, Lewis J, McPartlin M, Powell HR (1990) J Chem Soc, Chem Commun, 110
197. Bailey PJ, Johnson BFG, Lewis J, McPartlin M, Powell HR (1989) J Chem Soc, Chem Commun, 1513
198. Bailey PJ, Duer MJ, Johnson BFG, Lewis J, Conole G, McPartlin M, Powell HR, Anson CE (1990) J Organomet Chem, 383: 441
199. Drake SR, Johnson BFG, Lewis J (1989) J Chem Soc, Dalton Trans, 505
200. Charalambous E, Gade LH, Johnson BFG, Kotch T, Lees AJ, Lewis J, McPartlin M (1990) Angew Chem Int Ed Engl, 29: 1137
201. Zanello P (1990) Stereochemistry of Organometallic and Inorganic Compounds, Bernal I, (ed) Elsevier, Amsterdam; Vol 4, 181

Electronic Structure and Bonding in Actinyl Ions

R. G. Denning

Inorganic Chemistry Laboratory, South Parks Road, Oxford, OX1 3QR, United Kingdom

The actinyl ions exhibit an unusually robust covalent bond which has a profound influence on their chemistry. Their electronic structure has been unravelled by the use of a variety of optical measurements and by photoelectron spectroscopy, which together establish the composition and role of the valence orbitals. The experimental energy level scheme can be compared with the results of molecular orbital calculations of varying degrees of sophistication, the most successful being SCF calculations incorporating the effects of relativity. An important contribution to the bonding comes from the pseudo-core 6p shell, which is important in determining the linearity of the ions. A new concept, the Inverse *trans* Influence, is introduced in this context. There is some evidence that d-orbital π-bonding is more important than σ-bonding. The remarkable strength of the actinyl bond can be attributed to the presence in the valence shell of both f and d metal orbitals, giving each actinide-oxygen bond a formal bond order of three.

1	Introduction		217
	1.1	Properties of the Actinyl Ions	217
	1.2	Objectives of this Review	219
2	Experimental Electronic Structure		221
	2.1	Background	221
	2.2	Recent Single Crystal Optical Spectroscopy	222
	2.3	Excited State Parity	225
	2.4	Angular Momentum in the First Excited State	225
	2.5	Properties of Higher Energy Excited States	228
	2.6	The Tetragonal Equatorial Field	231
	2.7	Other Equatorial Fields	231
	2.8	Relative Energies of the Non-Bonding f-Orbitals	232
	2.9	Other f-Orbital Energies	235
	2.10	Relative Configuration Energies	237
	2.11	Two-Photon Spectroscopy	238
	2.12	Excited State Absorption	240
	2.13	Conclusions from Optical Measurements	244
	2.14	Filled Orbital Energies	244
	2.15	Summary of Experimental Energy Level Data	246
3	Theoretical Studies		249
	3.1	Background and Orbital Overlap	249
	3.2	Self-Consistent Field Methods	250
		3.2.1 Introduction	250
		3.2.2 The Multiple-Scattering Method	251
		3.2.3 The Discrete Variational Method	254
		3.2.4 The Pseudo-Potential Method	257
		3.2.5 Summary of SCF Results	258
	3.3	Extended Hückel Calculations	259
	3.4	Orbital Energies and Ionisation Potentials	260

4 Chemical Bonding 262
  4.1 Effective Charge on Uranium. 262
  4.2 Linear Geometry 264
  4.3 The Inverse *trans* Influence 265
  4.4 Vibrational Interaction Force Constants 268
  4.5 The U–O Bond and The Equatorial Field 269

5 Conclusions 271

6 References 273

1 Introduction

1.1 Properties of the Actinyl Ions

Compared with the lanthanides or the transition metals, the actinide elements introduce a striking array of novel chemical features, displayed most clearly in the chemistry of uranium. There is the variety of oxidation state, and to some extent the chemical diversity, typical of transition metals in the same periodic group, but physical properties which show that the valence electrons occupy f-orbitals in the manner of the lanthanides. This raises the question of the nature of the chemical bond in the compounds of these elements. The configuration of the uranium atom in the gas phase is $f^3 ds^2$, so it is natural to ask whether there are special characteristics of the bonding that reflect the presence of both f and d valence orbitals.

To answer this question we should choose simple covalent compounds and examine their electronic structure. Although uranocene, $U(C_8H_8)_2$, is an interesting example which has been examined both experimentally and theoretically [1], the covalent bond is seen at its simplest in the dioxo cations MO_2^{2+}. It is hard to over-emphasize the stability of this chemical unit, which is common to all the elements from uranium to americium [2]. Some of these elements exhibit the same unit in oxidation states other than six, e.g., UO_2^+, NpO_2^+, NpO_2^{3+}. For uranium(VI) there is very little chemistry which does not contain the UO_2^{2+} structural unit. Notable exceptions are UF_6, UCl_6, and the δ-phase of UO_3 which has the ReO_3 structure with the uranium atom at a perfectly octahedral oxide site [3], a geometry also found in the double perovskites $Ba_2MgW(U)O_6$ [4]. The octahedral oxide environment is, however, anomalous. The enthalpy of formation of δ-UO_3 [5], ΔH_f (298.15 K, s) = -1213.5 ± 1.25 kJ mol^{-1}, is about 10 kJ mol^{-1} less negative than that of the γ-phase [6]. This more stable phase has a much higher density, contains the uranyl grouping [7], and is obtained from the δ-phase by heating above 410 °C. The octahedral geometry in the double perovskites is presumably forced by the host tungsten lattice.

Uranium(VI) in UCl_6 is a strong oxidant, for example oxidising CH_2Cl_2 to chloroform [8], and decomposes to UCl_5 and chlorine at 120 °C. This property helps to explain the absence of sulphur analogues of the uranyl(VI) ion, but is much attenuated in uranyl compounds; the $UO_2^{2+}/U(IV)$ redox couple is only $+0.32$ V, and uranyl complexes can exist with such reducing ligands as the iodide ion [9].

The extraordinary stability of the UO_2^{2+} ion is also clear in its aqueous chemistry. If acid solutions are neutralised then, near pH 3.5, first dimers and then trimers are formed [10]. These have bridging hydroxide groups, but Raman spectroscopy shows that there is little change within the dioxo group as the condensation proceeds [11]. At pH 5 uranyl hydroxide is precipitated, the solid state structure confirming that the oxycation is intact [12].

Unlike the polyacids of vanadium, molybdenum and tungsten, uranium(VI) does not give a soluble oxyanion in alkaline solutions. Formal analogues of the metamolybdates and metatungstates can be prepared in the solid state, but there the resemblance ends. BaUO$_4$, which can be prepared by heating BaCO$_3$ and UO$_3$, has no structural similarity to BaWO$_4$, in which tetrahedral oxyanions are found. Instead uranyl ions are surrounded and linked together by a square array of four oxide ions, much further from uranium than the oxo-oxygen atoms, into continuous sheets [13]. This is the dominant structural theme in the solid state chemistry of uranium(VI) oxides, although the equatorial bonding can also be five coordinate as in γ-UO$_3$ [14]. Whereas the U–O distance within the uranyl group is typically 175 pm [15], equatorial oxygen distances are nearer 210 pm [14, 15] and similar to that found in the octahedral environment (208 pm) [3].

Actinyl ions are also unusual in that the dioxo unit is invariably close to linear, as illustrated in a survey of 180 structures obtained by diffraction methods [15]. Low symmetry environments seldom introduce more than a 5° deviation from linearity. This contrasts with the pattern in the d-block transition metal dioxo-cations which most frequently have internal angles closer to 90°. Two compounds with the formulation MO$_2$(Ph$_3$PO)$_2$Cl$_2$ (Fig. 1) illustrate this. When M is uranium the dioxo group has the *trans* configuration [16] but when M is Mo it is *cis*, with a O–Mo–O angle of 103° [17]. The metal–chlorine bonds, which are *cis* to the oxo-atoms in both compounds, have average lengths indicating that uranium is larger than molybdenum by 25 pm, nevertheless the

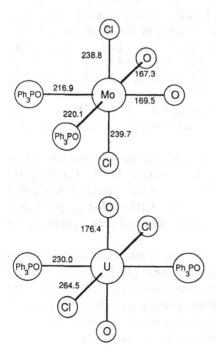

Fig. 1. Bond lengths, in pm, in complexes containing uranium and molybdenum dioxo groups

average metal-oxo distance is only 8 pm larger in the uranyl ion, confirming the remarkable shortness of the uranium oxygen bond.

The strength of this bond is established by thermodynamic data. A recent review [15] reports a mean U–O bond enthalpy, for dissociation to oxygen atoms, of 701 kJ mol^{-1} for UO_2^{2+}(g) and 710 kJ mol^{-1} for UO_2(g). Comparable values for some transition metal gaseous dioxides are: CO_2, 802; MoO_2, 587; WO_2, 635; RuO_2, 509; OsO_2, 770 kJ mol^{-1} [18]. The U–O bond is towards the top end of this range, and approaches the strength of the bond in carbon dioxide.

Not only is the uranyl ion thermodynamically robust, it is also kinetically inert. Experiments designed to measure the rate of isotopic oxygen exchange between the oxo atoms and water at room temperature, establish that the exchange half-life is greater than 40,000 hours [19]. This overall chemical stability accounts for an extensive coordination chemistry which is exploited, for example, in the solvent extraction separation processes used in the nuclear fuel cycle [20].

In the absence of chemical quenching, uranyl compounds have long luminescent lifetimes and high luminescent quantum efficiency [21]. Often, however, the excited state reacts chemically. The photochemistry of the ion, the most famous example of which is the uranyl oxalate actinometer, has generated an enormous body of work and been the subject of comprehensive reviews [22, 23]. It can occur both in solution and in the solid state. The most common reaction is the oxidation of organic substrates. Both the photochemistry and the remarkable properties of the covalent bond, demand a satisfactory interpretation in terms of the electronic structure.

1.2 Objectives of this Review

For an atom as heavy as uranium theoretical treatments of the molecular electronic structure are strongly dependent on simplifying assumptions so it is best to turn first to experiment for the basis of any analysis. This topic has been reviewed previously, but the interpretation of the experimental data has remained ambiguous and controversial [23]. It is the purpose of this review to bring the experimental position up to date, and to show that most of the uncertainties have now been resolved. The review is not exhaustive; I have selected only the most significant sources.

The electronic structure is primarily expressed through the relative energies of orbitals derived from the 5f, 6d, 7s and 7p valence shells on the uranium, and the 2s and 2p orbitals of the oxygen atoms, but some core orbitals on uranium may also participate. In uranium(IV) compounds the electronic spectra and magnetic susceptibilities are unequivocally those of the f^2 configuration suggesting that the 5f shell lies below the 6d shell; moreover it is well-known that the f–d separation increases with increasing oxidation state. With this in mind a tentative energy level scheme for the uranyl ion, which excludes the role of the

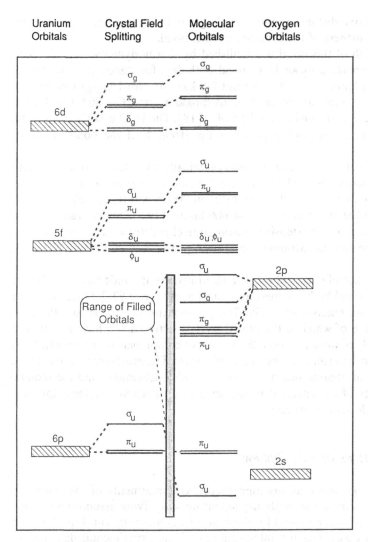

Fig. 2. A tentative energy level scheme for the uranyl ion, excluding spin-orbit components

spin-orbit interaction, is shown in Fig. 2. This diagram should prove useful in following the development of the electronic structural arguments which follow.

The ordering of the four filled, and nominally oxygen-based, valence orbitals in this figure is problematical, but is clearly an important indication of the nature of the bonding. This order should reflect the role of 5f and 6d actinide shells through the difference in the energy of the *gerade* and *ungerade* orbital sets as well as the relative significance of σ and π bonding. We will therefore begin by examining the experimental evidence for the ordering of the highest filled valence orbitals, which will in turn require an understanding of the order of the lowest energy empty orbitals.

2 Experimental Electronic Structure

2.1 Background

In 1957 Jørgensen, writing on the role of the spin-orbit interaction in ligand-to-metal charge-transfer transitions, made the first detailed attempt to understand the electronic excited states of the UO_2^{2+} ion [24]. In his view the optical spectrum of the uranyl ion was due to oxygen to uranium charge-transfer excitations. He pointed out that, for heavy metal atoms, the spin-orbit interaction should exceed the spin-correlation energy between the electron transferred to, and centred upon the metal atom, and the unpaired electron remaining on the ligand. In these circumstances there should be no effective spin selection rule and it would be most appropriate to apply the j–j coupling approximation to interpret the energies, (or in the nomenclature appropriate to the $D_{\infty h}$ group ω–ω coupling). Noting that the visible portion of the absorption spectrum of the UO_2^{2+} ion showed a number of bands with molar extinction coefficients of <10 mol^{-1} dm^3 cm^{-1} Jørgensen argued that, in the absence of a spin selection rule, the weakness of the spectrum was due to a *parity* selection rule, implying that the electron was excited to the metal 5f-shell from a *ungerade* orbital. By analogy with carbon dioxide he anticipated that the overlap of the oxygen π_u orbitals would be less than that of the σ_u, so the highest filled ligand orbitals would be the weakly bonding π_u. From this viewpoint the spectrum should be interpreted in terms of the states arising from a π_u to f excitation.

In 1961 a new hypothesis, now known to be incorrect, was introduced by McGlynn and Smith [25], based on the results of their overlap calculations. The highest filled orbital was attributed π_g symmetry, so that the lowest energy allowed electronic excitations would be to states from the configuration $\pi_g^3 \delta_u$, which would have Π_u and Φ_u orbital symmetry. Conservation of angular momentum requires that only the former should be optically excited, and when the spin-correlation and spin-orbit interactions are included this should provide $^3\Pi_{0u}$, $^3\Pi_{1u}$, $^3\Pi_{2u}$ and $^1\Pi_u$ states. Within this model the low intensity of the spectrum is explained by the spin selection rule, so the apparent presence of three transitions in the aqueous solution spectrum, on which McGlynn and Smith based their analysis, was attributed to the three spin-orbit components of the triplet state. The central of these three spectral regions is the most intense, as expected from the role of the spin-orbit interaction, which provides only the $^3\Pi_{1u}$ state with intensity from the $^1\Pi_u$ state.

McGlynn and Smith's interpretation quickly gained widespread currency on account of its simplicity and the familiarity of the singlet and triplet classification in use by photochemists working with light atom systems [22]. It was shown to be invalid by observations made in 1975, but is firmly embedded in the literature and even now is occasionally quoted by authors unfamiliar with recent progress. With hindsight it is unfortunate that McGlynn and Smith appear to have overlooked the significance of Jørgensen's work, and its implication that low spectral intensities were unlikely to be attributable to a spin selection rule.

An important advance was made in 1971 [26]. Görller-Walrand and VanQuickenborne compared the optical spectra of a large number of compounds containing the uranyl ion in different equatorial ligand field symmetries. They observed that the intensities are enhanced in non-centric environments, implying that the transitions are intrinsically parity forbidden, thereby confirming the main assumption of Jørgensen's analysis.

However, as we shall see, the key to the electronic structure lies in polarised crystal spectroscopy. At low temperatures many hundreds of sharp lines are observed in the electronic absorption spectra, with equivalent detail in the luminescence spectra. During the Second World War, as part of the Manhattan Project, in a largely unknown and pioneering feat of single crystal spectroscopy, measurements were made at liquid hydrogen temperatures for several uranyl compounds, the results being published by Dieke and Duncan in a detailed monograph in 1948 [27]. It is interesting that the objective of this work was to develop an optical method of isotope separation based on the selectivity implicit in the sharpness of the spectrum and the efficient photochemistry of the uranyl ion [28], and to this end a number of measurements were made on isotopic species. The application of lasers to isotope separation is now well established, but the experimental effort put into the spectroscopy of uranium, well before the invention of the laser, is a striking testimony to the range and scale of this wartime enterprise to develop a nuclear weapon. While the data of Dieke and Duncan and their coworkers, reveal the extraordinary complexity of the spectrum at low temperatures, their observations were not easy to interpret at that time or even subsequently, partly because the intensities were recorded photographically and partly because the paucity of crystal structure determinations did not allow the polarisation of the spectra to be related to molecular axes.

2.2 Recent Single Crystal Optical Spectroscopy

A series of optical experiments over the last 15 years has now clarified the position and I will select the data which show the nature of the excitations most clearly. The best information comes from the spectrum of $Cs_2UO_2Cl_4$ [29].

This material crystallizes in the monoclinic space group $C_{2/m}$. [30] The primitive unit cell contains only one molecule so there are no factor-group (Davydov) splittings. The orientation of the molecular axes with respect to the crystal axes and faces, shown in Fig. 3, is easy to establish. The space group requires C_{2h} point symmetry at the uranium atom; it is a centrosymmetric site. Figure 4 shows the intimidating level of detail in the visible and near UV spectrum, at 4.2 K, under modest resolution. Nevertheless we can see here (a) the low intensities first noted by Jørgensen, with their implication of a parity selection rule, (b) long progressions, well documented by Dieke and Duncan, in the UO_2 symmetric stretching mode at about 710 cm^{-1}, much lowered from the ground state frequency of about 830 cm^{-1}, and (c) sufficient detail to indicate

Electronic Structure and Bonding in Actinyl Ions

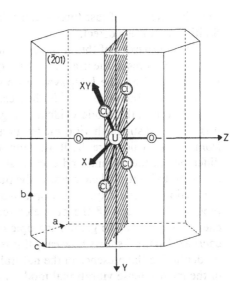

Fig. 3. Crystallographic axes, crystal habit and molecular orientation in $Cs_2UO_2Cl_4$; from Ref. [31]

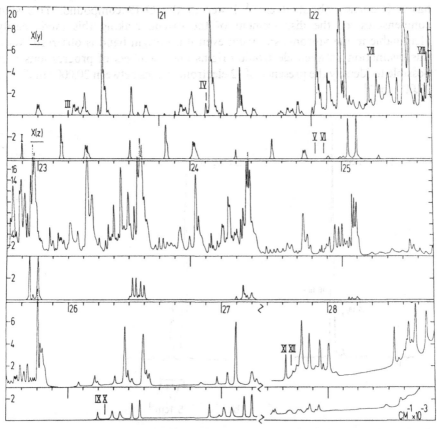

Fig. 4. Single crystal polarised absorption spectrum of $Cs_2UO_2Cl_4$ at 4.2 K; from Ref. [29]

that the presence of just three electronic states, as proposed by McGlynn and Smith, cannot be correct.

The complexity of this spectrum can be tackled effectively by replacing the oxygen-16 atoms in the uranyl ion by oxygen-18 [29]. This substitution has two main effects. First the UO_2 symmetric stretching frequency drops, from about 710 cm^{-1}, to about 675 cm^{-1} in the electronic excited states. As a result the long progressions in this mode, evident in Fig. 4, contract on the energy scale, so that it is easy to distinguish vibronic features associated with their higher members from those which are built on alternative electronic origins. Second, the difference in the ground and excited state oxygen frequencies appears in their zero-point vibrational energies, so the pure electronic transition energy becomes a function of the oxygen isotopic mass. When both oxygen-16 atoms are replaced by oxygen-18 the pure electronic excitation shifts 10 cm^{-1} to *higher* energy because of the greater isotopic decrease in the ground state zero-point energies. Figure 5 shows how easy this is to observe; indeed there is no difficulty in identifying the presence of the natural abundance of oxygen-18 (0.20%). One of the most intense vibrational modes in the spectrum is the UO_2 bend, which has a frequency about 10 cm^{-1} lower in the O-18 compound. This shift compensates for the displacement of the origin, making this mode easily identifiable in the vibronic spectrum even if no origin band is observed. With this distinction between electronic origins and members of progressions it is possible to identify the presence of 12 electronic states between 20,000 cm^{-1} and 29,000 cm^{-1}.

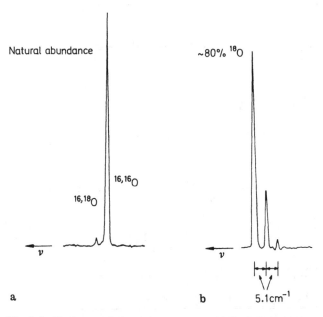

Fig. 5a,b. Single crystal absorption spectra at 4.2 K of $Cs_2UO_2Cl_4$ near 20,095 cm^{-1}, (**a**) natural isotopic abundances and (**b**) containing ~80% oxygen-18; from Ref. [125]

2.3 Excited State Parity

The parity of the electronic excited states is easily found. With a closed shell ground state of even parity and a centrosymmetric point symmetry, *ungerade* excited states are electric-dipole allowed, whereas *gerade* excited states can only be magnetic-dipole or electric-quadrupole allowed. Figure 6 shows six polarisations covering the first 900 cm^{-1} of the optical spectrum [31]. The notation shows the propagation axis in capital letters and the electric vector polarisation axis in parentheses. Comparing the $X(y)$ and $Z(y)$ experiments, much of the spectrum is identical but the lowest energy feature at $\sim 20,095$ cm^{-1}, labelled II, is only present in $Z(y)$. The electric field vector is the same in both cases so this band must be magnetic-dipole allowed, the transition being induced by the magnetic component of the radiation field along the molecular x-axis. As expected the band only appears in one other polarisation, $Y(z)$. The figure also reveals another magnetic-dipole allowed transition 1.6 cm^{-1} lower in energy labelled I, which appears only when the magnetic component of the field is in the y direction. These are the first two bands in the absorption spectrum and they also appear as luminescence origins; it follows that the pure electronic transitions are parity-conserving and that the first excited states are *gerade*. Similar polarisation experiments confirm that all the observed origins below 29,000 cm^{-1} correspond to excited states of even parity.

2.4 Angular Momentum in the First Excited State

Because the first two electronic origins are allowed by the x and y magnetic components of the radiation field it is a good guess, on the basis of the selection rules in $D_{\infty h}$, that the transition involves a change in the quantum number, Ω, describing the total angular momentum around the UO_2 axis, of one unit, and that the excitation should be described as a Σ_g^+ ($\Omega=0$) to $\Pi_g(\Omega=1)$ transition. Notice that with a large spin-orbit interaction it is prudent to avoid the use of spin angular momentum quantum numbers. However, a cylindrical symmetry representation is not sufficient because the almost tetragonal ligand field of the four chloride ions can, for example, cause a superposition of Π_g and Φ_g cylindrical field states.

By measuring the magnetic moment in the excited state parallel to the UO_2 axis one can decide which of these two cylindrical field parent states is dominant. The observations are shown in Fig. 7 [29]. The two components of the first excited state occur in orthogonal polarisations separated by 1.6 cm^{-1}, and the Zeeman effect, which is second-order in the absence of degeneracy, is too small in a 5T field to give a detectable shift, although the other transition displayed in this figure, occurring near 26,200 cm^{-1}, shows rather large Zeeman shifts. Nevertheless the superposition of the components induced by the field is clearly visible in the figure through the mixing of their polarisations, and allows the

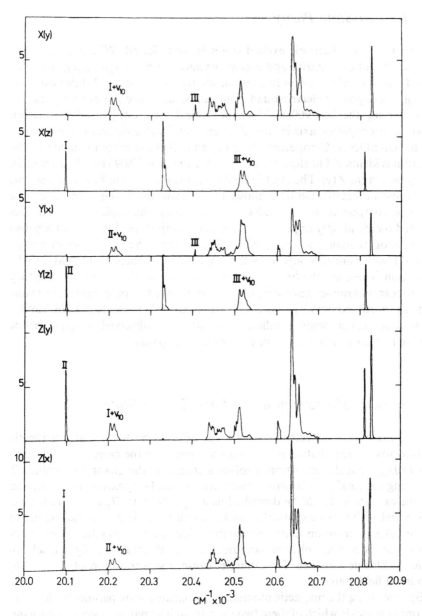

Fig. 6. Single crystal polarised absorption at 4.2 K of $Cs_2UO_2Cl_4$ in six polarisations. The polarisation notation is explained in the text; from Ref. [31]

magnitude of the matrix elements of the magnetic moment to be obtained. They give the state a moment of 0.16 μ_B.

There is no combination of spin and orbital angular momentum for a Φ_g state, where $\Omega = 3$, which can have such a small magnetic moment, so it follows

Electronic Structure and Bonding in Actinyl Ions

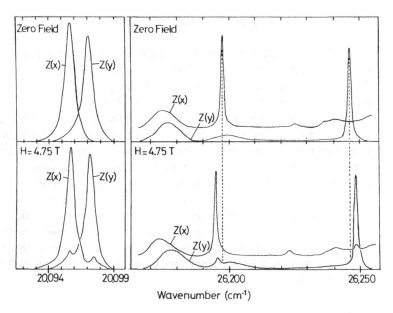

Fig. 7. The effect of a 4.75 T magnetic field applied parallel to the O–U–O axis on two polarisations of the spectrum of $Cs_2UO_2Cl_4$ at 4.2 K; from Ref. [29]

that the cylindrical field parent state must be predominantly Π_g [29]. There are not many configurations which can generate such a state, because the spin angular momentum of the two unpaired electrons must create a magnetic moment which cancels the orbital angular momentum, so the magnetic quantum numbers have the values, $M_A = 2$ and $M_\Sigma = -1$. (Λ and Σ are the $D_{\infty h}$ analogues of L and S in spherical symmetry.)

We now make the assumption that the lowest unoccupied orbitals are f_ϕ or f_δ, where the subscript encodes the absolute value of m_λ. The remaining f-orbitals are anti-bonding with respect to the oxygen and so are assumed to be too high in energy to be of significance. There are then only two elementary excitations consistent with the small magnetic moment. Either a $\sigma_u(\lambda=0)$ electron is excited to $\delta_u(\lambda=2)$ orbital to give a wavefunction, in the ω–ω basis, described by $|\bar{\sigma}_u \bar{\delta}_{u2+}\rangle$, where the barred functions imply $m_\sigma = -1/2$ and the subscripts describe the parity and, where required, the m_λ-value, or a $\pi_u(\lambda=1)$ electron is excited to a $\phi_u(\lambda=3)$ orbital and the wavefunction is $|\bar{\pi}_{u+1} \pi_{u-1} \bar{\pi}_{u-1} \bar{\phi}_{u+3}\rangle$. It is easy to verify that both these functions have no net magnetic moment along the UO_2 axis.

It is the ambiguity between these two configurations and the implications for the highest filled molecular orbital which has dominated the discussion of the significance of the electronic spectrum in recent years [23]. To make a decision between these possibilities it is necessary to consider the properties of some of the other, higher energy, excited states.

2.5 Properties of Higher Energy Excited States

Figure 8 illustrates the pattern of the excited states with correlation diagrams for the two configurations suggested by the Zeeman effect measurements [32]. In these diagrams the abscissa represents the magnitude of the electron-electron repulsion energy and shows the evolution between the ω–ω limit and the Russell-Saunders limit. Notice that there are twice as many states arising from the $\pi^3\phi$ configuration as from the $\sigma\delta$ configuration, but that both configurations span much the same energy range. The main difference is a doubling of the number of states in the $\pi^3\phi$ configuration, caused by the spin-orbit interaction in the π_u orbital, which should be mostly oxygen-centred. The magnitude of this splitting is difficult to estimate because it depends on the covalency in the uranium oxygen π-bond, but even if this is large, it cannot be much greater than 1000 cm^{-1}. Both the δ_u and ϕ_u f-orbitals are non-bonding with respect to oxygen so, whichever orbitals the excitation occurs from, there should be additional states nearby in energy, formed either from the $\sigma_u\phi_u$ or from the $\pi_u^3\delta_u$ configuration.

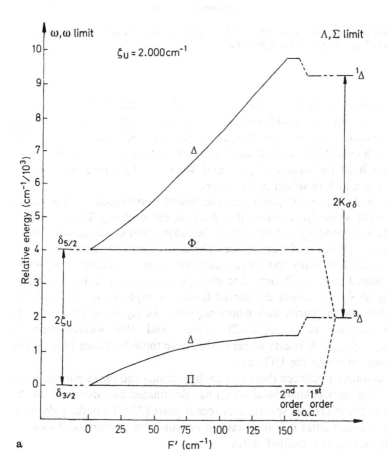

Electronic Structure and Bonding in Actinyl Ions

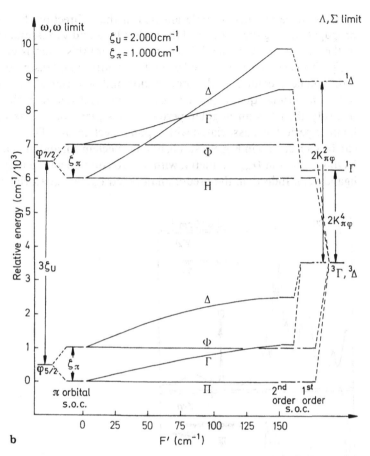

Fig. 8a,b. Correlation diagrams for the states arising from (a) the $\sigma\delta$ configuration and (b) the $\pi^3\phi$ configuration; from Ref. [32]

We shall see in due course that there is sufficient evidence to exclude the presence of states derived from excitations to the antibonding f-π_u^* orbitals. So if a σ_u electron is excited to either a δ_u or a ϕ_u orbital we anticipate a total of *eight* excited states in $D_{\infty h}$ symmetry or *sixteen* states in the actual C_{2h} site symmetry. On the other hand if a π_u electron is excited to the same orbitals there should be *sixteen* $D_{\infty h}$ excited states or *thirty-two* C_{2h} states, within much the same spectral range. Thus, a distinction should be possible on the basis of the *density* of excited states in the 9000 cm^{-1} range for which sharp detail is observed in the spectrum.

Figure 8 suggests that the second excited state should have $\Delta_g(\Omega=2)$ symmetry if the configuration is $\sigma\delta$, but $\Gamma_g(\Omega=4)$ if it is $\pi^3\phi$. Neither of these states is accessible through the magnetic dipole transition mechanism. However when $\Delta\Omega=2$ the transition is allowed by the electric-quadrupole mechanism. In Fig. 6 a band near 20,405 cm^{-1}, labelled III, in the $X(y)$ and $Y(x)$ polarisations appears to be magnetic-dipole allowed in the z direction, but when the light propagates along the bisector of the molecular x and y axes, and is still polarised in the xy plane (i.e. in the $XY(xy)$ polarisation) this band is effectively absent [31]. This is what is anticipated when the quadrupolar tensor component of the radiation field, defined by the propagation and polarisation vectors, is projected onto a molecular quadrupolar transition moment tensor in the xy plane. The intensity then has an angular period in this plane of $\pi/2$ as opposed to the familiar period of π associated with dipolar mechanisms. This periodicity, which can be seen clearly in Fig. 9, establishes the mechanism and indicates a transition having $\Delta\Omega = 2$ in $D_{\infty h}$ symmetry, with its two components split by the equatorial ligand field. Indeed, in the predominantly tetragonal equatorial environment of

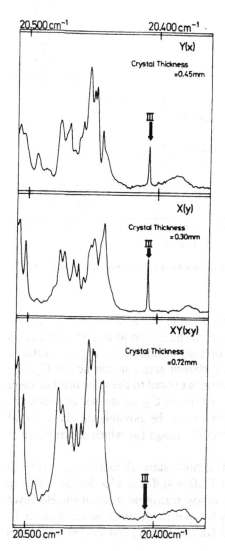

Fig. 9. Polarised absorption spectra of three different polarisations in the spectrum of $Cs_2UO_2Cl_4$ at 4.2 K, showing the quadrupolar intensity of the feature at 20,405 cm^{-1}; from Ref. [31]

the chloride ions only those states with $\varDelta_g(\Omega=2)$ symmetry in $D_{\infty h}$ would be expected to be split by the local ligand field.

While these observations favour the choice of the $\sigma\delta$ configuration, it nevertheless remains possible that when the full Hamiltonian for the states derived from the $\pi^3\phi$ configuration, and configurations that interact with it, is diagonalised, the order of the states suggested in Fig. 8 might be modified in such a way as to be consistent with the observations. For this reason it is important to establish the symmetries and energies of as many of the excited states as possible, preferably in more than one equatorial ligand symmetry, and to test these against a comprehensive Hamiltonian.

2.6 The Tetragonal Equatorial Field

The presence of a tetragonal field splitting, of the order of 500 cm^{-1}, in a $\varDelta_g(\Omega=2)$ state has a direct bearing on the nature of the excited configuration. Görller-Walrand and VanQuickenborne first realised that the strong axial ligand field, together with the spin-orbit interaction placed constraints on the ability of the equatorial ligands to resolve the degeneracy of the cylindrical field states [33]. In the isolated ion these states form conjugate pairs whose angular momentum is described by the magnetic quantum numbers M_Ω. This angular momentum is composed of a contribution from both the excited f-electron and the hole in the highest filled-orbital. In this representation any equatorial field perturbation of lower than cylindrical symmetry is off-diagonal, and to resolve the degeneracy it must superpose or mix the components of a conjugate pair. The two possible configurations $\sigma\delta$ and $\pi^3\phi$ respond differently in this regard. Because the perturbation is a one-electron operator, the splitting of the f-electron degeneracy is denied, to first order in the perturbation, by the orthogonality of the conjugate one-electron functions describing a hole with orbital angular momentum, as is the case for the $\pi^3\phi$ configuration. On the other hand, in the case of the $\sigma\delta$ configuration where the hole has no orbital angular momentum, it is possible to construct states in the Russell-Saunders basis where the equatorial field resolves the cylindrical field degeneracy in first-order [32, 33]. So, the observation of significant equatorial field splittings is, in itself, good evidence for the role of the $\sigma\delta$ configuration.

2.7 Other Equatorial Fields

There is a general similarity between the optical spectra of almost all uranyl compounds in the near UV, so the correct choice of excited configurations should explain the spectroscopic properties in other ligand environments. This proposition can be tested by examining the polarised crystal spectrum of CsUO$_2$(NO$_3$)$_3$ at low temperatures [34]. In this material the UO$_2$ group is surrounded by an approximately hexagonal group of six oxygen atoms from the three bidentate nitrate groups. There are now two molecules per unit cell but their UO_2 axes are parallel, so that simple polarisation measurements are again possible. The UO_2 axis is also coincident with the uniaxial direction of the trigonal space-group permitting the observation of magnetic circular dichroism (MCD) and allowing the measurement of the magnetic moment of the first excited state. This is found to be 0.01 μ_B. The Zeeman splitting of the second excited state establishes that it is degenerate with a magnetic moment of 1.1 μ_B. In hexagonal or trigonal equatorial fields only states with $\Omega = 3$ are split, so the first two states appear, from their degeneracy, their polarisation [34], and their magnetic moments to correspond to cylindrical field states with $\Omega = 1$ and $\Omega = 2$ respectively, exactly the same pattern found in $Cs_2UO_2Cl_4$.

The complex nitrate spectrum can be partially analysed by using oxygen-18 shifts [34], in the manner described in Section 2.2. But there is an additional useful feature. The vibronic structure includes internal modes of the nitrate ions. These modes have nitrogen-15 shifts which identify them unambiguously [35]. The frequencies of these modes in coordination compounds are well-established so the optical frequencies can be used to determine the energy of the pure electronic transition to which they are coupled, while the polarisation of the vibronic spectrum constrains the choice of the electronic symmetry.

When these techniques are applied systematically it is possible to locate the position of *ten* electronic excited states in the spectrum of $CsUO_2(NO_3)_3$ (counting both components of degenerate states) although only seven of these have observable origin transitions. For $Cs_2UO_2Cl_4$ *twelve* electronic states are identifiable, of which seven have observable origins.

2.8 Relative Energies of the Non-Bonding f-Orbitals

It should also be possible to distinguish the merits of the two hypotheses if the ordering of the δ and ϕ components of the f-orbitals, which are non-bonding with respect to the oxygen atoms, were able to be established independently. A natural, although indirect, way to do this is to identify the ground state of the NpO_2^{2+} ion, which has the f^1 configuration. Magnetic susceptibility [36] and electron spin resonance [37] data are available for the compounds with trigonal equatorial fields, $NaNpO_2(CH_3COO)_3$ and $RbNpO_2(NO_3)_3$. Accurate measurements can also be made using the very large Zeeman splittings in the spectrum of $CsNpO_2(NO_3)_3$ [38], an example of which is shown in Fig. 10. The temperature dependence easily distinguishes the ground and excited state splittings and gives ground state g-values of $g_\perp = 3.34\ \mu_B$ and $g_\perp = 0.2\ \mu_B$, in excellent agreement with the ESR data. The lowest spin-orbit coupled state from the f_δ orbital, $\Delta_{3/2u}$, has a first order value for $g_\parallel$ of 2.0 μ_B, while $\Phi_{5/2u}$, from the f_ϕ orbital has $g_\parallel = 4.0\ \mu_B$, suggesting that the ground state is predominantly f_ϕ. When the equatorial field and the second-order spin-orbit interactions are included the calculated g-value for $\Phi_{5/2u}$ is very close to that observed. It was the

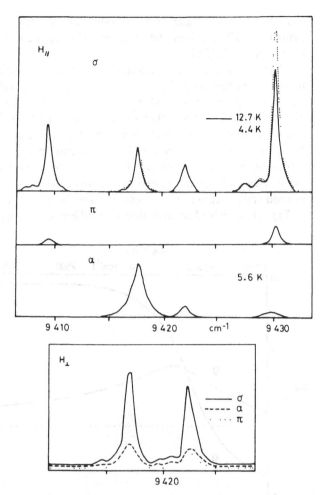

Fig. 10. The effect of a 5.0 T magnetic field on the polarised absorption spectrum of CsNpO$_2$(NO$_3$)$_3$. In zero field only a single absorption occurs at 9420.2 cm^{-1}; from Ref. [38]

ESR determination [37] of the ground state magnetic moments which provided the first direct evidence that the valence shell contains an *f*-electron.

The properties of the ground state, however, do not in themselves determine the lowest orbital energy because the spin-orbit components of f_ϕ orbital are separated by $3\mathscr{G}$, approximately 6600 cm^{-1}, whereas those of the f_δ orbital are separated by only $2\mathscr{G}$, or 4400 cm^{-1}. A further complication is caused by the second order spin-orbit interaction between $\delta_{5/2}$ and $\phi_{5/2}$. In addition the trigonal equatorial field separates the components of f_ϕ by another 3000 cm^{-1}. It is therefore possible for the *centre of gravity* of the states arising from the f_δ configuration to lie below those from the f_ϕ configuration, despite the dominant role of the latter in the ground state. Careful spectroscopic work can locate all four states from these two orbitals and identify them by their magnetic properties. The result is unambiguous, placing the f_ϕ orbital energy 1700 cm^{-1} *above* that of f_δ [38].

Turning now to the tetragonal ligand field, the optical Zeeman effect is also useful in establishing the ground state of Cs$_2$NpO$_2$Cl$_4$ [38]. The measured values are g$_\parallel$ = 1.38 μ_B and g$_\perp$ = 1.32 μ_B. At first sight this almost isotropic g-value is hard to comprehend in a strong axial field. But on closer analysis it can be seen to arise naturally from the accidental degeneracy of the $\Delta_{3/2u}$ and $\Phi_{5/2u}$ states, which allows angular momentum perpendicular to the cylindrical field axis; a feature which will be obvious to those familiar with the angular momentum raising and lowering operators. The energies of the states and the ground state g-values as a function of the f-orbital energy difference are shown in Fig. 11, which illustrates that this difference is very tightly constrained by

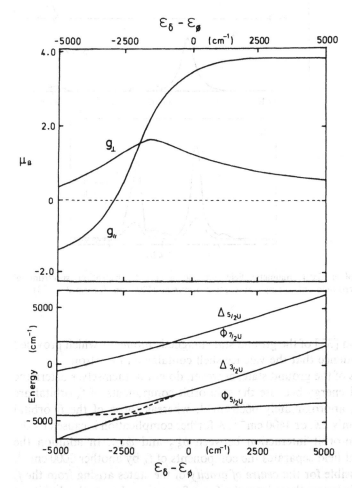

Fig. 11. The energies lowest energy f-electron states and ground state g-values in Cs$_2$NpO$_2$Cl$_4$ as a function of the difference in the δ and ϕ orbital energies; from Ref. [38]

the observed g-values. Indeed these measurements establish that f_ϕ is 1960 ± 30 cm^{-1} above f_δ in this compound. Despite the large difference in the equatorial field in the chloride and nitrate compounds, the separation between the δ and ϕ orbitals seems almost unchanged, and is presumably dominated by the role of the oxo-ligands. For what it is worth, in an electrostatic model the fourth-order tensor components of the potential from both the equatorial and axial ligands raise the energy of f_ϕ above f_δ.

Although this result is an important guide to the relative ordering of the f-electron configurations in the excited states of the uranyl ion, it is not conclusive because the configuration energies are composed of both virtual orbital eigenvalues, which correspond to the neptunyl f-orbital energies, and the interaction energy between the excited f electron and the hole in the valence shell, which can only be estimated theoretically and will be discussed in due course.

2.9 Other f-Orbital Energies

The spectrum of the neptunyl ion can also be used to locate other f-orbital components, and this is important in identifying the magnitude of the f_σ and f_π interactions with oxygen. Figure 12 shows a portion of the single crystal absorption and MCD spectrum of NpO$_2^{2+}$ in the trigonal environment of CsUO$_2$(NO$_3$)$_3$ [38]. The features labelled VI belong to a state with a different Huang-Rhys factor to those labelled VII. The latter state exhibits a large reduction in the M–O stretching frequency, from 860 cm^{-1} to $\sim$750 cm^{-1}, compared to the ground state, similar in magnitude to the reductions found in uranyl ion excited states. In the other state, labelled VI, this frequency is 812 cm^{-1}, indicating a much smaller weakening of the Np–O bond. There is a second state with the same characteristics, at higher energy, as expected for the two spin orbit components, $\Pi_{1/2u}$ and $\Pi_{3/2u}$. Zeeman effect measurements, and the sign of the MCD confirm the identity of these states, and place the f_π orbital about 14,000 cm^{-1} above f_ϕ [38].

Much the same result is obtained for the neptunyl ion doped as an impurity in Cs$_2$UO$_2$Cl$_4$. The transition to the state derived from the f_σ orbital is not observed and must be covered by intense charge-transfer transitions at higher energy. However, the large number of observables, including g-values and magnetic-dipole intensities, help to constrain the possible energy of this unobserved state (within the limitations of a one-centre model), placing it more than 40,000 cm^{-1} above f_ϕ.

The Huang-Rhys parameter for the $f_{\delta,\phi}$–f_π transitions in the neptunyl ion is about 0.8, indicating a significant Np–O bond lengthening in the excited states derived from the f_π configuration. In contrast there is no evidence of a progression in the NpO$_2$ stretching frequency in transitions between the states of the f_δ and f_ϕ configurations, which are non-bonding with respect to oxygen.

The reduction of the NpO$_2$ frequency from 802 cm^{-1} in the ground state to 765 cm^{-1} shows that the Np–O bond is also weakened when the electron

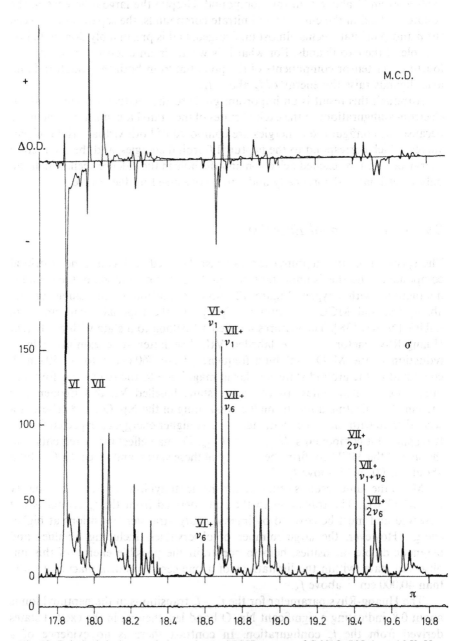

Fig. 12. Polarised single crystal absorption and MCD spectra of CsNpO$_2$(NO$_3$)$_3$ at 4.2 K. Upper portion MCD, middle portion σ, and lower portion π polarisations; from Ref. [38]

occupies the f_π orbital, so this orbital interacts significantly with the oxygen p_π orbitals. The $f_\pi - f_\phi$ separation (which reflects the anti-bonding character of the f_π orbitals) is 14,000 cm^{-1} in CsNpO$_2$(NO$_3$)$_3$ and 12,800 cm^{-1} in Cs$_2$NpO$_2$Cl$_4$, in line with their relative Np–O bond strengths implied by their ground state symmetric stretching frequencies, which are 860 and 802 cm^{-1} respectively.

2.10 Relative Configuration Energies

Although there is a small possibility of ambiguity in the order of the $\sigma\delta$ and $\pi^3\phi$ excited configuration energies in the uranyl ion, the first excited charge-transfer configurations of the neptunyl ion offer an independent test of the relative order of the highest filled oxygen orbitals. In this case the excited valence electron joins the f-electron to give either a σf^2 configuration or a $\pi^3 f^2$ configuration. Calculations of the magnitude of the repulsion between the f electrons, confirm the expectation from Hund's second rule that the $\gamma\delta\phi$ configurations, where γ is either σ or π, are more stable by about 4000 cm^{-1} than either of the configurations, $\gamma\delta\delta$ or $\gamma\phi\phi$ [39]. This is a sufficiently large difference to ensure that either the $\sigma\delta\phi$ or the $\pi^3\delta\phi$ configuration will provide the first excited state regardless of small differences between the f_δ and f_ϕ energies. So, in contrast to the uranyl ion, the properties of the first excited state are determined by the nature of the excited valence electron, and we can ignore the uncertainty arising from the ordering of the non-bonding f-orbitals. If a σ-electron is excited the first excited state should be $^4H(\Lambda=5)_{7/2u}$, but if a π-electron is excited the first excited state may either be $^4\Gamma(\Lambda=4)_{5/2u}$ or $^4I(\Lambda=6)_{9/2u}$, both states having the same electron–electron repulsion energy.

A clear test of these alternatives may be made by measuring the $g_\parallel$ values for the first excited states either by Zeeman measurements on Cs$_2$NpO$_2$Cl$_4$, or by MCD on CsNpO$_2$(NO$_3$)$_3$ [39]. These g-values carry a sign, defining the relative energy in the magnetic field of the M_Γ components within the appropriate double group representations. These signs can be experimentally determined from the sign of the MCD for CsNpO$_2$(NO$_3$)$_3$ or through the magnitude of the Zeeman splitting of vibronic sidebands in Cs$_2$NpO$_2$Cl$_4$. The results are collected in Table 1.

The outcome is unambiguous. Only the $\sigma\delta\phi$ configuration gives the correct sign and magnitude for the g-values in either case. The quantitative agreement

Table 1. $g_\parallel$-values for excited states of the neptunyl ion

Symmetry	Observed	Calculated		
		$^4H_{7/2u}(\sigma\delta\phi)$	$^4\Gamma_{5/2u}(\pi_+\pi_-{}^2\delta\phi)$	$^4I_{9/2u}(\pi_+{}^2\pi_-\delta\phi)$
D_{3h}	+3.36	+4.1	−2.1	+6.1
D_{4h}	−3.8	−4.1	+2.1	+6.1

between the observed and calculated g-values can be improved by including the equatorial field in the model Hamiltonian, and this calculation is partially successful in locating the energies of the first five excited states in both compounds [39]. This result provides evidence that the highest filled valence orbital has σ_u symmetry.

2.11 Two-Photon Spectroscopy

In Sect. 2.5, I pointed out that the $\sigma_u \gamma_u$ and $\pi_u^3 \gamma_u$ configurations differ mainly in the density of their excited states. The former should create 16 states, the latter 32. Polarised absorption spectroscopy only detects 12 excited states, so both configurations remain possible. The cynic can claim that many more states are present but are undetected. This argument can never be completely overturned, but can be undermined by increasing the number of observables characterising the states, and by reducing the probability that some states are not detected. Two-photon absorption (TPA) is very helpful in this regard.

In centro-symmetric systems TPA spectroscopy obeys selection rules which are complementary to those operating in electric-dipole allowed one-photon spectroscopy. For intrinsically parity-conserving electronic transitions, as in the present case, the spectrum is greatly simplified by the removal of vibronic or 'false' origins, and is dominated by pure electronic transitions and progressions in totally symmetric modes [40]. Those modes which do couple, and have even parity, cannot be dipolar, so that any lattice dispersion is associated with the coupling of higher order multipole moments. Such interactions fall off sharply with distance, and so exhibit a small lattice dispersion, a feature which can sharpen the vibronic spectrum significantly compared to one-photon absorption (OPA). TPA selection rules also give access to those pure electronic transitions which are not magnetic-dipole allowed, while the variety of possible polarisation experiments can be very helpful in defining electronic symmetry. Finally, the low absorption cross-section of TPA permits the study of regions of the spectrum of pure single crystals where the optical density in (OPA) is too great to permit the transmission of light.

All these characteristics can be seen in Fig. 13 where a portion of the TPA polarised spectrum of $Cs_2UO_2Cl_4$ is compared with a single-photon polarised spectrum [40]. The spectrum is particularly easy to interpret, and in its entirety confirms the location of the 12 states previously proposed from the analysis of the one-photon spectrum. In particular strong *origin* bands are found where previously the existence of the electronic excited states had to be inferred from an analysis of the vibronic structure. But in addition, TPA locates two further excited states in the near ultraviolet which were previously unknown, making a total of 14 excited states.

More importantly, the high contrast ratio, which arises from the sharpness of the spectrum, taken with the more permissive regime of TPA selection rules, sharply reduces the probability that there are undetected states in this spectral

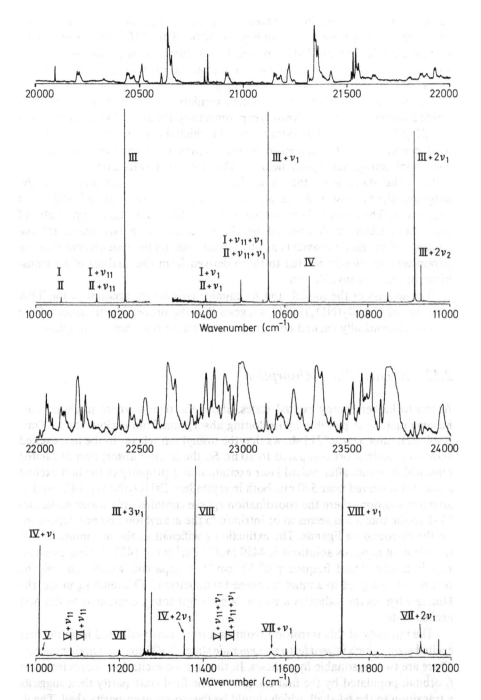

Fig. 13. A portion of the single photon absorption and TPA spectra of $Cs_2UO_2Cl_4$ at 4.2 K. Energy scales for the two experiments are matched in paired segments; from Ref. [40]

region. This point can be emphasised by examining the fine detail of the spectrum on an expanded scale. In Fig. 14 the band at 21,105 cm^{-1} is due to the natural abundance (0.075%) of oxygen-17 substituted uranyl ions. All the fine structure at this level of sensitivity in the spectrum is assignable to known vibrational modes, or to acoustic phonons.

To illustrate the information content available from all these experiments Table 2 summarises what is known experimentally about the 14 excited states in Cs$_2$UO$_2$Cl$_4$. Figure 15 illustrates the way in which the free ion states in the ω–ω limit evolve under the influence of the electron-electron repulsion and the equatorial tetragonal ligand field. In this empirical calculation, which also includes the states from the $\sigma\pi$ configuration which are not shown on the diagram, eleven observed energies have been fitted to a model with nine parameters. The result is, however, strongly constrained, by (a) the symmetry of each state, which is determined by the experimental polarisations, (b) the inclusion of magnetic moments as observables, and (c) the requirement that the parameters are closely related to those derived from the analysis of f–f transitions in the neptunyl(VI) ion.

The success of the model, and its equally satisfactory account of the TPA spectrum of CsUO$_2$(NO$_3$)$_3$ [41], suggests that the proper identification of the lowest electronically excited configurations must be regarded as complete.

2.12 Excited State Absorption

Access to higher energy excited states, above the region where single crystals become opaque is possible by exploiting absorption from excited states. Time resolved luminescence [21] shows that the uranyl ion relaxes to the first excited state on a scale short compared to 10 ns. So the intense absorption of excited ions, which occurs after pulsed laser excitation, is a property of the first excited state. It is observed near 570 nm, both in crystalline CsUO$_2$(NO$_3$)$_3$ [42], and in aqueous solution where the coordination sphere contains only water molecules [43], so the transition seems to be intrinsic to the uranyl ion and not dependent on the surrounding ligands. The extinction coefficient at the maximum of this transition in aqueous solutions is 4450 mol^{-1} dm^3 cm^{-1} [43]. A clear progression in a vibrational frequency of 570 cm^{-1} is apparent, which can only be realistically assigned to a much weakened symmetric U–O stretching mode; the Huang-Rhys factor indicates a major bond-lengthening, compared to the first excited state.

The intensity of this transition requires it to be parity-allowed from the first excited state, which is *gerade*, to an *ungerade* state 17,500 cm^{-1} to higher energy. There are two reasonable hypotheses. In the first the excitation occurs from the f_δ orbital, populated by the first excitation. The final state parity then suggests a transition to the 6d-shell, which should be the closest even-parity shell. The d_δ orbital is non-bonding with respect to oxygen and ought to be the lowest energy component of this shell, so this proposal is inconsistent with the drop in the

Electronic Structure and Bonding in Actinyl Ions

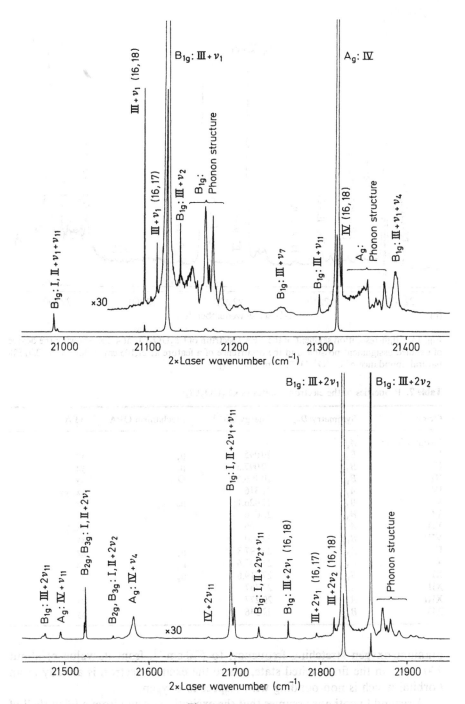

Fig. 14. (a) (*Contd.*)

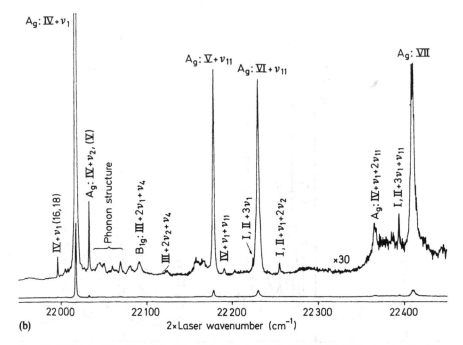

Fig. 14. High-resolution segment of the TPA spectrum of $Cs_2UO_2Cl_4$ at 4.2 K, showing the degree of detailed assignment possible and the occurrence of a feature at 21,105 cm^{-1} due to the 0.075% natural abundance of $U^{17}O^{16}O^{2+}$; from Ref. [40]

Table 2. Properties of the electronic states of $Cs_2UO_2Cl_4$

Origin	Symmetry(D_{2h})	Energy/cm^{-1}	Mechanism OPA	TPA
Ground State	A_g	0		
I	B_{2g}	20 095.7	μ_y	xz
II	B_{3g}	20 097.3	μ_x	yz
III	B_{1g}	20 406.5	Θ_{xy}	xy
IV	A_g	21 316		xx, yy
V	B_{2g}	22 026.1	μ_y	xz
VI	B_{3g}	22 076		
VII	A_g	22 406		xx, yy
VIII	B_{1g}	22 750		xy
IX	B_{2g}	26 197.3	μ_y	xz
X	B_{3g}	26 247.6	μ_x	yz
XI	B_{1g}	27 719.6	μ_z	xy
XII	A_g	27 757		xx, yy
XIII	A_g	29 277		xx
XIV	B_{1g}	29 546		xy

uranium oxygen stretching frequency, to 570 cm^{-1}, from its value of about 700 cm^{-1} in the first excited state, where the excited electron is already in an f-orbital which is non-bonding with respect to oxygen.

A second hypothesis assumes that the excitation occurs from a filled shell of even parity to fill the hole in the σ_u orbital caused by the first excitation. This

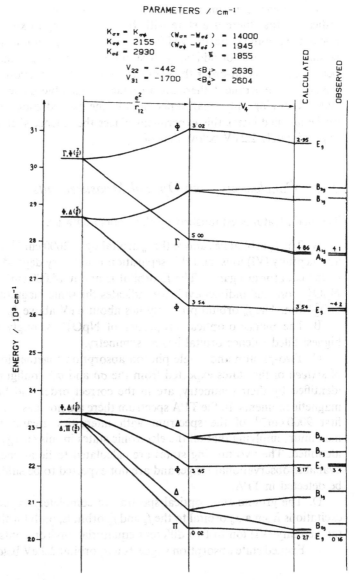

Fig. 15. Calculated and observed excited state energy levels of $Cs_2UO_2Cl_4$. The symmetries indicated at the right of the diagram are in D_{4h}. The numbers associated with the levels are magnetic moments in Bohr magnetons; from Ref. [40]

excitation may occur from either a σ_g or a π_g shell. Both are bonding with respect to the metal 6d orbitals, and so the excited state should exhibit a significant lowering in the U–O bond stretching frequency. Such a transition should be polarised perpendicular to the uranyl axis for an excitation from the π_g shell but parallel to the axis for excitation from the σ_g shell.

Preliminary experiments [44] show parallel polarisation for this absorption, indicating that there is a state with Π_u symmetry approximately 37,000 cm^{-1} above the ground state, derived from the $\sigma_g \delta_u$ configuration. Transitions to this configuration are forbidden, in first order, from the ground state, although permitted in second-order through the spin-orbit interaction. The excited state absorption experiment therefore provides a valuable device for locating this state. Assuming only small differences in the excited state relaxation energies (to be justified later), this experiment places the valence shell σ_g orbital energy approximately 2.2 eV below that of σ_u.

2.13 Conclusions from Optical Measurements

The main features established by the optical data are:

A. The f_ϕ orbital lies above the f_δ orbital by ~ 2000 cm^{-1} in both uranyl(VI) and neptunyl(VI) ions, and this separation is not very dependent on the nature of the equatorial ligands. The f_π orbital is about 14,000 cm^{-1} above f_ϕ in the NpO$_2^{2+}$ ion, and indirect evidence indicates the same magnitude of gap for the UO$_2^{2+}$ ion. The f_σ orbital probably lies about 6 eV above f_δ.

B. The magneto-optical properties of NpO$_2^{2+}$ strongly suggest that the highest filled valence orbital has σ_u symmetry.

C. Two-photon and single photon absorption together locate *fourteen* out of *sixteen* of the states expected from the $\sigma\delta$ and $\sigma\phi$ configurations. They are identified by their symmetry, are in the correct order and have the expected magnetic moments. In the TPA spectrum there are no unassigned features in the first 9000 cm^{-1} of the spectrum with intensities greater than 1% of the maximum, implying that all the electronic states in this energy range have been identified. The two missing states are calculated to lie at energies beyond the available observational range, and are not expected to be sufficiently intense to be detected in TPA.

D. The ground state optical spectra are completely explicable in terms of excitations from a σ_u orbital to the f_δ and f_ϕ orbitals, both for the uranyl ion and the neptunyl(VI) ion in two different equatorial environments.

E. Excited state absorption suggests a σ_g orbital 2.2 eV below the σ_u orbital.

2.14 Filled Orbital Energies

Turning now to the more tightly bound filled orbitals we examine the evidence from photoelectron spectroscopy. By comparing the ultra-violet excited photoelectron spectrum (UPES) of volatile uranyl beta-diketonate complexes [45] with the spectrum of complexes of the same ligands with other metals, two sharp ionisations are observed, at 10.30 and 10.80 eV, which seem to be assignable to uranyl-oxygen ionisations. The lower of these ionisations increases in relative intensity when observed using He(II) radiation, which may be compatible with greater 5f character. The remaining uranyl-oxygen bands are seriously masked by ligand ionisations, but it is suggested that two other bands are located at 9.65 and 11.40 eV. If this is correct the whole span of filled orbitals is 1.75 eV. Support for this view comes from the high temperature PES of gaseous UO_2, where valence shell ionisations are identified at 9.87 and 10.31 eV [46]. The first ionisation, of the non-bonding f-electrons, is known from electron-impact measurements to be at 5.4 eV [47] but is not observed in the PES.

In a study which has proved vital to our understanding of the bonding, Carnall and his coworkers have examined the XPES of a large number of solid uranyl compounds, mostly containing oxide and fluoride ligands [48]. In all these compounds a 'valence' region is observed, centred near 5 eV relative to the Fermi level, with a width of about 3 eV, covering the ionisation of the twelve valence electrons nominally associated with the 2p shell on oxygen, but no substructure is resolvable.

The most informative part of this spectrum is the ionisation from the more strongly bound 6p shell. The two spin-orbit components of this shell, whose mean binding energy is ~ 20 eV, are separated by ~ 13 eV, a magnitude readily explained by the large penetration of this orbital. This can be illustrated by the calculated value of $\langle r^{-3} \rangle$, with which the spin-orbit coupling constant scales. In the neutral atom relativistic (Dirac-Fock) SCF calculations [49] predict this quantity to be 14 times larger than for the 5f shell. The upper atomic component, $6p_{3/2}$, is clearly split, by up to 5 eV, in the XPES spectrum, and to an extent which correlates with variations in the U–O bond length. This is shown by the results for thirteen compounds [48], collected in Fig. 16. When it is realised that

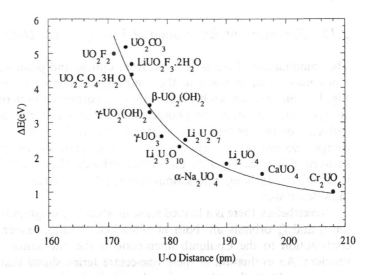

Fig. 16. XPES measurements of the crystal-field splitting of the U $6p_{3/2}$ level for a series of uranium compounds plotted versus U–O separation; from Ref. [48]

the radius at which the $6p_{3/2}$ wavefunction has a maximum amplitude (83 pm) is considerably greater than the equivalent radius in the $5f_{5/2}$ orbital (52 pm) [49] (see Fig. 19), it is clear that the $6p_{3/2}$ shell must play an important role in the uranium oxygen bonding, where typical bond lengths are 170 pm.

A recent study of the XPES spectrum of $Cs_2UO_2Cl_4$ [50], is useful because it can be compared directly with the comprehensive optical studies of this species described in the preceding sections. It shows a $6p_{3/2}$ splitting of 4.4 eV, and a valence region centred at 5.6 eV with a FWHM of 3.2 eV, partially covered on the low ionisation energy side by the chloride ionisation which, by comparison with CsCl, contributes a width of 1.4 eV.

Conventionally the $6p$ shell would be regarded as part of the core, but the XPES evidence indicates that this is a serious misconception. It has been pointed out by Pyykkö [51], that the neptunium nuclear quadrupole coupling in neptunyl(VI) compounds, which can be studied by ESR and Mössbauer spectroscopy, is most probably dominated by the charge asymmetry associated with the covalency in this $6p_{3/2}$ shell. The coupling is consistent in sign and magnitude with the removal of 0.17 electron from the axial direction as expected if the $6p_z$ orbital participates in the valency. The large magnitude of the coupling is due to the dependence of the nuclear quadrupole interaction on $\langle r^{-3} \rangle$, and is difficult to explain in terms of charge asymmetry in the $5f$ or $6d$ orbitals. Unfortunately not all the compounds for which the quadrupole coupling has been measured [52] are chemically or structurally well characterised, but for the three that are, $RbNpO_2(NO_3)_3$, $NaNpO_2(CH_3COO)_3$, and $K_3NpO_2F_5$, there is a good correlation between the magnitude of the quadrupole splitting and the bond length inferred from vibrational data. So it is clear that this pseudo-core orbital plays a significant part in the uranyl bonding.

2.15 Summary of Experimental Energy Level Data

The combination of the deductions from optical and photo-electron spectroscopy may be summarised in the energy level diagram of the uranyl ion in Fig. 17. This diagram is schematic in that it incorporates both orbital ionisation energies obtained from the photoelectron spectra and excited state energies derived from the optical data. It is important to realise that optical excited state energies do not necessarily give a useful indication of eigenvalue differences between either the filled or the virtual orbitals; they are influenced by the Coulomb and exchange interactions between the excited electron and the hole in the valence shell.

Nevertheless, there is a limited sense in which the diagram is useful. Because the f_δ and f_ϕ orbitals are both non-bonding towards oxygen, the dominant contribution to the Coulomb interaction is the one-centre contribution on uranium. An evaluation of these one-centre terms shows that the correction needed to obtain the relative virtual orbital energies from the configuration energies is small compared to the separation between them [31].

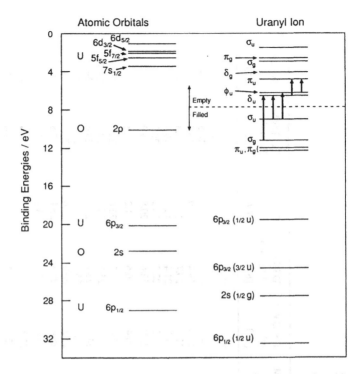

Fig. 17. Summary of energy levels of the uranyl ion in $Cs_2UO_2Cl_4$. For clarity the spin-orbit components are included for the pseudo-core orbitals, but not for the valence shell or empty orbitals. Atomic orbital energies are taken from Ref. [67]. *Arrows* connect levels whose position has been established by optical spectroscopy

Differences in the electron-hole interaction energy can lead to a change in the order of excited states as compared with the ionisation energies of the filled orbitals from which the excitations occur, and is well-known in the UPES of transition-metal acetylacetonates [53], where the lowest energy ionisation occurs from ligand orbitals, but the lowest energy optical excitations are the *d-d* transitions. In the present system the lowest energy optical excited states are apparently due to excitations from a uranyl based σ_u orbital and cover a range of 1.1 eV, but the first ionisation in $Cs_2UO_2Cl_4$ appears to be from a chloride ion orbital [50], despite the absence of chloride to uranium charge-transfer transitions within the initial portion of the optical spectrum. Taken with the poor resolution and ambiguity in the valence-band PES this means that not much is definite about the relative orbital energies in this region apart from the range of about 3.5 eV which they span, and the implication, from the excited state absorption, that the σ_g orbital lies about 2.2 eV below σ_u.

It is not possible to position the π_g and π_u components of the valence shell reliably, although the polarisation of the excited state absorption suggests that the π_g orbital is well below σ_g, while the absence of any excited states at-

Table 3. SCF eigenvalues/eV

Method		MS-Xα	Rel-MS	Rel-MS	Rel-MS+SO	DS-DVM ion	DS-DVM ion	DS-DVM +ligands[‡]	DS-DVM Opt. TS	Rel-DVM Perturb	RECP
Reference		[59]	[61]	[62]	[64]	[66]	[67]	[67]	[67]	[72]	[73]
(6d)	δ_{3g}	15.5	—	—	—	18.6	—	—	—	—	—
(5f)	σ_{1u}	13.8	—	—	—	—	—	—	—	—	—
	π_{3u}	18.3	21.1	17.1	16.9	20.4	18.6	19.4	17.9	16.5	—
	π_{1u}	18.3	21.2	17.1	17.3	20.4	21.2	20.7	19.0	16.9	—
	ϕ_{7u}	20.9	23.1	19.1	18.7	21.0	21.2	21.4	19.7	19.3	—
	δ_{5u}	21.2	23.1	19.1	18.7	21.3	22.2	22.2	20.5	19.4	—
	ϕ_{5u}	20.9	24.4	19.1	19.7	21.0	22.0	22.6	20.8	20.2	—
	δ_{3u}	21.2	23.9	19.1	19.4	21.3				19.9	—
Gap		1.3	0.3	4.0	2.0	2.4	0.4	0.6	1.2	1.8	—
(2p)	σ_{1u}	22.9	25.5	23.1	22.8	23.7	22.4	23.2	22.0	22.0	29.5
	σ_{1g}	23.7	26.3	24.9	24.6	24.0	23.5	25.6	24.7	22.6	28.7
	π_{3g}	22.5	24.6	23.5	23.1	24.0	24.5	25.7	24.7	23.6	29.5
	π_{1g}	22.5	24.6	23.6	23.2	24.5	24.6	25.8	24.9	23.7	29.5
	π_{3u}	23.3	25.6	23.5	23.2	24.7	25.2	26.0	24.6	23.2	29.4
	π_{1u}	23.3	26.5	23.6	23.7	25.2	25.4	26.4	25.1	23.5	29.4
Span		1.2	1.9	1.8	1.9	1.5	3.0	3.2	3.1	1.7	0.8
(6p) σ_{1u}		29.0	33.9	—	31.2	31.6	31.5	32.8	—	30.4	—
Charge on U		~2.0	—	—	—	2.5	2.6	3.2	—	2.0	2.54
1st IP		—	—	27.1	—	~28.4	—	—	—	27.3	—

All eigenvalues are negative. 'Gap' indicates the separation between filled and empty orbitals. 'Span' is the range of energies within the filled valence set. Orbital subscripts are twice the value of Ω.
[‡] These values have been shifted arbitrarily to match the free ion values, see Ref. [67].

tributable to the $\pi_u^3\phi$ configuration in the optical spectra implies that the π_u orbital is also near the lower end of the range of valence shell energies.

Figure 17 has been constructed as follows. A value for the first ionisation potential is derived from the photoelectron spectrum in the solid state, using a method to be described in Section 3.4. Orbitals whose relative energies are determined from the optical spectroscopy of both NpO_2^{2+} and UO_2^{2+} are designated by transition arrows. The position of the d_δ orbital is taken from theoretical calculations, see Table 3, while the remaining d-orbitals are assumed to be antibonding to a degree determined by their overlap.

The core ionisation energies are obtained from the XPES data [48] and are used to locate the orbitals drawn from the uranium $6p$ shell and oxygen $2s$ shell. They have been corrected to be consistent with the estimate of the first ionisation potential in the solid state. We shall see that these orbitals are strongly mixed. Apart from the chemical sensitivity of the '$6p_{3/2}$' splitting there is no experimental means of identifying the core orbital symmetries, so their labelling relies on the results of the SCF calculations to be discussed in subsequent sections.

3 Theoretical Studies

3.1 Background and Orbital Overlap

The uranyl ion has proved an important test-bed for the application of new theoretical techniques to molecules containing heavy atoms, where relativistic contributions to the energy are important. Pyykkö has compiled a recent bibliography [54]. The theoretical techniques which have been applied are reviewed below. Their limitations, and the dominant features of the bonding that they reveal are also emphasized.

The first major attempt to quantify the bonding in the uranyl ion was made by Belford and Belford [55] who used uranium non-relativistic Hartree-Fock $5f$ and $6d$ wavefunctions, in the U^{5+} ion, to estimate the magnitude of their overlap with the σ and π combinations of the oxygen $2p$ orbitals. The overlap integrals for the f-orbitals are small at a bond length of 167 pm. The Belfords pointed out that this is especially so for the f_σ orbital because the position of its angular nodes gives rise to extensive overlap cancellation in regions of different phase. Using non-relativistic SCF atomic orbitals the overlap integrals at 167 pm are f_σ, 0.066; f_π, 0.075; d_σ, 0.175; d_π, 0.269. This pattern, in which π overlaps exceed σ overlaps, is quite the opposite of that expected for p-orbitals, suggesting that familiar notions about the relative importance of σ and π bonding should be avoided.

This point was picked up by Newman [56] who, in an important early contribution, showed that the use of atomic *relativistic* SCF wavefunctions for

U^{5+}, produces an expansion of the f-orbitals and to a lesser extent the d-orbitals; this is the result of the increased shielding caused primarily by the contraction of s and p core orbitals, for which there are large stabilising relativistic energy corrections. Orbitals with many radial nodes have high curvature in the region close to the nucleus, where the potential is large, implying high velocities and large mass-velocity corrections. As a result of this expansion, the f-orbital oxygen-$2p$ overlap integrals increase relative to those of the d-orbitals. The overlap integrals calculated by Newman, at a U–O distance of 167 pm, are f_σ, 0.081; f_π, 0.133; d_σ, 0.132; d_π, 0.315. On the basis of overlap alone, the order of the occupied valence orbitals in increasing energy would be $\pi_g^4 < \pi_u^4 < \sigma_g^2 < \sigma_u^2$. This qualitative prediction is close to the results of the best SCF calculations but, as we shall see, there are other important factors which determine this ordering.

More recent quantitative work, involving the actual calculation of orbital energies, may be divided into self-consistent field calculations and Extended Hückel calculations.

3.2 Self-Consistent Field Methods

3.2.1 Introduction

The methodology is often identified by technical labels, so here I will try to identify the main characteristics of each type of calculation. Hartree-Fock calculations use a non-relativistic one-electron Hamiltonian, while Dirac-Fock calculations solve the relativistic (Dirac) form of the one-electron Hamiltonian. The eigenfunctions in the latter case are four-component states giving both positive energy eigenvalues describing the spinor states for the electron, and negative energy positron states. The calculations described here all contain Slater's name in their description to indicate that a *local* exchange potential has been substituted in place of the application of the non-local exchange operator. To be specific we write for the non-relativistic one electron Hamiltonian:

$$\hat{h} = -1/2 \nabla^2 + V(\mathbf{r}) \qquad (1)$$

where $V(\mathbf{r})$ represents the molecular potential, which is the sum of Coulomb and exchange terms:

$$V(\mathbf{r}) = V_c(\mathbf{r}) + V_x(\mathbf{r}) \qquad (2)$$

The Coulomb potential contains nuclear and electronic parts:

$$V_c(\mathbf{r}) = \sum_a \frac{Z_a}{|\mathbf{r} - \mathbf{R}_a|} + \int \frac{d\mathbf{r}' \rho(\mathbf{r}')}{|\mathbf{r} - \mathbf{r}'|} \qquad (3)$$

where $\rho(\mathbf{r}')$ is the molecular charge density. The local exchange potential is defined by

$$V_x(\mathbf{r}) = 3\alpha[(3/8\pi)\rho(\mathbf{r})]^{1/3} \tag{4}$$

where α is a constant with a value normally chosen such that $2/3 < \alpha < 1$. The approximation in Eq. (4), due to Slater, implies that the potential is locally dependent only on $\rho(\mathbf{r})$. This so-called X-α approximation and methods for choosing α to satisfy the virial theorem have been closely studied and give excellent agreement with full Hartree-Fock calculations, where this simplification is not made.

For an atom the evaluation of the potentials in Eq. (2) is achieved straightforwardly, using trial wavefunctions, by integrating the charge density radially. Once the potentials are known the wavefunctions are obtained by numerical integration of the radial part of the Schrödinger equation, and the process is continued iteratively to self-consistency. In molecules the loss of spherical symmetry makes the procedure much more difficult, and in particular for molecules containing heavy atoms the number of matrix elements which must be evaluated becomes prohibitive. Calculations on the uranyl ion have employed three different approaches to circumvent this difficulty.

3.2.2 The Multiple-Scattering Method

In this method, due to Johnson [57], the molecule is delineated by atomic spheres, centred on the nuclei and tangentially in contact, within which the potential, which includes contributions from charges outside the sphere, is spherically averaged. In general, this amounts to expanding the potential due to other atoms in spherical harmonics about each atomic centre. All terms in the expansion, except the first (spherical) one are ignored. The neglect of convergent higher order multipolar components is more damaging when the number of neighbouring atoms is small, as in the triatomic uranyl ion, where the first term dropped on the uranium atom is quadrupolar. On the other hand, in octahedral transition metal complexes the first term to be dropped is hexadecapolar, corresponding to a potential which is intuitively closer to spherical.

It is obvious that spheres in contact in this way can enclose only a portion of electron density because we know that the chemical bond involves significant overlap of atomic charges. To obtain the correct asymptotic behaviour for the wavefunctions in the regions outside the spheres the potential is assumed constant within the confines of an outer sphere centred on the centre of the molecule; beyond this the potential is again assumed spherically symmetrical—the whole pattern being described as a 'muffin-tin' potential. In the constant potential region the solutions of the Schrödinger equation are the same as those used to describe the scattering of waves. These functions are then required to join smoothly in amplitude and slope with the usual atomic type functions within the spheres. The eigenfunctions obtained in this way are not linear combinations of atomic orbitals, but do reflect the symmetry properties of the Hamiltonian. Similarly, any population analysis is confined to partitioning between the different regions, and it is not easy to assign effective charges to individual atoms.

It is important to realise that the MS-Xα method is not an *ab initio* method. In hetero-atomic molecules a choice of the relative size of atomic spheres has to be made, as well as the values of the α parameter to use in the various regions. These cannot be variationally determined. Despite these difficulties the method is surprisingly successful in practice. Considerable improvement is possible by allowing the atom centred spheres to overlap [58].

The first application of the MS-Xα method to the uranyl ion used a non-relativistic method [59]. The calculation starts with the charge densities derived from the U^{2+} ion and neutral oxygen atoms. This choice can be justified by examining the results of relativistic atomic Hartree-Fock calculations as a function of the charge on the uranium ion [60], the results of which are shown in Fig. 18. The average of the XPES ionisation energies of the $6p$ electrons in $Cs_2UO_2Cl_4$ measured by Teterin et al. [50] and corrected to the vacuum level by a method which we describe later, is 25.8 eV. In due course we will show that the Madelung and polarisation corrections in this material are of the order of 20 eV, so that the ionisation energy of the $6p$ electrons in the isolated uranyl ion is predicted to be ~ 45 eV, or 1.68 a.u. This figure compares well with the calculated $6p$ orbital energy of 1.55 a.u. for U^{2+} [60].

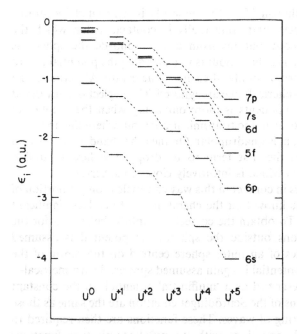

Fig. 18. Variation of the orbital energies of uranium, in atomic units, with charge, as determined by a Hartree-Fock relativistic calculation; from Ref. [60]

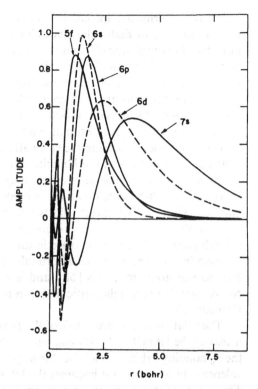

Fig. 19. Amplitude of the wave-functions of the neutral uranium atom as a function of radius, in atomic units; from Ref. [60]

For the purposes of the MS-Xα calculation the core was assumed frozen, and on uranium only the 6s, 6p, 5f, 6d and 7s orbitals were considered. (Fig. 19, which also comes from Ref. [60], shows why, at a typical metal–oxygen distance of 175 pm (3.3 a.u.) it is necessary to consider the charge distribution in all these orbitals.) The eigenvalues from this calculation are included in the first column of Table 3. Despite its neglect of relativistic effects, this approach describes the essential pattern of the bonding orbitals. However, the HOMO-LUMO gap is underestimated, and the HOMO, described as π_g, is not consistent with the experimental symmetry.

The MS-Xα method was extended by Yang et al. to include relativistic effects starting from the Dirac equations [61] but this calculation could not be iterated to self-consistency for technical reasons. The results, in column 2 of Table 3 are rather poor, in that the HOMO-LUMO gap is far too small, and the highest filled levels are again derived from π_g.

The deficiencies of this procedure have been carefully analysed by Boring and Wood [62] who worked with an approximate treatment of the Dirac equations, due to Cowan and Griffith [63]. In this method the spin-orbit operator is omitted from the one-electron Hamiltonian but the mass-velocity and Darwin terms are included. Because the trace of the spin-orbit coupling matrix vanishes in first-order over closed shells, there is no significant error introduced in the charge distribution associated with this approximation, at any rate for core electrons. Using this simplification, the normal MS-Xα-SCF methodology is applicable and the spin-orbit interaction can be introduced as a perturbation taken to first or higher order, following convergence. Armed with this technique, Boring and Wood were able to identify the influence of the relativistic part of the Hamiltonian, and to illustrate how the results of Yang et al. arise from not performing the calculation self-consistently with respect to the relativistic component of the Hamiltonian.

Briefly, the relativistic corrections may be divided into two contributions, 'direct' effects which stabilise all the orbitals, but which are only large for the 7s and 6p orbitals, and 'indirect' effects which arise from the contraction accompanying this stabilisation. The increased shielding from the contracted core orbitals causes a large net increase in the energy of the f-orbitals and a smaller increase in the energy of the 6d-orbitals. This is the same result as found by Newman in uranium atoms [56], and is also clearly displayed as a graphical comparison between relativistic and non-relativistic properties, in the work of Desclaux [49].

The relativistic upwards shift of the f-orbitals within the uranyl ion greatly increases the HOMO-LUMO gap but also has the effect of raising the energy of the σ_u bonding orbital, which has the largest f-orbital contribution amongst the valence orbitals, so that it becomes the HOMO, in agreement with experiment. The results of this work are in column 3 of Table 3. To complete the analysis Wood, Boring and Woodruff [64], extended the method by introducing the spin-orbit perturbation following SCF convergence, and diagonalising it within three separate blocks of valence orbitals, which crudely correspond to (a) the uranium 6p and oxygen 2s set, (b) the oxygen 2p set and (c) the uranium 5f-set (these being virtual orbitals). The perturbation is calculated *ab initio* from the SCF functions. Their results appear in column 4 of Table 3. There is an inevitable weakness in this method because the large splitting (~ 13 eV) of the 6p shell, which we know to be involved in valency, cannot be incorporated during the SCF convergence.

3.2.3 The Discrete Variational Method

This is a numerical method of evaluating the many integrals required by an SCF calculation. In the work of Rosen, Ellis, Adachi and Averill [65] the MOs are expanded as LCAOs in the normal way and a variational procedure is used to determine the LCAO coefficients. However, the matrix elements are evaluated, using orbitals defined numerically rather than analytically, over a grid of sample points, using an appropriate weighing function at each point. The Coulomb potential is greatly simplified by expressing the molecular charge density as a sum of atomic orbital charge densities, which in turn are determined by atomic orbital populations derived from a Mulliken partitioning of the overlap densities. The technique is called the self-consistent charge approximation (SCC). When this methodology is incorporated with the full Dirac form of the one-electron Hamiltonian, using the local exchange potential approximation, the procedure is described as the Dirac-Slater Discrete Variational Method [66].

This is a genuine *ab initio* method, although approximate in the sense described above. There are no disposable parameters. Results for the free molecular ion [66, 67] appear in columns 5 and 6 of Table 3. Column 7 gives results for the more realistic case where the ion is surrounded by an array of six equatorial charged atoms, with which, in the interests of simplicity, no orbital overlap is permitted [67]. Their Coulomb potential raises the energy of the uranium-centred orbitals compared to the oxygen-centred orbitals. The main effect is an increase in the HOMO-LUMO gap, because the LUMOs are the non-bonding f-orbitals which are directed toward the equator of the ion and are strongly influenced by this equatorial potential. For the same reason there is also an increase in the separation between the highest filled orbital, which has appreciable $5f$ and $6p$ content, and the lower lying valence orbitals, which have more oxygen character. Nevertheless, the gap calculated from the figures of column 7 is only 0.60 eV, as compared with an optical excitation energy of 2.5 eV. Following Walch and Ellis [67], the figures in column 7 of Table 3 have been arbitrarily shifted down in energy to allow a comparison with the free molecular ion.

Walch and Ellis also report the results of an SCF calculation which can be compared with optical measurements; they use the 'transition state' method due to Slater [68]. The calculation is carried out with one-half electron removed from the HOMO and added to the LUMO. The results are shown in Column 8 of Table 3. They predict an optical threshold energy of 1.2 eV, still rather small compared to experiment. This discrepancy may be a consequence of using the Mulliken population analysis, which has well-known disadvantages [69], as an integral part of the energy calculation, or inadequacies in the basis set, acknowledged by the authors [70], whose high energy components are limited to the $5f$, $6d$ and $7s$ orbitals of the neutral uranium atom. In U^{2+} the $7p$ shell is only about 5 eV above the $7s$ shell [60] and its inclusion in the basis should be most significant in the σ_u HOMO, and so would be expected to increase the 'gap' energy.

Using the SCC method, a charge density analysis gives the uranium a charge of $+3.2$ and the oxygen atoms a charge of -0.6 each. The polarity of the bond may be overestimated by this procedure because a decrease in positive charge on the metal would raise the f-orbital energies relative to those of the oxygen orbitals, increasing the HOMO-LUMO gap closer to that which is observed. On the other hand, as we shall see, the magnitude of the formal charge on the uranium atom is consistent with the inference from experiment.

A related approach is due to Snijders, Baerends and Ros [71], who also use the discrete variational method for determining the wave-functions. However the Coulomb potential is handled more exactly by fitting the molecular charge density as an expansion in Slater-type functions on every atom. The expansion coefficients are chosen to minimise the difference between the exact and simulated density. This device again reduces the calculation load in deriving the Coulomb potential. On the other hand, this method is quite different in that the relativistic effects are introduced as a perturbation, using a SCF basis from the non-relativistic Hamiltonian.

The method works as follows. The mass velocity, Darwin and spin-orbit coupling operators are applied as a perturbation on the non-relativistic molecular wave-functions. The redistribution of charge is then used to compute revised Coulomb and exchange potentials. The corrections to the non-relativistic potentials are then included as part of the relativistic perturbation. This correction is split into a core correction, and a valence electron correction. The former is taken from atomic calculations, and a frozen core approximation is applied, while the latter is determined self-consistently. In this way the valence electrons are subject to the 'direct' influence of the relativistic Hamiltonian and the 'indirect' effects arising from the potential correction terms, which of course mainly arise from the core contraction.

Once the perturbation Hamiltonian has been determined in this way it is applied, in first order only, to the non-relativistic eigenfunctions. The procedure works very well for atoms, with good agreement between the perturbation treatment and full Dirac-Fock-Slater calculations, and also gives good results for both I_2 and HgI_2. The applications to the uranyl ion [72] are particularly interesting, being the first SCF attempt to include the valence interactions with the equatorial ligands explicitly. Energies were obtained for both the free ion and the tetragonal complex ion $[UO_2F_4]^{2-}$; the eigenvalues for the free ion appear in columns 9 of Table 3. The results are quite successful in that they predict a realistic HOMO-LUMO gap of about 1.8 eV. The description of the significant factors is the same as that provided by Boring and Wood, but is made explicit by the perturbation method in breaking down the effect of the contributions to the perturbation Hamiltonian. The dominant term is the correction to the core potential for the 5f-electrons, which provides an upward shift of 4.8 eV in the virtual f-orbitals and 3.2 eV in the HOMO.

There is, however, a disturbing prediction from this approach which highlights a weakness. The f_δ orbitals span b_{1u} and b_{2u} representations of the D_{4h} point group describing the $[UO_2F_4]^{2-}$ ion. In the non-relativistic calculation, the former is non-bonding by symmetry, whereas the latter interacts with the fluorine p_π-orbitals. As expected the b_{2u} orbital is found about 0.49 eV above the b_{1u} orbital. However when the relativistic perturbation is added, the dilution of the anti-bonding b_{2u} orbital with fluorine 2p character combined with the large f-orbital core-potential correction causes a much smaller upward shift of this orbital than of the b_{1u} non-bonding orbital, which is 100% 5f in composition. This leads to the alarming prediction that the non-bonding component of this pair of orbitals is now *higher* in energy than the anti-bonding component. This result is not in agreement with the experimental sign of the tetragonal field splitting [40], nor with intuition.

The origin of this anomaly can be found in the order of application of the equatorial bonding and relativistic components of the Hamiltonian. With a relativistic core correction of about 5 eV and an equatorial bonding interaction computed to be about 0.5 eV (actually the experimental splitting of the f_δ orbitals is closer to 0.05 eV), it is inappropriate to incorporate the latter interaction first, and subsequently to apply the relativistic perturbation in first order. Indeed with relativistic corrections of this magnitude it must be seriously questioned whether the proper charge distribution is determined at all by the non-relativistic eigenfunctions. While care is taken to correctly determine the magnitude of the perturbation corrections, their application in first order to the non-relativistic basis, does not allow any consequential redistribution of charge.

The calculation of De Kock et al. gives the HOMO 64% fluorine $2p$ character. By implication the first optical excited states would largely involve charge-transfer from fluorine to uranium. There are two reasons why this must be wrong. First, the experimental energy of the first excited states is only very weakly dependent on the chemical nature of the equatorial ligands, and second the Huang-Rhys factor for the excitation of the totally symmetric stretching mode of the four chloride ions in $Cs_2UO_2Cl_4$ is known to be very small [40], implying little or no change in metal–halogen bond length accompanying transitions from the HOMO.

If the upwards relativistic shift of the $5f$-orbitals were to be invoked prior to the inclusion of the equatorial valence interaction, the fluorine–uranium bond would be expected to be much more polar, both the uranium f-orbitals and the HOMO being raised in relation to the fluorine orbitals. A polar description of the equatorial bonding is in much better accord with the chemical properties of the equatorial ligands, and agrees more closely with the observed magnitude of the equatorial field splitting, which is an order of magnitude smaller than suggested by the calculations.

The perturbative HFS method described above has also been applied to interpret the UPES spectrum of gaseous linear UO_2 [46]. Here the experimental evidence is rather weak; electron impact gives a first ionisation from the f-shell at 5.4 eV, and the photoelectron spectrum locates two ionisations at 9.87 and 10.31 eV. Much of spectrum is obscured by features due to UO. As usual, with PES data lacking angular resolution, the assignments are tentative. The perturbative-HFS calculations, which use the transition state procedure, predict f-electron ionisation at 5.1 eV and six ionisations in the range 8.3 to 10.3 eV.

3.2.4 The Pseudo-Potential Method

Here a pseudo-potential is used to represent the core-valence interaction. Thereby the SCF procedure is reduced in scale to encompass only the valence electrons. The development of the method is described in an application to the uranium atom [60]. The procedure is as follows. First a set of valence pseudo-orbitals is formed from a linear combination of atomic orbitals, with coefficients chosen to optimise these orbitals with respect to those computed from the Hartree-Fock relativistic (HFR) procedure of Cowan and Griffith [63] by applying three criteria. They are required to be nodeless, have the minimum number of spatial undulations, and to fit the SCF functions as closely as possible. The dominant component in these expansions is the actual valence orbital being simulated; for example, in the $5f$ pseudo-orbital the atomic $5f$ function is combined with a small contribution from a $4f$ function in such a way as to remove the radial node. Once the pseudo-orbitals are defined, an effective core potential can be defined for each type of valence orbital. These potentials incorporate the effect of the relativistic core contraction on the energies of the valence orbitals, and represent the electrostatic influence and orthogonality requirements of the core electrons. To allow efficient evaluation of the matrix elements of these effective core potentials they are represented analytically as a linear combination of Gaussian functions. Similarly, the pseudo-orbitals are represented by Gaussian function expansions containing up to four components. Once the core potential is defined the valence electron calculation need only consider the outermost pseudo-core $6s$ and $6p$ and valence orbitals $5f$, $6d$ and $7s$ on uranium and the oxygen $2s$ and $2p$ orbitals. Such a calculation is amenable to the variational all-electron SCF methods based on Gaussian orbitals which have been applied to molecules containing light atoms.

The economy of effort in the relativistic effective core potential method enables the calculation of orbital and total energies as a function of bond angle and bond length. Wadt has made such a study for the uranyl ion and for the isoelectronic ThO_2 molecule [73], which is apparently non-linear in the gas phase [74]. This is an *ab initio* procedure. However, because the radial properties of the valence orbitals have been tampered with at small radii, they cannot be used for the computation of the spin-orbit interaction, which depends on $\langle r^{-3} \rangle$, in the valence orbitals. The orbital energies presented by Wadt therefore exclude the spin-orbit interaction. His results for the linear geometry appear in column 10 of Table 3. The total span of the valence orbital energies is only 0.8 eV and the HOMO is calculated to have σ_g symmetry. Both these results are in poor agreement with experiment, although the optimum bond length of 167 pm and the linear geometry are satisfactorily predicted. It may be that the overlap integrals are improperly computed when radially nodeless orbitals are used, and it is undoubtedly wrong to incorporate the spin-orbit interaction as a post-SCF correction, when this interaction splits the $6p$ pseudo-core orbitals by ~ 13 eV.

3.2.5 E. Summary of SCF Results

The most successful truly *ab initio* calculation is the Dirac-Slater Discrete Variational Method of Walch and Ellis [67]. This handles the relativistic part of the Hamiltonian more rigorously than other approaches, and illustrates the importance of the equatorial ligands in determining the energy of the first optical transitions. Furthermore, the use of an optical transition state calculation makes a comparison with the data of Sect. 2 worthwhile. The identity of the HOMO is correctly anticipated, while the second highest filled MO, σ_g, is correctly placed 2.7 eV lower, comparing well with the experimental value of 2.2 eV. The total span of the filled 'valence' orbitals, is predicted to be 3.1 eV in close agreement with the FWHM of 3.2 eV indicated by the photoelectron data in Sect. 2. 14. On the other hand the first optical transition energy, 1.2 eV, is about half the experimental value, a defect which may be related to an inadequacy in the basis set.

The splitting of the pseudo-core $6p$ orbitals is also satisfactorily dealt with. It is from this calculation that the orbital labels in Fig. 17 are drawn. It is particularly striking that the highest energy component of the '$6p_{3/2}$' shell has $\omega = 1/2$. Naively this could be taken to imply that the shell is subject to a crystal field perturbation from the oxygen atoms, which raises $6p_z$ in relation to $6p_{x,y}$, because the valence interaction with the oxygen $2p$ shell would lead to a splitting of the opposite sign, Fig. 2. However, the calculation shows that the oxygen $2s$ orbitals, which have a similar initial energy, have a strong role in determining this pattern, their σ_u combination raising the energy of the $\omega = 1/2$ component of the $6p_{3/2}$ shell.

Some workers have calculated the effective charge on the uranium atom and, where available, their values, ranging from +2.0 to +3.2, are included in Table 3. In Sect. 4.1 we will examine some experimental evidence for the magnitude of this charge. Some calculations evaluate orbital ionisation potentials as well as eigenvalues. Where they are reported the first ionisation potentials are included in Table 3: in Sect. 3.4 an attempt will be made to compare them with measured values.

3.3 Extended Hückel Calculations

The simplest of these is due to Hoffmann et al. [75]. This is an ordinary Extended Hückel non-relativistic calculation using Slater-type orbitals, which are of the double-zeta variety for the uranium $6d$ and $5f$ orbitals. However the orbital exponents and energies are matched to the Dirac-Fock atomic functions of Desclaux [49] so that the main features of the relativistic orbitals are approximately introduced. The main purpose of this calculation was to investigate the angular dependence of the total energy as a function of metal–ligand bond angles, and this will be examined in Sect. 4.2. Despite this simplicity, the valence orbitals are found to have approximately the correct span of ~ 4 eV, with the σ_u HOMO orbital well separated from the remaining orbitals.

A more ambitious Extended Hückel Treatment has been developed by Pyykkö and Lohr [76] allowing the inclusion of relativistic effects with refined orbital exponents. In its most recent form the method, known as REX, does not rely on empirical parameters but rather derives its parameterisation from *ab initio* atomic calculations, and allows for charge-consistency [51]. The importance of charge-consistency in EHT calculations was pointed out long ago by a number of groups, the concept being developed extensively by Fenske et al. [77]. The diagonal orbital energies are the main parameters in the calculation. These are a strong function of the formal charge on the atom carrying the orbital and, through the Coulomb interaction, the charge on the other atoms. As the valence interaction redistributes charge, the diagonal energies are recomputed and the procedure is repeated to achieve charge consistency.

In the work of Larsson and Pyykkö [51] the energy dependence of the atomic orbitals is defined by a polynomial fit of the charge dependence of the energy of any one AO on the population of the other AOs, as derived from Dirac-Fock calculations [49]. Orbital populations are obtained by the Mulliken approximation. The orbital energies are corrected for the Coulomb interaction of the total charge on other atoms. The wavefunctions are of double-zeta variety, the radial parameters being optimised separately for the different spinors.

This method does not permit any change in the radial parameters of the orbitals in the course of the calculation, but this approximation receives some justification from the weak charge dependence of $\langle r \rangle$ determined from the atomic HFR calculations of Hay et al. [60], as shown in Fig. 20.

The simplicity of the method allows it to be used to describe quite complex species such as $[UO_2Cl_4]^{2-}$ and $[UO_2F_4]^{2-}$. The results are, however, in rather poor agreement with experiment. The HOMO is correctly identified but the HOMO-LUMO gap is underestimated at ~ 1 eV and the uranyl valence orbitals span almost 10 eV, which is far too large.

3.4 Orbital Energies and Ionisation Potentials

SCF calculations provide absolute orbital energies for the HOMO. In Table 3 these are seen to vary from 22.0 to 28.7 eV. It is possible to make an approximate comparison of these values with the XPES ionisation energies. Walch and Ellis

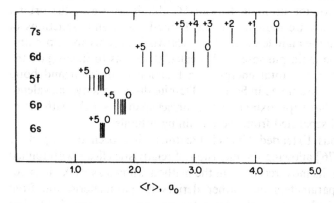

Fig. 20. Variation of $\langle r \rangle$, in atomic units, for the orbitals of uranium as a function of charge, from Ref. [60]

[67] have investigated the one-electron binding energies using the transition state procedure [68]. The result is a rigid downward shift of all the occupied energy levels of ~6 eV so, using the ground state eigenvalue of the HOMO, 22.4 eV, calculated by these workers, the first ionisation energy of the isolated ion is predicted to be ~28.4 eV. A similar calculation by Wood, Boring and Woodruff [64], using their relativistic MS technique, gives a rigid shift, of about 4.2 eV, between the SCF eigenvalues and the binding energies. With their figures the HOMO is calculated to have a binding energy of 27.1 eV.

We can compare this theoretical ionisation potential with the experimental value [50] of 5.6 eV for a crystal of $Cs_2UO_2Cl_4$, taking care to correct the measurements to the vacuum potential rather than the Fermi level of the spectrometer. For an insulating solid the Fermi level is poorly defined but the measured values may be satisfactorily referenced by aligning the cesium and chlorine ionisations with those in CsCl also measured by the same workers. For example, Teterin et al. report the Cl $3p$ ionisation at 4.5 eV in CsCl, very close to their value for the same ionisation in $Cs_2UO_2Cl_4$. The *absolute* binding energies (relative to the vacuum level) of the valence electrons in CsCl and other alkali halides have been carefully studied by Citrin and Thomas [78] who find an experimental value of 10.1 eV for the Cl $3p$ ionisation and a theoretical value of 8.0 eV. They believe their experimental value to be too large by about 1.0 eV on account of sample charging, so we assume a true value of 9.1 eV. The Cl $3s,3p$ and Cs $5s,5p$ ionisations in CsCl reported by Teterin et al. are on average 4.8 eV less than the experimental values of Citrin and Thomas. Comparing all these results, the absolute binding energy of the valence band, or uranyl, electrons in $Cs_2UO_2Cl_4$, is $(5.6+4.8-1.0)$ eV, or 9.4 eV. This figure is in good agreement with the gas-phase UPS data of Fragala et al. [45], see Sect. 2.14, where Fermi level corrections are absent.

In a crystalline lattice the binding energies are effected by the Madelung potential [78]. The potential at uranium can be calculated in $Cs_2UO_2Cl_4$ from the known atom positions [30], assuming unit charges on all the ions [79]. The UO_2^{2+} ion can be treated as a point 2+ charge, no significant change in the potential being introduced by this approximation. Convergence of the lattice potential is effectively complete over a sphere of radius 3 nm and has the value 18.9 eV. On ionisation from a solid the final state is stabilised relative to the initial state by the polarisation of the surrounding medium. The appropriate corrections [78] have been calculated for alkali halides and for chlorides are ~1 eV. Using the theoretical value [67] for the isolated ion of 28.4 eV, we therefore predict a first ionisation energy of 8.5 eV in good agreement with the experimental value. The lower theoretical value, 27.12 eV, from Ref. [64], would be consistent with the experimental result if allowance is made for a small degree of covalency in the uranium chlorine bond, which would reduce the Madelung potential.

The absolute energy values obtained from the Hückel calculations are quite different, with the HOMO of the *isolated* uranyl ion lying near -12.0 eV. The difficulty lies in the proper choice of the initial atomic orbital energies. The most usual practice is to start with neutral atom orbital energies, and this appears to be the starting point in Refs. [75] and [76]. The molecular eigenvalues are therefore more appropriate to the neutral UO_2 species than the uranyl ion, although the population analysis is of course undertaken with the correct number of electrons. Even the self-consistent charge calculation of Larsson and Pyykkö [51] which allows for the rescaling of the orbital energies as charge is redistributed within the molecule obtains a similar energy for the HOMO, but it is unclear what initial orbital energies these workers have introduced.

4 Chemical Bonding

4.1 Effective Charge on Uranium

The effective charge on the uranium atom can be assessed both by theory and by experiment. The theoretical techniques rely on some charge partitioning scheme such as the Mulliken population analysis. Perhaps the best calculation available assigns the uranium, in uranyl complexes, a charge of $+3.2$ [67], but because this neglects any equatorial ligand overlap, it is likely to be too high. Experimentally one may examine the apparent radius of the uranium atom using crystallographic data, determine the solution stability of complexes with ligands bound in the equatorial positions, or use a spectroscopic measure of the electron–ligand interaction, in each case making a comparison with reference compounds. Interpolation provides an effective oxidation state, which should be comparable with the theoretical estimates which neglect equatorial ligand overlap. The actual charge will of course be smaller.

To interpret the uranium atom radius we simplify the argument by considering only chloride ligands. Table 4 shows the uranium–chlorine distances for a number of well-defined compounds containing uranium and chlorine. Where possible the examples have regular octahedral geometry, but all the simple halides have larger coordination numbers, with variable distances within the first coordination sphere; for these the quoted distances are the shortest ones. Table 4 also includes the frequency of the totally symmetric uranium–chlorine stretching mode, for centrosymmetric coordination, or the highest U–Cl frequency in other cases. There is a good inverse correlation between this frequency and the bond length. Comparing the oxo-compounds with the reference compounds, the data in the Table strongly suggest that the effective charge on the uranium atom in $Cs_2UO_2Cl_4$ is of the order of $+3.5$, which implies an effective charge of -0.75 on each oxygen atom. In the monoxo species $[UOCl_5]^-$ the uranium–chlorine bond lengths give an apparent charge of about $+4.5$ consistent with the smaller influence of a single oxo ligand, and this requires a charge of -0.5 on the oxygen atom.

Table 4. Bond lengths and vibrational frequencies as a function of oxidation state

Ox. State	Compound	R_{U-Cl}/pm	v_{sym}/cm^{-1}	Ref.
VI	UCl$_6$	241	367	[98, 99]
V	UCl$_5$	244	324/367	[100, 101]
	(PPh$_3$CH$_2$Ph)UCl$_6$	249.1		[102]
	CsUCl$_6$		348	[99]
	(Me$_4$N)UCl$_6$		345	[99]
	(Ph$_3$PO)UCl$_5$	248.8(cis)		[103]
IV	UCl$_4$	264.4	311	[104, 105]
	Cs$_2$UCl$_6$	262.1	308	[104, 99]
	(PEtPh$_3$)$_2$UCl$_6$	262.4		[102]
	(Me$_4$N)$_2$UCl$_6$		296	[99]
III	UCl$_3$	293		[104]
	Cs$_2$NaUCl$_6$	273(5)	276*	[106, 107]
	RbUCl$_4 \cdot$ 5H$_2$O		232	[108]
	Sum of ionic radii	285		[109]
"VI"	Cs$_2$UO$_2$Cl$_4$	266.9†	264	[28, 29]
	(Et$_4$N)$_2$UO$_2$Cl$_4$	266.8		[110]
"V"	(PPh$_4$)UOCl$_5$	253.6(cis)	293	[111]

* This value is taken from Cs$_2$NaCeCl$_6$, whose lattice parameter is 1094.6 pm [112] and is almost the same as that for the uranium analogue, 1093.5 pm [107].
† This is not the bond length reported in Ref. [29], which contains computational errors, but is derived from the atomic coordinates reported therein.

A second approach, due to Choppin, is based on the stability of fluoride complexes in aqueous solution. The free energy change obtained from the stability constant of UO$_2$F$^+$ in 1.0 mol dm^{-3} aqueous NaClO$_4$, is compared to those for model systems such as ThF^{3+}, NdF^{2+}, and CaF$^+$, using an extended Born equation whose validity has been tested on other actinide systems [80]. This measurement gives an effective charge of 3.2 ± 0.1 to the uranium atom in the uranyl ion. Other actinide atom charges determined in the same way are, NpO$_2^{2+}$, 3.0 ± 0.1; PuO$_2^{2+}$, 2.9 ± 0.1, and NpO$_2^+$, 2.2 ± 0.1. These results follow the expected trend of increasing ligand-to-metal charge transfer as the effective nuclear charge of the metal increases.

The influence of the oxo-ligands on the apparent charge on uranium can also be estimated from a comparison of ligand field parameters. Table 5 shows the magnitudes of the crystal-field radial parameters, $\langle B_4 \rangle$ and $\langle B_6 \rangle$, obtained from

Table 5. Crystal field parameters/cm^{-1} for actinide compounds

Compound	UCl$_3$	UCl$_6^{2-}$	UCl$_6^{2-}$	UCl$_6^-$	Cs$_2$UO$_2$Cl$_4$(expt)
References	[104, 113]	[114]	[115]	[116]	[40]
$\langle B_4 \rangle$	484	2132	2060	3425	
$\langle B_6 \rangle$	606	1322	1822	949	
$\Delta E(f_\delta)$	32	641	417	1464	442

spectroscopic and crystal structure data for some chloride containing compounds in oxidation states III, IV and V. These values may be used to predict the magnitude of the splitting of the f_δ orbitals, as a function of oxidation state, caused by four equatorial chloride ions. The experimental value for $Cs_2UO_2Cl_4$ is also included in Table 5. In the point charge model the magnitude of this splitting is determined by opposed contributions from $\langle B_4 \rangle$ and $\langle B_6 \rangle$ [32], and is therefore sensitive to their relative values. This feature and the limitations of the model make it difficult to take the results as more than a guide to the apparent charge. Nevertheless, there appears to be a good correlation between the magnitude of the splitting and formal oxidation state, suggesting a value near $+4$ for the uranium atom in $Cs_2UO_2Cl_4$.

The unanimity between the experimental methods and the theoretical estimates is impressive. They agree on a uranium charge of about $+3.2$, although there is no provision for quantifying the charge transfer from the equatorial ligands.

4.2 Linear Geometry

A number of authors have discussed the preference of the uranyl ion for linear geometry. As is often the case, a naive calculation provides the most insight. In their non-relativistic Hückel calculation Tatsumi and Hoffmann [75] examine the role of the various valence and pseudo-core orbitals as a function of bond angle displaying their results in the form of Walsh diagrams. If the pseudo-core $6p$ orbitals are excluded from the basis the total energy is very weakly dependent on the OUO bond angle, and the calculation suggests a bent geometry, but inclusion of the $6p$ set in the basis provides a strong energetic preference for the linear molecule. The explanation works in the following way. The $6p$ shell is the highest filled uranium core orbital, and as confirmed by the XPES, it interacts primarily with the σ_u combination of oxygen $2p$ and $2s$ orbitals. The $2p$ energy is thus raised so that it better matches the energy of the f_σ orbital with which it bonds. It is this interaction which ensures that the σ_u valence orbital is the HOMO. On the other hand, in the bent geometry, where the upward displacement caused by the $6p$ interaction has approximately half the magnitude [81], the bonding interaction with the f-orbitals is correspondingly small, so that this geometry has less stability. The Extended Hückel calculations of Larsson and Pyykkö [51] reach the same conclusion.

The relativistic effective core potential calculations of Wadt [73] examine the equilibrium bond angles in the UO_2^{2+} ion and the iso-electronic ThO_2 molecule for which matrix isolation studies indicate a bond angle of $122° \pm 2$ [74]. This calculation has some weaknesses in failing to determine the correct order of filled MOs (Sect. 3.2.4.), but is sufficiently economical to allow the investigation of the total energy as a function of bond length and bond angle. The calculation succeeds in predicting a sensible value, of 164 pm, for the equilibrium bond length in the isolated uranyl ion (equatorial ligands would cause an expansion towards the distance of 175 pm typical of real compounds), and an equilibrium bond angle of 118° for the ThO$_2$ molecule.

The analysis of this geometry difference depends on the relative energies of the 5f and 6d orbitals in the two species. The lower charge on the thorium atom results in 6d orbital energies which are lower than those of the 5f shell, so that the former are dominant in determining the bonding and geometry in ThO$_2$. For this species the bonding in then closer to the situation for the MoO$_2^{2+}$ ion, investigated by Tatsumi and Hoffmann [75], where only d-orbitals are available, the *cis* geometry being favoured by allowing the maximum use of the empty d-orbitals. On the other hand in the uranyl ion the f-orbitals now lie well below the 6d in energy and dominate the geometry.

Wadt points out that the 6p-2p interaction should be even more significant in ThO$_2$ than in the uranyl ion, because of the higher energy of the 6p shell, and argues that this cannot therefore be a principal cause of the linear bonding in the uranyl ion, otherwise ThO$_2$ would also be linear. He therefore attributes the linearity of the uranyl ion to the angular properties of the 5f orbitals, which he believes to stabilise the linear geometry. His argument is not consistent with the role of the f-shell as investigated by Tatsumi and Hoffmann [75], who give Walsh diagrams showing that, as long as the 6p shell is ignored, the inclusion of the f-orbitals in the basis actually stabilises the bent geometry. If, as argued by Wadt, the f-orbital energies in ThO$_2$ are so high as to be ineffective in the bonding, it may be possible to dismiss the role of the 6p-2p interaction in facilitating the f-orbital bonding in ThO$_2$, but it is wrong to extrapolate this dismissal to the uranium case, where the relative energies are quite different. There is scope here for a more detailed analysis of geometry using an SCF calculation which is more successful in describing the orbital energies.

Despite the importance of the interaction between the uranium 6p and oxygen 2p valence orbitals, described by Jørgensen as 'pushing from below' [23], it is important to realise that the dominant σ_u interaction is bonding. This is shown by the change in OUO stretching frequency, from 832 cm^{-1} to 714 cm^{-1} in Cs$_2$UO$_2$Cl$_4$ when a valence electron is excited from the σ_u orbital to those f-orbitals which are non-bonding with respect to oxygen, while the Huang-Rhys factor for these transitions suggests an expansion of the equilibrium bond length from 176 pm in the ground state to 183 pm [82] in the excited state.

4.3 The Inverse trans Influence

The well-known *trans* effect, manifest in the substitution kinetics of d-block elements, has a ground-state analogue called the *trans* Influence, in which tightly bound ligands appear to selectively weaken the bonds to ligands *trans* to their own position [83]. The main kinetic consequences appear in the substitution of square planar complexes. Strongly bound ligands such as the oxo-group, might be expected to exert a similar influence, although they are not represented amongst the class of compounds for which kinetic studies have been made.

Table 6. Bond lengths/pm in halide and pseudo-halide complexes

Compound	Ox. State	R_{M-X} cis	R_{M-X} trans	% diff.	Ref.
K_2NbOF_5	V, d^0	184	206	+12.0	[117]
$K_2MoOF_5 \cdot H_2O$	V, d^1	188	202	+7.4	[118]
$(AsPh_4)_2NbOCl_5$	V, d^0	240.2	255.5	+6.4	[119]
$(PPh_3CH_2Ph)WSCl_5$	VI, d^0	230.4	246.1	+6.8	[120]
$(AsPh_4)_2NbO(NCS)_5$	V, d^0	209	227	+8.6	[121]
$(AsPh_4)_2ReN(NCS)_5$	VI, d^1	202.1	230.7	+14.2	[122]
K_2OsNCl_5	VI, d^2	236.2	260.5	+10.2	[123]
$(PPh_4)UOCl_5$	VI, f^0	253.6	243.3	−4.0	[111]
$(Et_4N)_2PaOCl_5$	V, f^0	264	242	−8.3	[124]

A collection of metal–halogen and pseudo–halogen bond lengths, illustrating this influence, is shown in Table 6. In all the *d*-block compounds the oxo-, thio- and nitrido-groups induce an obvious lengthening, by 5–10%, in the *trans* metal–halogen bond by comparison with the *cis* halogens. On the other hand, in the formally similar $[UOCl_5]^-$ and $[PaOCl_5]^-$ ions the *trans* M–Cl bond is considerably *shorter* than the *cis* M–Cl bonds, suggesting the operation of what can be described as an Inverse *trans* Influence. This simple idea explains naturally why two strongly bound ligands, as in the dioxo compounds, are arranged mutually *trans* to one another in actinide chemistry where their inverse *trans* influence is cooperative. On the other hand, in *d*-block chemistry the most stable configuration for two ligands with a strong *trans* influence will be *cis*.

The *trans* U–Cl bond shortening in $[UOCl_5]^-$ has been interpreted by De Wet and du Preez [84] as a lengthening of the *cis* bond lengths by the direct electrostatic influence of the oxo group, which is, of course closer to the *cis* positions. If such an argument is correct it should apply in the *d*-block metal mono-oxo complexes, so the data in Table 6 make it untenable. Our discussion of effective atomic charges, shows that the formal charge on the oxygen atoms in the oxo-group is in the range −0.5 to −0.75, a result which, on an electrostatic argument, should have the opposite effect to that proposed by de Wet and du Preez.

There is a naive, but pleasingly simple view of the source of the *trans* influence for metal ions with a formally empty valence shell. We might treat the effect of the ligand, viewed as an anion, as a simple electrostatic perturbation acting on the core electrons. The polarisation of the core by a single tightly-bound ligand can be expressed in terms of dipolar, quadrupolar and higher order multipole moments. However, the predominant moment will be influenced, through the denominator in the perturbation energy expression, by the relative energies of excited states corresponding to the excitation of core electrons into the valence shell.

On this basis, if the highest filled core orbitals have *opposite* parity to the lowest energy valence shell orbitals, then the dominant component of the polarisation will be *dipolar*, leading to an accumulation of negative charge in the *trans* position, thereby forcing the second ligand into the *cis* position. This is the case for early transition metals, where the core orbitals are *p*-type and the valence orbitals are *d*-type. Good examples are the *cis* geometry of the dioxo cations of V(V), Mo(VI) and W(VI). The same factor may contribute to the bent geometry of BaF_2 in the gas-phase if the predominant valence shell contributions to the polarisation involve 6s and 5d shells.

On the other hand, where the core shell and valence shell share the *same* parity the leading component of the polarisation is likely to be *quadrupolar*, with an accumulation of electronic charge in the *cis* positions. Under these circumstances the second ligand should be most stable in the *trans* position, corresponding to the inverse *trans* influence. Thus when the *p*-shell provides the highest energy core orbitals, and the lowest energy valence orbitals are *f*-type, as in the actinyl ions, the *trans* geometry is favoured, but if the *d*-shell is lower in energy than the *f*-shell, as in thorium, then the *cis* geometry of gaseous ThO_2 is to be expected.

An inverse *trans* influence is, of course, also anticipated for the early B-metals, where the core is *d*-type and the valence shell is *s*-type, and will contribute to the dominance of linear two-fold coordination in the chemistry of Ag(I), Au(I) and Hg(II). A nice example, with some technological importance, of the difference between A and B-metals comes from the crystal structures of $KTiOPO_4$ [85], and $KSnOPO_4$ [86]. Both structures are very similar, the Ti and Sn atoms being in a distorted octahedron of oxygen atoms. The metal atoms and the oxo-oxygen atoms alternate in zig-zag chains through the lattice. There are two crystallographically distinct molecules per unit cell. In the titanium compound the oxo-group atoms form long and short bonds to each titanium, the short bonds being 13.0% and 11.9% less than the mean Ti–O distance in the octahedron, and the long bonds being 9.0% and 6.3% greater than that mean, showing the operation of the *trans*-influence. In the tin compound all the tin-oxo bonds are almost equal in length, and are 3.6, 3.8, 4.2 and 4.6% less than the mean Sn–O bond distance in the octahedron, as expected for an inverse *trans*-influence. The asymmetry in the titanium compound appears to be the key feature in determining its high second-order (non-linear) optical polarisability, which is exploited in frequency-doubling crystals for laser systems.

We can interpret the *trans* geometry of OsO_2^{2+} [87], which has the d^2 configuration, in a similar way. Here the first oxo-ligand acts, in the first order of perturbation theory, to induce a quadrupolar charge distribution in the valence shell, as the *d*-electron pair occupies either the xy or x^2-y^2 components in response to a ligand on the z axis; the second ligand must then follow in the *trans* position.

This simple model depends on the properties of a deformable core, and is only likely to be significant for heavier, polarisable metals. In addition, there are the directional influences of the valence orbital overlaps, which have been widely explored, and, in general, favour *cis* geometry in the *d*-block. For example, the Extended Hückel calculations of Tatsumi and Hoffmann show the dominant role of 4*d*-π orbitals in stabilising the *cis* geometry in MoO_2^{2+} [75]. In this geometry the four oxygen π-type lone pairs can interact with three vacant d-orbitals, whereas in the *trans* geometry there are only two d-orbitals which can serve this purpose.

The main difference in the actinides is the inclusion of the pseudo-core $6p$-orbitals within the valence set. In this way the properties of the deformable core are incorporated within a molecular orbital scheme. The short metal–oxygen bond implies the interaction of the O-$2p$ orbital, not only with the empty $5f$-orbitals but also with the filled uranium $6p$-orbitals. The overall effect is a transfer of charge from the $6p_\sigma$ and O-$2p$ orbitals to the uranium f_σ. According to the relativistic SCF calculations of Walch and Ellis [67] the loss of charge from the $6p_\sigma$ orbital at the equilibrium distance in the uranyl ion is between 0.1 [67] and 0.2 electrons [66], a value consistent with the estimate of 0.17 electron given by Larsson and Pyykkö [51], based on the experimental sign and magnitude of the nuclear quadrupole coupling. However, as shown in Figs. 19 and 20 the radial extent of the $6p$ orbitals is much larger than the $5f$; the average value of the electron radius is some 25 pm greater. The bond-shortening of the *trans* uranium–chlorine in $[UOCl_5]^-$ can then be seen to be a consequence of an oblate charge distribution in the $6p$ shell, the charge being specifically reduced in the *trans* direction through the $6p$-$5f$ hybridization induced by the oxo-group.

4.4 Vibrational Interaction Force Constants

There should be a dynamic analogue of the inverse *trans* influence expressed through the interaction force constants for linear triatomic species. Many years ago Bader [88] pointed out that the interaction force constants for the linear actinyl ions of U, Np, Pu and Am, were negative, whereas most linear triatomic molecules have positive interaction force constants. For example, the symmetric stretching mode frequency is lower than that of the asymmetric mode in the $[UO_2Cl_4]^{2-}$ [29], but the reverse is true for $[OsO_2Cl_4]^{2-}$ [87]. In Bader's work the interaction force constant is related to the square of the linear electron–phonon (or vibronic) coupling. (The stereochemical consequences of strong interactions of this type have been elaborated by Burdett [89] using the same formalism as Bader, although he prefers to describe them as an expression of the second-order Jahn-Teller effect.)

In linear centrosymmetric triatomics a negative interaction force constant corresponds to a lowering of the force constant for the totally symmetric mode. This is the dynamic analogue of the inverse *trans* influence, because the bond shortening is cooperative. Conversely, a positive interaction force constant implies the dynamic analogue of the normal *trans* influence. For closed shell systems the electronic perturbation provided by the totally symmetric mode links the ground state with low-lying electronic excited states of the same Σ_g^+ electronic symmetry, whereas that due to the asymmetric mode involves Σ_u^+ excited states. In each case the only significant states are those that influence the strength of the metal–ligand bond.

Both carbon dioxide and the dioxo-osmium cation have positive interaction force constants, implying a softening of the *ungerade* stretching mode as a result of the interaction with excited states of Σ_u^+ electronic symmetry. In carbon dioxide the HOMO has π_g symmetry and the LUMO with π_u symmetry is strongly anti-bonding; they are linked by the vibronic coupling of the asymmetric mode. In the OsO_2^{2+} ion we may ignore the two valence shell d-electrons which are non-bonding with respect to oxygen, occupying orbitals of δ symmetry for which there are no counterparts accessible via the vibronic perturbation. The highest energy oxygen-based MO should have π_u symmetry, the *gerade* ligand based orbitals being depressed by their interaction with the metal d-orbitals, and the significant excited states are those in which a π_u electron is excited to the anti-bonding d_π set. The dominant vibronic perturbation again requires the asymmetric mode.

On the other hand, in the uranyl ion, the HOMO has σ_u symmetry and the first excited states are all from the *ungerade* f-orbital manifold. The filled *gerade* orbitals are at least 2 eV lower in energy, see Sect. 2.15, so the σ_g to σ_u excitations occur at higher energies than σ_u to σ_u excitations. On this basis, Σ_g^+ excited states should dominate the vibronic coupling, leading to the negative interaction force constant implicit in the vibrational frequencies.

The experimental interaction force constants become more negative across the actinyl series, the values being UO_2^{2+}, -30; NpO_2^{2+}, -54; PuO_2^{2+}, -89; and AmO_2^{2+}, -166 Nm^{-1} [88]. This trend is consistent with the decreasing energy gap between the filled *ungerade* valence orbitals and the f-orbitals, as shown by the energy of the first valence shell excitation, which occurs at 2.5 eV in UO_2^{2+}, but at 1.6 eV in NpO_2^{2+} [39]. This difference reflects the greater effective nuclear charge, and more oxidising properties of the later actinyl ions.

4.5 The U–O Bond and The Equatorial Field

The O–U–O vibrational frequencies in uranyl compounds span a wide range. In view of the imprecision in the location of the oxygen atoms by X-ray diffraction due to the large scattering amplitude of uranium, many authors have attempted to find a relationship between the U–O vibrational frequencies, or the force constants derived from them, and the bond length. It has been the practice to analyse the U–O stretching modes with a force field containing only a bond stretching force constant and an interaction force constant, and to relate the bond length to the vibrational data through a version of Badger's rule [48, 90].

Bond length correlations with either force constants or with the symmetric or antisymmetric stretching frequency have recently been reviewed [91]. A survey of 27 uranyl compounds gives a satisfactory correlation between the symmetric stretching frequency and the bond lengths determined by diffraction methods with the form:

$$R_{U-O} = (10650 \, v^{-2/3}) + 57.5 \tag{5}$$

where R_{U-O} is in picometres and the frequency is expressed in cm^{-1}. The symmetrical stretching frequency, usually observed in the Raman or luminescence spectra provides a reasonable indication of the relative length of the O–U–O bond, because this mode is, to a good approximation, kinematically decoupled from the equatorial ligand vibrations. Equation (5) therefore allows us to relate the O–U–O bond length, estimated in this way, to the nature of the equatorial ligands.

Figure 21 shows that there is a useful correlation between this bond length and the energy of the first electronic excited state. By implication, the shorter the uranium–oxygen bond the higher is the energy of the first electronic transition. The same Figure shows that there is a relationship between the U–O bond length and the formal charge of the equatorial ligand atoms. For example, if the nitrate ion is taken as a covalent unit with a charge of $-1/3$ on each oxygen atom, then in $CsUO_2(NO_3)_3$ the uranyl ion is exposed to an equatorial charge of -2. In this way one can obtain the following equatorial charges: $NaUO_2(CH_3COO)_3$, -3; $Cs_2UO_2Cl_4$, -4; $K_4UO_2(CO_3)_3$, -4; $K_3UO_2F_5$, -5; and $BaUO_4$, -8. This

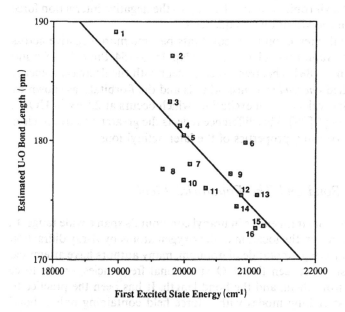

Fig. 21. Relationship between the U–O bond length, derived from Eq. (5) using the UO_2 symmetric stretching frequency, and the energy of the first electronic excited state. Data points and their sources are as follows:

1.	$BaUO_4$	[126]	9.	$UO_2(NH(CH_2COO)_2)_2$	[127]
2.	UO_2MoO_4	[128]	10.	$Cs_3UO_2(SCN)_5$	[129]
3.	$Cs_3UO_2F_5$	[130]	11.	$\alpha\text{-}UO_2SO_4, 3.5H_2O$	[125]
4.	$Rb_3UO_2F_5$	[130]	12.	$RbUO_2(CH_3COO)_3$	[27]
5.	$K_3UO_2F_5$	[130]	13.	$NaUO_2(CH_3COO)_3$	[131]
6.	$K_4UO_2(CO_3)_3$	[132, 27]	14.	$K_2UO_2(NO_3)_4$	[133]
7.	$Cs_2UO_2Cl_4$	[29]	15.	$RbUO_2(NO_3)_3$	[134, 27]
8.	$Cs_2UO_2Br_4$	[135]	16.	$CsUO_2(NO_3)_3$	[34]

rather coarse description of the equatorial ligand field nevertheless correlates well with the ground state vibrational frequencies. It is possible to rationalise the position of $Cs_3UO_2(SCN)_5$ by noting the reduced equatorial charge accompanying delocalisation within the thiocyanate ion. The dependence on the counter-cation size in the pentafluoro complexes also fits this model, the smaller cations lowering the effective negative charge in the equatorial plane.

Using a more covalent description, there is a relationship between the basicity of the equatorial ligands and the length of the U–O bond. For example, in the survey of bond distances of Ref. [15] it is clear that the longest oxo-bonds occur when the equatorial ligands are oxide ions, but long bonds of ~180 pm also occur with ligands such as the peroxide ion [92] and the hydroxylamide ion [93]. A number of compounds with the latter ligand have U–O distances determined with the high precision of neutron diffraction. All these ligands can be regarded as strong bases and therefore strong donors; in contrast, anions of strong acids, such as the nitrate ion, form complexes with the shortest uranium–oxo bonds.

These observations can be reconciled with the bonding scheme in the following way. Increasing the charge, or donor orbital energy, in the equatorial region, with the U–O bond length frozen, raises the energy of the metal orbitals in relation to those on oxygen in the manner illustrated by the work of Walch and Ellis [67], so that the $5f$-$2p$ bonding interaction decreases, while the $6p$-$2p$ antibonding interaction increases. The U–O bond will therefore lengthen, as implied by the vibrational frequencies. This lengthening in turn weakens the σ_u bonding interaction so that the energy of HOMO is raised in relation to the non-bonding f-orbitals and the first optical excitations shift to lower energy.

5 Conclusions

Spectroscopic evidence, augmented by the results of SCF calculations, establishes the pattern of energy levels shown in Fig. 17. We notice that the pattern of the filled MOs is consistent with the magnitudes of the overlap integrals which are larger for the $6d$-orbitals than for the $5f$ orbitals and larger for their π components than their σ components. However, the large gap between the σ_u bonding orbitals and the remaining components of the valence set is determined largely by the participation of the pseudocore $6p$ shell, whose significance in the bonding scheme is established by the XPES data.

The f-orbitals reflect this pattern, because, despite the small magnitude of the f_σ overlap integral, this orbital appears to be more than 4 eV above the energy of the non-bonding f-orbitals, compared with the f_π orbitals at only 1.5 eV. So the spectroscopic evidence implies that the σ-bonding of the f-orbitals is enhanced through the role of the $6p$ shell, which may be thought of as pushing the oxygen $2p$-orbitals from below so that a better energy match is obtained with the $5f$ set.

In alternative language, the bonding is enhanced by the formation of 6p-5f hybrids, in much the same manner as the linear bonding in the post transition metals is enhanced by the formation of s-d hybrids. While the energy gap between the $6p_{3/2}$ shell and the $5f_{5/2}$ shell in the uranium atom (~ 17 eV) [49] is much larger than that between the 6s and 5d shells in mercury (4.5 eV) [94] the overlap of the uranium 6p shell becomes important because of its large average radius. This radius and the matching energy of the oxygen 2s orbitals requires that the latter are also incorporated in the bonding scheme.

By comparison the uranium 6s shell is almost 30 eV more tightly bound so that the *gerade* bonding interactions are determined primarily by the properties of the oxygen p-orbitals and uranium 6d orbitals. Although the highest filled *gerade* orbital has σ_g symmetry as expected from the magnitude of the overlap integrals, the low value of the U–O stretching frequency in the state derived from exciting this electron implies that it is strongly bonding. We can therefore infer that the π_g bonding orbitals, which appear from the excited state absorption spectrum to be of lower energy, are probably even more significant than σ_g orbitals in their contribution to the bond. Finally the position and antibonding character of the f_π orbitals shows that the π_u valence orbitals also make a significant contribution to the bonding.

Taken together, these observations signify a role for both the 5f and 6d shells in the bonding, and show that all twelve electrons in the valence shell participate. This description gives each uranium–oxygen linkage a formal bond order of three. From this point of view it is the availability of metal valence orbitals of both parities which make the actinyl bond so strikingly different from metal-oxo bonds in the transition series, where only the metal d-shell plays a dominant role.

The 6p shell is critical in determining the stability of the linear geometry, which is characteristic of the actinyl ions, and contrasts with d-block oxy-cations. The linearity is a consequence of the 6p-5f hybridization of the valence orbitals. These same electronic features account for the strengthening of the bonds to ligands *trans* to the oxo group in actinide compounds which I have described as the Inverse *trans* Influence. The equivalent effect in the d-block, called the *trans* influence, leads to a bond weakening. These directional features of the bonding have a clear counterpart in determining the sign of the interaction force constants.

Finally, in the light of what is known about the electronic structure, it is worth commenting on the long luminescence lifetimes and extensive photochemistry of the uranyl ion. The lifetime of uranyl ion luminescence at low temperatures in solids is close to that determined by the radiative relaxation rate, which can be estimated from the absorption intensity. Non-radiative relaxation mechanisms are therefore very inefficient. We may trace this to the nature of the first excited state, which has Π_g symmetry in the isolated ion, and to the magnitude of the energy gap separating it from the ground state.

In the standard treatment of non-radiative relaxation [95] the rate depends on the presence of a *promoting* mode which links the ground and excited states through the vibronic perturbation, and an *accepting* mode, many quanta of which are created by the internal conversion of the electronic excitation energy. The accepting mode corresponds to the vibrational coordinate describing the difference in the ground and excited state equilibrium geometry, which in this case is the UO_2 symmetric stretch with a frequency of $\sim 850\ cm^{-1}$. The rate is dependent on the inverse power of the number of quanta which are required to bridge the electronic energy gap. For the uranyl ion 24 quanta of the accepting mode need to be excited simultaneously, so that this is a very inefficient mechanism.

The symmetry of the first excited state requires the promoting mode to be of π_g symmetry. There are no internal modes of the uranyl ion with this symmetry so the relaxation relies on the vibronic perturbation of the equatorial ligand modes with the analogous symmetry. As we have seen, even the static influence of these equatorial ligands is small, so the promoting modes are also inefficient.

The photochemistry of the uranyl ion is predominantly that of the photo-oxidation of a variety of substrates. The mechanistic aspects of this process have been very thoroughly discussed [23] and for organic substrates the primary mechanism appears to be hydrogen atom abstraction to generate the species UO_2H^{2+}. This is easily understood, in terms of our knowledge of the first excited state, as a process in which the hole in the σ_u orbital acts as the source of the oxidation. The unpaired electron in this orbital has strong oxygen $2p_z$ character, and is sterically accessible to chemical attack by the hydrogen atom along the direction of the O–U axis. By this route the excited state of the uranyl ion can evolve directly into the ground state of the U(V) species $UO(OH)^{2+}$, which is formally analogous to U(V) mono-oxo species such as $UOCl_5^{2-}$ which have been studied both by Ryan and by Selbin and his co-workers [96] and have an electronic structure related to that of the NpO_2^{2+} ion. The role of UO_2H^{2+} in exciplex formation has been discussed by Marcantonatos [97].

Acknowledgements. I would like to thank Professor H. U. Güdel for an invitation to deliver a series of lectures in Swiss Universities. This review has evolved from those lectures. I am also deeply grateful to Professor C. K. Jørgensen for many valuable discussions over a number of years, to Dr J. P. Day for originating my interest in this subject, and to Professor Peter Day, for encouragement and advice. I am particularly indebted to my colleagues, Dr. T. R. Snellgrove, Dr. D. R. Woodwark, Dr. J. O. W. Norris, Dr. P. J. Stone, Dr. J. R. G. Thorne, Dr. D. I. Grimley, T. J. Barker and I. D. Morrison for their essential contributions to the development and analysis of much of the optical work described in the review.

6 References

1. Streitweiser A (1979) in: Marks TJ, Fischer RD (eds) Organometallics of the *f*-elements, Reidel, Dordrecht
2. Katz JJ, Seaborg GT, Morss LR (1986) The chemistry of the actinide elements, 2nd edn, Chapman and Hall
3. Wait E (1955) J Inorg Nucl Chem 1: 309; Weller MT, Dickens PG, Penny DJ (1988) Polyhedron 7: 243

4. Bleijenberg KC (1980) Struct and Bonding, 42: 97; Blasse G (1976) Struct and Bonding 26: 43; Blasse G (1980) *ibid* 42: 1
5. Dickens PG, Lawrence SD, Penny DJ, Powell AV (1989) Sol Stat Ionics 32/33: 77
6. Cordfunke EHP, Ouweltjes W (1981) J Chem Thermodyn 13: 187
7. Siegel S, Hoekstra HR (1971) Inorg Nucl Chem Lett 7: 455
8. Kolitsch W, Müller U (1979) Z Anorg Allgem Chem 410: 21
9. Day JP, Venanzi LM (1966) J Chem Soc (A) 1363
10. Sylva RN, Davidson MR (1979) J Chem Soc Dalton Trans 3: 465
11. Toth LM, Begun GM (1981) J Phys Chem 85: 547
12. Taylor JC (1971) Acta Cryst B 27: 1088
13. Reis AH, Hoekstra HR, Gebert E, Peterson SW (1976) J Inorg Nucl Chem 38: 1481
14. Greaves C, Fender BEF, Fender (1972) Acta Cryst B28: 3609
15. Denning RG (1983) In: Gmelin Handbook, Uranium A6: 31
16. Bombieri G, Forsellini E, Day JP, Azeez WI (1978) J Chem Soc Dalton Trans 677
17. Butcher RJ, Penfold BR, Sinn E (1979) J Chem Soc Dalton Trans 668
18. Glidewell C (1977) Inorg Chim Acta 24: 149
19. Gordon G, Taube H (1961) J Inorg Nucl Chem 19: 189
20. Findlay JR, Glover KM, Jenkins IL, Large NR, Marples JAC, Potter PE, Sutcliffe PW (1977) In: Thompson R (ed) The modern inorganic chemicals industry, The Chemical Society, Spec Pub 31, London
21. Thorne JRG, Denning RG, Barker TJ, Grimley DI (1985) J Lumin 34: 147; Sugitani Y, Nomura H, Nagashima K (1980) Bull Chem Soc Japan 53: 2677
22. Rabinowitch E, Belford RL (1964) Spectroscopy and photochemistry of uranyl compounds, Mcmillan, New York; Burrows HD, Kemp TJ (1974) Chem Soc Rev 3: 139
23. Jørgensen CK, Reisfeld R (1982) Struct and Bonding 50: 121
24. Jørgensen CK (1957) Acta Chem Scand 11: 166
25. McGlynn SP, Smith JK (1961) J Mol Spect 6: 164
26. Gorller-Walrand C, VanQuickenborne LG (1972) Spectrochim Acta A 28: 257; *idem* (1971) J Chem Phys 54: 4178
27. Dieke GH, Duncan ABF (1949) Spectroscopic properties of uranium compounds, McGraw-Hill, New York
28. McClure DS (private communication)
29. Denning RG, Snellgrove TR, Woodwark DR (1976) Mol Phys 32: 419
30. Hall D, Rae AD, Walters TN (1966) Acta Cryst 20: 160
31. Denning RG, Snellgrove TR, Woodwark DR (1975) Mol Phys 30: 1819
32. Denning RG, Snellgrove TR, Woodwark DR (1979) Mol Phys 37: 1109
33. Gorller-Walrand C, VanQuickenborne LG (1972) J Chem Phys 57: 1436
34. Denning RG, Foster DNP, Snellgrove TR, Woodwark DR (1979) Mol Phys 37: 1089
35. Denning RG, Short IG, Woodwark DR (1980) Mol Phys 39: 1281
36. Gruen DM, Hutchinson CA (1954) J Chem Phys 22: 386; McGlynn SP, Smith JK (1961) J Mol Spectrosc 6: 188
37. Bleaney B, Llewellyn PM, Pryce MHL, Hall GR (1954) Phil Mag 45: 992; Bleaney B (1955) Disc Faraday Soc 19: 112
38. Denning RG, Norris JOW, Brown D (1982) Mol Phys 46: 287
39. Denning RG, Norris JOW, Brown D (1982) Mol Phys 46: 325
40. Barker TJ, Denning RG, Thorne JRG (1987) Inorg Chem 26: 1721
41. Barker TJ (1990) D Phil Thesis, Oxford
42. Abramova IN, Abramov AP, Tolstoy NA (1969) Opt Spectrosc 27: 293
43. Allen DM, Burrows HD, Cox A, Hill RJ, Kemp TJ, Stone TJ (1973) J Chem Soc Chem Comm 59; Sergeeva G, Chibisov A, Levshin L, Karyakin A (1974) J Chem Soc Chem Comm 159; Burrows HD (1990) Inorg Chem 29: 1549
44. Denning RG, Morrison ID (unpublished)
45. Fragala I, Condorelli G, Tondello A, Cassol A (1978) Inorg Chem 17: 3175
46. Allen GC, Baerends EJ, Vernooijs P, Dyke JM, Ellis AM, Feher M, Morris A (1988) J Chem Phys 89: 5363
47. Rauh EG, Ackerman RJ (1974) J Chem Phys 60: 1396
48. Veal BW, Lam DJ, Carnall WT, Hoekstra HR (1975) Phys Rev B12: 5651
49. Desclaux JP (1973) At Data Nucl Data Tables 12: 311
50. Teterin YA, Baev AS, Mashirov LG, Suglobov DN (1984) Dokl (Phys Chem) 277: 131

51. Larsson S, Pyykkö P (1986) Chem Phys 101: 355
52. Dunlap BD, Kalvius GM (1974) in: Freeman AJ, Darby JB (eds) The actinides, electronic structure and related properties, Vol I, Academic Press, New York
53. Evans S, Hamnett A, Orchard AF (1972) J Coord Chem 2: 57
54. Pyykkö P (1987) Inorg Chim Acta 139: 243
55. Belford RL, Belford G (1961) J Chem Phys 34: 1330
56. Newman JB (1965) J Chem Phys 43: 1691
57. Johnson KH (1973) Adv Quant Chem 7: 143
58. Rösch N, Klemperer WG, Johnson KH (1973) Chem Phys Lett 23: 149
59. Boring M, Wood JH, Moscowitz JW (1975) J Chem Phys 63: 638
60. Hay PJ, Wadt WR, Kahn LR, Raffenetti RC, Phillips DH (1979) J Chem Phys 71: 1767
61. Yang CY, Johnson KH, Horsley JA (1978) J Chem Phys 68: 1000
62. Boring M, Wood JH (1979) J Chem Phys 71: 392
63. Cowan RD, Griffin DC (1976) J Opt Soc Am 66: 1010
64. Wood JH, Boring M, Woodruff SB (1981) J Chem Phys 74: 5225
65. Rosen A, Ellis DE, Adachi H, Averill FW (1976) J Chem Phys 65: 3629
66. Ellis DE, Rosen A, Walch PF (1975) Int J Quantum Chem Symp 9: 351
67. Walch PF, Ellis DE (1976) J Chem Phys 65: 2387
68. Slater JC (1974) The self-consistent field for molecules and solids, McGraw-Hill, New York
69. Roby KR (1974) Mol Phys 47: 81
70. Ellis DE (private communication, quoted in Ref. [64])
71. Snijders JG, Baerends EJ, Ros P (1979) Mol Phys 38: 1909
72. DeKock RL, Baerends EJ, Boerrigter PM, Snijders JG (1984) Chem Phys Lett 105: 308
73. Wadt WR (1981) J Am Chem Soc 103: 6053
74. Gabelnik SD, Reedy GT, Chasonov MD (1974) Chem Phys 60: 1167
75. Tatsumi K, Hoffmann R (1980) Inorg Chem 19: 2656
76. Pyykkö P, Lohr LL (1981) Inorg Chem 20: 1950; Pyykkö P, Laaksonen L (1984) J Phys Chem 88: 4892
77. Fenske RF, DeKock RL (1970) Inorg Chem 9: 1053
78. Citrin PH, Thomas TD (1972) J Chem Phys 57: 4446
79. Denning RG (unpublished)
80. Choppin GR, Rao LF (1984) Radiochim Acta 37: 143
81. Burdett JK (1980) Molecular shapes, Wiley, New York, p 90
82. Woodwark DR (1977) D Phil Thesis, Oxford
83. Nugent WA, Mayer JA (1988) Metal ligand multiple bonds, Wiley, New York, p 156; Appleton TG, Clark HC, Manzer LE (1973) Coord Chem Rev 10: 335
84. De Wet JF, Du Preez JGH (1978) J Chem Soc Dalton Trans 592
85. Tordjman I, Masse R, Guitel JC (1974) Z Kristall 139: 103
86. Phillips MLF, Harrison WTA, Stucky GD (1990) Inorg Chem 29: 3245; Thomas PA, Glazer AM, Watts BE (1990) Acta Cryst B46: 333
87. Griffith WP (1962) J Chem Soc 3248; Griffith WP (1964) J Chem Soc 245; Griffith WP (1969) J Chem Soc (A) 211
88. Bader RFW (1960) Mol Phys 3: 137
89. Burdett JK (1980) Molecular shapes. Wiley, New York, p 64
90. Jones LH (1959) Spectrochim Acta 15: 409
91. Bartlett JR, Cooney RP (1989) J Mol Struct 193: 295
92. Alcock NW (1968) J Chem Soc (A) 1588
93. Adrian HWW, Van Tets A (1978) Acta Cryst B34: 88; Adrian HWW, Van Tets A (1977) Acta Cryst B33: 2997
94. Moore CE (1958) Circ US Nat Bur Stand, Vol III
95. Jortner J, Rice SA, Hochstrasser RM (1969) Adv Photochem 7: 149; Riseberg LA, Moos HW (1968) Phys Rev 174: 429
96. Ryan JL (1971) J Inorg Nucl Chem 33: 153; Selbin J, Ballhausen CJ, Durrett DG (1972) Inorg Chem 11: 510
97. Marcantonatos MD (1978) Inorg Chim Acta 26: 41
98. Taylor JC, Wilson PW (1974) Acta Cryst B30: 1481
99. Shamir J, Silberstein A, Ferraro J, Choca M (1975) J Inorg Nucl Chem 37: 1429
100. Müller U, Kolitsch W (1974) Z Anorg Chem 410: 32
101. Kolitsch W, Müller U (1975) Z Anorg Chem 418: 235

102. De Wet JF, Caira MR, Gellatly BJ (1978) Acta Cryst B34: 1121
103. Bombieri G, Brown D, Mealli C (1976) J Chem Soc Dalton Trans 2025
104. Schleid T, Meyer G, Morss LR (1987) J Less-Common Met 132: 69
105. Brown D (1966) J Chem Soc (A) 766
106. Aurov NA, Volkov VA, Chirkst DE (1983) Radiokhimiya 25: 366
107. Amberger H-D, Rosenbauer GG, Fischer RD (1976) Mol Phys 32: 1291
108. Barnard R, Bullock JI, Gellatly BJ, Larkworthy LF (1972) J Chem Soc Dalton Trans 1932
109. Shannon RD, Prewitt CT (1969) Acta Cryst B25: 787
110. Bois C, Dao N-Q, Rodier N (1976) Acta Cryst B32: 1541
111. Bagnall KW, du Preez JGH, Gellatly BJ, Holloway JH (1975) J Chem Soc Dalton Trans 1963
112. Furrer A, Güdel HU, Darriet J (1985) J Less-Common Met 111: 223
113. Crosswhite HM, Crosswhite H, Carnall WT, Paszek AP (1980) J Chem Phys 72: 5103
114. Wagner W, Edelstein N, Whittaker B, Brown D (1977) Inorg Chem 16: 1021
115. Johnston DR, Satten RA, Schreiber CL, Wong EY (1976) J Chem Phys 44: 3141
116. Edelstein N, Brown D, Whittaker B (1974) Inorg Chem 13: 563
117. Pinsker GZ (1967) Sov Phys Crystall 11: 634
118. Grosjean D, Weiss R (1967) Bull Chem Soc France 3054
119. Müller U, Lorenz I (1980) Z Anorg Allg Chem 463: 110
120. Drew MGB, Fowles GWA, Page EM, Rice DA (1981) J Chem Soc Dalton Trans 2409
121. Kamenar B, Prout CK (1970) J Chem Soc (A) 2379
122. Carrondo MAA de CT, Shahir R, Skapsi AC (1978) J Chem Soc Dalton Trans 844
123. Bright D, Ibers JA (1967) Inorg Chem 8: 709
124. Brown D, Reynolds CT, Moseley PT (1972) J Chem Soc Dalton Trans 857
125. Denning RG, Ironside CN, Thorne JRG, Woodwark DR (1981) Mol Phys 44: 209
126. Lam RUE'T, Blasse G (1980) J Inorg Nucl Chem 42: 1377
127. Brittain HG, Perry DL (1981) J Phys Chem 85: 3073
128. Smit WMA, Blasse G (1984) J Lumin 31: 114
129. Softley TP (1981) Part II Thesis, Oxford
130. Flint CD, Tanner PA (1981) Mol Phys 43: 933
131. Denning RG, Ironside CN, Snellgrove TR, Stone PJ (1982) J Chem Soc Dalton Trans 1691
132. Brittain HG, Tsao L, Perry DL (1984) J Lumin 29: 285; Anderson A, Cheih C, Irish DE, Tong JPK (1980) Can J Chem 58: 1651
133. Leung AF, Tsang KK (1979) J Phys Chem Solids 40: 1093; McGlynn SP, Smith JK, Neely WC (1961) J Chem Phys 35: 105
134. Flint CD, Sharma P, Tanner PA (1982) J Phys Chem 86: 1921
135. Flint CD, Tanner PA (1981) Inorg Chem 20: 4405

A New Approach to Structural Description of Complex Polyhedra Containing Polychalcogenide Anions

M. Evain* and R. Brec

Laboratoire de Chimie des Solides, I.M.N., Université de Nantes, 44072 Nantes Cedex 03, France

Because of the occurrence of polyanionic groups, in particular of $(X_2)^{2-}$ pairs, that are found along with the monoatomic ligands, rather complex coordination groups $(M(X_i)_m X_n)$ (with $i = 2$ in most cases) are encountered in transition metal chalcogenide families. In order to facilitate the classification and the description of most phases of those families, a general method to generate and to classify $(M(X_i)_m X_n)$ polyhedra is introduced. With this method it is possible to visualize the links between coordination groups, either directly through common coordination anions or indirectly through polyanionic bridges. It is shown how the proposed procedure simplifies the understanding of the structural array of numerous complex compounds, in particular in the organo-halogeno-chalcogenides, halogeno-chalcogenides and chalcogeno-phosphates of transition metals. Furthermore, the way the developed method allows comparative studies and offers the possibility of conceiving new phases is exposed.

1	Introduction	278
2	The Change of Coordination	278
	2.1 The Different Classes of Polyhedra	278
	2.2 The Alpha, Beta, and Gamma Transformations	279
	2.3 Symbolization	281
3	Structural Arrangements Based on Identical Polyhedra: Homopolyhedral Construction I[i]I	282
	3.1 The F Polyhedron	283
	3.2 Th F1, F2, and F'2 Polyhedron	285
	3.3 The G1 Polyhedron	286
	3.4 The G2 and G'2 Polyhedra	287
	3.5 The G3 Polyhedron	288
	3.6 The H2 and H'2 Polyhedra	289
	3.7 The H3 Polyhedron	298
	3.8 The H4 Polyhedron	301
4	Structural Arrangements Based on Different Polyhedra: Heteropolyhedral Construction I[i]I'	302
	4.1 The F[b]G1 Hetero-Construction	302
	4.2 The F[b]H2 Hetero-Construction	303
	4.3 The H3[d2]H4 Hetero-Construction	304
5	Conclusion	305
6	References	306

* Author to whom correspondence is to be addressed

1 Introduction

Many halogenides and chalcogenides of transition elements form, from complex (MX_n) coordination groups, large or infinite arrangements through anion sharing or by means of polyanionic bridges. Thus the structural analysis and the description of such phases usually start from particular features (e.g. symmetry, distortion...) of the basic (MX_n) constituting units. This approach applies as well to the portrayal of inorganic phases as to the representation of organometallic materials that exhibit halogen and chalcogen ligands.

Some complication takes place in the case of chalcogenide phases with occurrence, along side the classical S^{2-} groups, of polysulfide or polyselenide entities as chelating ligands $((S_2)^{2-}$ or $(Se_2)^{2-}$ for instance). Indeed, in that case, we are dealing with complex coordination groups $(M(X_i)_m X_n)$ $((M(X_2)_m X_n)$ if we limit ourselves to dianionic X_2 entities) that are not taken into account in the usual classification of regular or semi-regular polyhedra.

In order to rationalize the studies of transition metal chalcogenides with irregular $(M(X_2)_m X_n)$ polyhedra, we develop in this article a method that allows the generation and the classification of these polyhedra (many examples of constitutive units are given). Furthermore, we propose a molecular approach to describe the way these polyhedra can interbond in the crystal network. We also give evidence that, in view of the proposed classification, many structures apparently very different from one another – chosen in the M–X, M–X–Y and M–P–X systems (X = chalcogen, Y = halogen) and in some organometallic phases – can very logically and easily be analyzed and compared. Finally we suggest, from some particular combinations, potential occurrence of new phases.

2 The Change of Coordination

2.1 The Different Classes of Polyhedra

Among the many coordination polyhedra (tetrahedron, octahedron, trigonal prism...) encountered in chemistry, and more specifically in inorganic chemistry, we consider a classification in three families. The first one includes regular polyhedra, that is, polyhedra with edges of the same length and with faces of the same type (octahedra and tetrahedra for example). The second family corresponds to polyhedra with edges of the same length but without any face type restriction; these polyhedra are referred to as semi-regular polyhedra and include, for instance, trigonal prisms and antiprisms. The remaining polyhedra, which present no regularity, constitute the third family. The lack of regularity occurs, in particular, when ligands become polyanionic.

In the following two sections, we will present a method that allows one to derive $(M(X_i)_m X_n)$ units of the last group (irregular polyhedra) from units of the first two groups (regular and semiregular polyhedra). We will also label each polyhedron according to the following notation:

– a capital letter will represent the number of apexes X (coordination number): A for one apex, B for two apexes, C for three apexes ... (for example the octahedron will be labelled F). Capital I and J will be used as generic letters.

– a succession of digits *mno*... will indicate the number of polyanionic ligands in the growing order of condensation; that is, m for the (X_2) pairs, n for the number of (X_3) trios.... Zero labels after the last non zero digit will be omitted.

For instance, the polyhedra (MX_6), $(M(X_2)X_4)$, and $(M(X_2)(X_3)X)$ will be labelled F, F1 and F11, respectively.

To simplify the discussion, we will limit the study to $(M(X_i)_m X_n)$ units with i = 2 and 2m + n ⩽ 8), that is, to A*m* – H*m* units, m ⩽ 4.

2.2 The Alpha, Beta, and Gamma Transformations

From classical regular or semi-regular polyhedra, various types of coordination transformation can be imagined (within our limitation i = 2, vide supra). Those changes belong to two categories: one that corresponds to a preservation of the metal-coordination number and the other that implies its modification by one or several units. The alpha change – α or internal pairing (IP) – relates to the first category and is based on the formation of an (X_2) pair from two unpaired X corners of the original polyhedron:

$$X + X \xrightarrow[\text{IP}]{\alpha} (X_2)$$

Such a pairing is illustrated in Table 1. For the parent polyhedron $(M(X_2)_m X_n)$ this corresponds to:

$$(M(X_2)_m X_n) \xrightarrow[\text{IP}]{\alpha} (M(X_2)_{m+1} X_{n-2}) \quad n \geq 2$$

Let us illustrate this coordination modification by considering the modification of the pristine (MX_6) octahedron. The making of a (X_2) pair leads to F1 according to:

$$F\ (MX_6) \xrightarrow[\text{IP}]{\alpha} F1\ (M(X_2)X_4)$$

Next, F1 generates F2 and F'2:

$$F1\ (M(X_2)X_4) \xrightarrow{\alpha} F2,\ F'2\ (M(X_2)_2 X_2)$$

Only two representatives of the $(M(X_2)_2 X_2)$ polyhedron, F2 and F'2, have been considered.

Table 1. Generation of complex polyhedra: Application of the internal pairing (IP) and external pairing (EP) operations (α and β, respectively) on the F octahedron

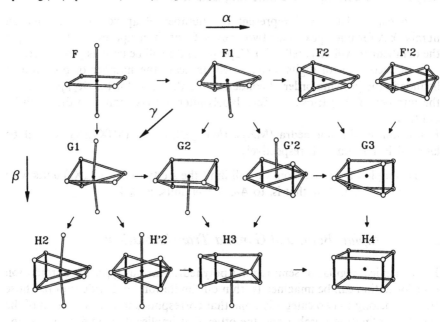

The beta modification – β or external pairing (EP) – corresponds to a change of the coordination number with a substitution of an X apex by a polyanionic group (X_2) according to:

$$(M(X_2)_m X_n) + X \xrightarrow[EP]{\beta} (M(X_2)_{m+1} X_{n-1}) \qquad n \geq 1$$

For instance, we have (see Table 1):

$$F(MX_6) + X \xrightarrow[EP]{\beta} G1(M(X_2)X_5)$$

$$G1(M(X_2)X_5) + X \xrightarrow[EP]{\beta} H2, H'2 (M(X_2)_2 X_4)$$

The operation chain $F \to G1 \to H2 \ldots$ stops at L6 (the $M(X_2)_6$ polyhedron) when all the apexes have been transformed.

The last modification we define is the gamma transformation (γ) that consists in the addition of a single apex X to the coordination sphere of a given polyhedron. Such a modification is illustrated in the following $F1 \to G1$ change:

$$F1(M(X_2)_m X_n) + X \xrightarrow{\gamma} G1(M(X_2)_m X_{n+1})$$

Structural Description of Complex Polyhedra Containing Polychalcogenide Anions   281

The extra X apex contributes to a coordination-number increase of one unit without any pairing (i.e., the number of pairs remains unchanged).

Similar transformations can be easily imagined for i > 2 (mutation of an (X_2) group in an (X_3) group or of an apex X in an (X_3) group . . .).

2.3 Symbolization

A given polyhedron I can be envisioned in many ways. For instance, an octahedron may be described as the (MX_6) group (**1a**) with six apexes surrounding the central atom, but it can also be seen as two X_3 triangles sandwiching M (**1b**) or as three opposite X_2 edges coordinating M (**1c**) or even as two apexes capping a (MX_4) unit (**1d**). Depending upon the type of connection involved in the molecular or crystalline framework, one may choose one or the other of the various descriptions. If it is rather easy to recognize an F unit in a given network and quite simple to give a comprehensible schematic representation of the corresponding structure, it seems to be somewhat more difficult to picture the F1, F2, F'2, G . . . polyhedra. To ease the understanding of the different structures hereinafter introduced, we propose a two-dimensional symbolization that simplifies the representation of the polyhedra and emphasizes their links throughout the crystalline network. For instance, with that symbolism, **1b**, **1c**, and **1d** reduce to **2a**, **2b** and **2c**, respectively. Only the essential features of each polyhedron are conserved, that is, the overall coordination and the presence or not of polyanionic groups. In Table 2 are presented possible symbolisms for the polyhedra already introduced in Table 1.

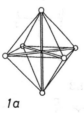
1a

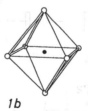
1b

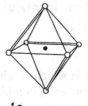
1c

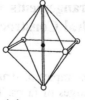
1d

2a

2b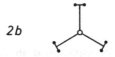

2c

Table 2. Schematic representation of the polyhedra introduced in Table 1. Central, metallic cations are represented by an *empty circle*. The surrounding anions appear either as *black disks*, for isolated apexes, or as *triangles* or *rectangles* for $(X_2)X$ and $(X_2)_2$ groups, respectively (in those units, the (X_2) pairs are underlined by a *thicker line*)

3 Structural Arrangements Based on Identical Polyhedra: Homopolyhedral Construction I[i]I

There exist many examples of structures in which octahedra are directly linked through vertexes, edges or faces. The number of combinations increases if one wants to link the octahedra indirectly through complex anionic groups, as found for example in the thiophosphates, arseniates, polysulfides....

The purpose of this section is 1) to characterize and symbolize each polyhedron of Table 1 with the help of an appropriate notation that allows the classification of the various homopolyhedra arrangements I[i]I and their representation in the network built from them, 2) to illustrate these arrangements from a few examples to demonstrate the generality and the usefulness of our systematic and, in the process, to reveal the internal coherence of structures apparently very different. Unknown I[i]I arrangements will be suggested.

A notation, in many respects similar to that used for the polyhedra, is applied to the labeling of the links:

– a lower case letter to indicate the number of shared apexes, a for one X apex in common, b for two .... Lower case i will be used as a generic link.

– a succession of numbers *mno*... to list the number of polyanionic groups (vide supra). Two groups $(MX(X_2)_2(X_3))$ with a $(X_2)X$ triangle link are therefore labelled H21[c1]H21.

3.1 The F Polyhedron

The three ways to combine F polyhedra (octahedron) two by two (F[i]F) are shown in 3. The first possibility 3a corresponds to a face sharing ([c] type connection), the second one 3b to an edge sharing ([b] type connection), and the third one 3c to an apex sharing ([a] type connection).

In the various compounds presenting direct combinations of F polyhedra, two have retained our attention: $Nb_2S_2Cl_4 \cdot 4(SC_4H_8)$ [1] and $V_2P_4S_{13}$ [2]. They belong to two families: the organochalcogeno-halogenides and the chalcogeno-phosphates of transition elements, in which many examples will be taken.

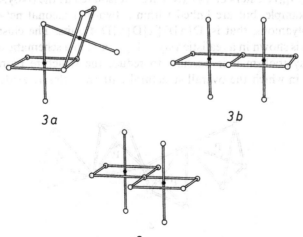

3a   3b

3c

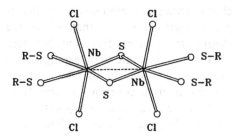

Fig. 1. Projection in an arbitrary direction of one molecule of $Nb_2S_2Cl_4 \cdot 4(SC_4H_8)$: Illustration of the F[b]F **3b** connection

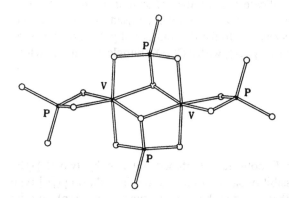

Fig. 2. $V_2P_4S_{16}$ structural fragment of (2D) $V_2P_4S_{13}$ structure (*open circles* stand for sulfur atoms): F[b]F **3b** junctions in a chalcogeno-phosphate of transition metal

In $Nb_2S_2Cl_4 \cdot 4(SC_4H_8)$, a molecular solid without polyanionic bonding, the (M_2X_{10}) cluster is a combination of two F octahedra resulting from the edge sharing F[b]F **3b** and characterized by a metal to metal bond (286.8 pm) between the two niobium (IV) ions through the common SS edge (Fig. 1). The chlorine atoms occupy the four privileged apexes of the symbolic representation introduced in **3b**. The large and bulky tetrahydrothiophen (SC_4H_8) groups are located at the four corners left available at the tip of the F octahedra.

The (V_2S_{10}) clusters of $V_2P_4S_{13}$ are not isolated as the (Nb_2X_{10}) are in the previous example, but are linked within a two-dimensional network through (P_4S_{13}) polyanions, that is, D[a]D[a]D[a]D **4** units. The cluster/polyanion connection is shown in a realistic way in Fig. 2 and in a schematic one in **5**. With the latter symbolism it is possible to reduce the structure representation to diagram **6**, in which the overall structural pattern is clearly evidenced.

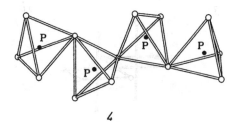

4

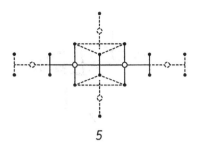

5

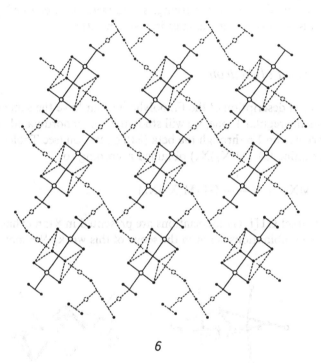

6

3.2 The F1, F2, and F'2 Polyhedra

F1, issued from F through the alpha (IC) operation, can be described as a monocapped polyhedron. The uncapped basic unit $(M(X_2)X_3)$ is the result of the opposition of two fragments: one triangle $(X_2)X$ and one edge 2X (see Tables 1 and 2). F1 polyhedra may be associated in various ways. In 7 is presented a F1[c1]F1 combination with occurrence of a common triangular face. There do not exist, at least within the three families under consideration, structural types that illustrate either this F1[c1]F1 connection or other F1[i]F1 connection types.

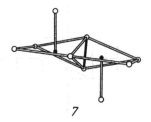

7

This lack of examples is verified again for the F2 and F′2 polyhedra (see Tables 1 and 2). We note here that F2 can be viewed either as a rectangle opposed to an edge or as two triangles arranged in an eclipsed position. F′2 corresponds to two facing triangles in a star position.

3.3 The G1 Polyhedron

After the brief description of the Fn polyhedra that have the same coordination number as the regular F one, we will study the heptacoordinated Gn polyhedra which derive from Fn through the beta (EP) operation (see Tables 1 and 2). Let us first examine G1 (M(X$_2$)X$_5$) obtained according to:

$$F(MX_6) + X \xrightarrow{\beta} G1(M(X_2)X_5)$$

A few direct G1[i]G1 associations are presented in **8** (an exhaustive catalog of possible combinations is not in the scope of this work). We may illustrate the

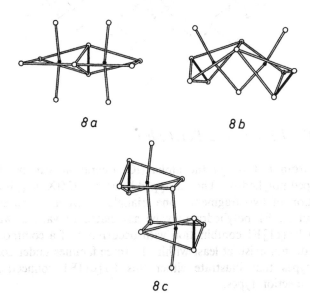

8a   *8b*

8c

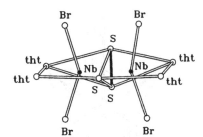

Fig. 3. Portrayal of the G1[c1]G1 link in an Nb$_2$S$_3$Cl$_4 \cdot$4(SC$_4$H$_8$) molecule

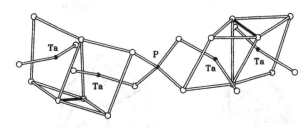

Fig. 4. Indirect connection of two G1[c1]G1 units in the three-dimensional phase Ta$_2$P$_2$S$_{11}$ (*open circles* stand for sulfur atoms)

G1[c1]G1 connection **8a** with examples which somewhat recall those of Sect. 3.1, e.g. with Nb$_2$S$_3$Cl$_4 \cdot$4(SC$_4$H$_8$) [3] and Ta$_2$P$_2$S$_{11}$ [4].

The extra sulfur of Nb$_2$S$_3$Cl$_4 \cdot$4(SC$_4$H$_8$) formula as compared to that of Nb$_2$S$_2$Cl$_4 \cdot$4(SC$_4$H$_8$) makes up a triangular (S$_2$)S group that connects the two G1 units (see Fig. 3) and substitutes the S$_2$ edge, shared by the two octahedra of Nb$_2$S$_2$Cl$_4 \cdot$4(SC$_4$H$_8$).

If we find isolated (M$_2$X$_{11}$) clusters in Nb$_2$S$_3$Cl$_4 \cdot$4(SC$_4$H$_8$), within Ta$_2$P$_2$S$_{11}$ the (Ta$_2$S$_{11}$) units are linked to one another through D (PS$_4$) tetrahedra (see Fig. 4) to constitute a double, tridimensional network with wide empty tunnels. This structure will be analyzed later along with those of TaPS$_6$ and Ta$_4$P$_4$S$_{29}$.

3.4 The G2 and G'2 Polyhedra

It is sometimes difficult to recognize a G2 or a G'2 polyhedron in a crystal network. For example, in the Mo$_3$S$_7$Cl$_4$ structure [5], as represented by the authors of the crystal determination (Fig. 5), what kind of environment does the molybdenum have in the cluster? (Mo(S$_2$)$_2$SCl$_2$) polyhedra are no other than G'2 (M(X$_2$)$_2$X$_3$) units (see Table 1) as can be clearly seen in **9**, obtained after rotation of one of the (Mo(S$_2$)$_2$SCl$_2$) units. What about the connection type between the G'2 polyhedra? They actually share c1 faces (G'2[c1]G'2 connection type) as pictured in **10**, a strong distortion allowing the formation of quasi-equilateral (Mo$_3$S$_7$Cl$_6$) triangular clusters that leads to relatively short

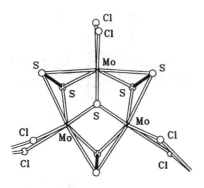

Fig. 5. Distorted G'2 polyhedra, connected through cl face-sharing, in the $Mo_3S_7Cl_6$ clusters of the $Mo_3S_7Cl_4$ framework (projection in an arbitrary direction)

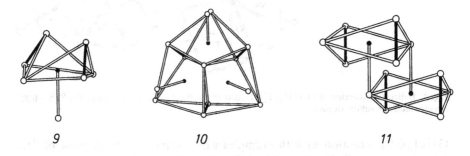

9   10   11

(274.1, 274.7, and 274.8 pm) intermetallic distances. This is not the only type of G'2[i]G'2 junction in this structure. The ($Mo_3S_7Cl_6$) clusters constitute infinite, zigzagging ($Mo_3S_7Cl_4$) rows, through the sharing of Cl–Cl edges (G'2[b]G'2 shown in 11).

Bibliographic searches have not disclosed any reports of G2 polyhedra (see Table 1). We will then proceed with other types of polyhedra.

3.5 The G3 Polyhedron

Let us consider again the ($Mo_3S_7Cl_6$) cluster as depicted in Fig. 5, keeping in mind that G3 (see Table 1) is obtained through the α(IP) operation from G2 or G'2, viz:

$$G2, G'2\,(M(X_2)_2X_3) \xrightarrow{\alpha} G3\,(M(X_2)_3X)$$

The substitution of the six chlorine atoms of the cluster by sulfur atoms allows the application of the α operation, leading to the (Mo_3S_{13}) group encountered in $Mo_3S_{13}(NH_4)_2$ [6] (see Fig. 6). In this case, it is quite easy to see the G3 surrounding 12 of the molybdenum atoms. The G3[c1]G3 interpolyhedral association gives the $(Mo_3S_{13})^{-2}$ anion 13 in which the Mo–Mo distances (271.9, 271.9 and 272.5 pm) are close to that found in $Mo_3S_7Cl_4$.

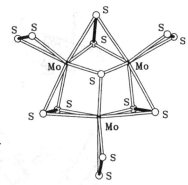

Fig. 6. Depiction of the (Mo_3S_{13}) groups of $Mo_3S_{13}(NH_4)_2$ that shows the G3 environment of molybdenum atoms

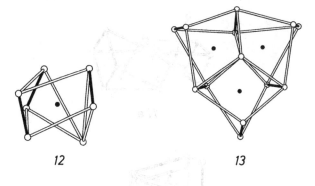

12   *13*

3.6 The H2 and H'2 Polyhedra

With the study of the Hm ($m = 2 - 4$) polyhedra (see Table 1), we are dealing with a particularly important domain, both for the understanding of the structures as well as for the interest raised by some of the phases built from these groups. The structural examples based on the Hm units are quite numerous (except for H'2 for which no structural illustration has been found yet) and the Hm[i]Hm combinations quite varied, sometimes even unexpected and surprising.

As well as for the F2 and G2 homologues, two arrangements can be considered for H2, namely, H2 and H'2 (see Tables 1 and 2). H2 is made up from the opposition of a rectangle $(X_2)_2$ and an edge X_2 with two capping apexes; H'2 is built upon the opposition of two triangles $(X_2)X$ with, once again, two capping apexes.

Possible H2[dm]H2 and H2[cm]H2 face sharing connections are gathered in **14** (we will not consider the junctions leading to (X_3) group formation that were previously excluded within the polyhedra). In **15** is illustrated a possible H2[b]H2 edge sharing.

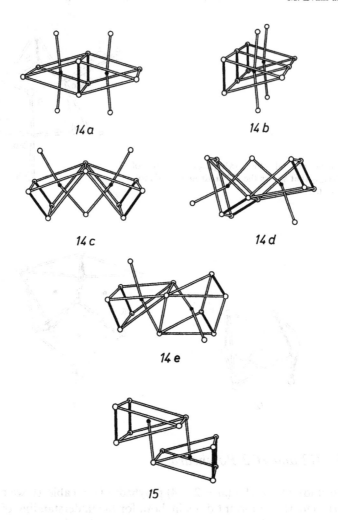

14a

14b

14c

14d

14e

15

In the $Nb_2X_nCl_4 \cdot 4(SC_4H_8)$ series (X = chalcogen), we have selected $Nb_2Se_4Cl_4 \cdot 4(SC_4H_8)$ [7] that corresponds to the **14a** association (H2[d2]H2 type) (see Fig. 7). This $Nb_2X_nCl_4 \cdot 4(SC_4H_8)$ family is remarkable in that it clearly illustrates the succession of β (EP) operations:

$$F(MX_6) \xrightarrow{\beta} G1(M(X_2)X_5) \xrightarrow{\beta} H2(M(X_2)_2X_4)$$

or:

$$NbX_2Cl_2 \cdot 2(SC_4H_8) \xrightarrow{\beta} Nb(X_2)XCl_2 \cdot 2(SC_4H_8)$$

$$Nb(X_2)XCl_2 \cdot 2(SC_4H_8) \xrightarrow{\beta} Nb(X_2)_2Cl_2 \cdot 2(SC_4H_8)$$

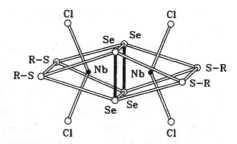

Fig. 7. The isolated H2[d2]H2 bicapped biprism of Nb$_2$Se$_4$Cl$_4$·4(SC$_4$H$_8$)

or in the dimeric form (see Figs 1 and 3):

$$Nb_2X_2Cl_4 \cdot 4(SC_4H_8) \xrightarrow{\beta} Nb_2(X_2)XCl_4 \cdot 4(SC_4H_8)$$

$$Nb_2(X_2)XCl_4 \cdot 4(SC_4H_8) \xrightarrow{\beta} Nb_2(X_2)_2Cl_4 \cdot 2(SC_4H_8)$$

The β operation or, more precisely, the inverse β transformation (β^{-1}: substitution of an (X$_2$) pair by a lone X atom) is not purely abstract; it corresponds in this case to the following chemical reaction [3]:

$$Nb_2S_3Cl_4 \cdot 4(SC_4H_8) \xrightarrow[\text{triphenylphosphine}]{\beta^{-1}} Nb_2S_2Cl_4 \cdot 4(SC_4H_8)$$

Numerous compounds in which the H2 polyhedra share simultaneously faces and edges can be found. This is in particular the case in ZrSe$_3$ [8] in which H2 units make infinite (ZrSe$_5$) chains (stacking of the H2[c1]H2 type shown in **14b**) laterally bonded through b connections (shown in **15**) to develop a two-dimensional network (Fig. 8). Several compounds (TiS$_3$, ZrS$_3$, ZrTe$_3$, HfS$_3$, HfSe$_3$ [9]) derive from the ZrSe$_3$ structural type. In particular, they all present a uniform metal to metal distance along the (MX$_5$) rows. NbS$_3$ [10] departs,

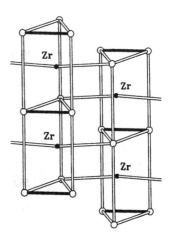

Fig. 8. The structure of ZrSe$_3$ projected along an arbitrary axis. Selenium atoms are shown as *open circles*

with alternative Nb–Nb short (303.7 pm) and long (369.3 pm) distances along the chains.

The creation of $(M_2(X_2)_2X_8)$ units through H2[d2]H2 association (shown in **14a**) favors the occurrence of metal to metal pairs (if allowed by the electron counting). Two types of direct connections between the $(M_2(X_2)_2X_8)$ units are essentially observed. In $Nb_3Se_5Cl_7$ [11], analyzed in Sect. 4.2, the inter-unit bonds take place through face sharings (H2[c]H2, **14c**) from which infinite (M_2X_9) rows are built (Fig. 9). The other association often observed is of the H2[b]H2 type (depicted in **15**). Since the $(M_2(X_2)_2X_8)$ group possesses eight different X_2 edges between two unpaired X corners, one in a capping position and the other in a non-capping position, several configurations are available, corresponding to the various H2[b]H2 arrangements. To be able to distinguish them from one another, an extra notation is needed. Let us add a U (up) label when the H2[b] connection is made up with a capping position and an apex situated above the plane defined by the central M atom and the capping atoms and a D (down) label in the other case as shown in **16**. If we exclude the participation of the same apex to two different junctions, we obtain, for two connections by $(M_2(X_2)_2X_8)$ group (one for each H2 polyhedron), the various

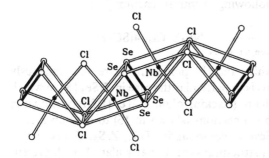

Fig. 9. Illustration of a H2[c]H2 combination in the Nb_2X_9 chains of $Nb_3Se_5Cl_7$

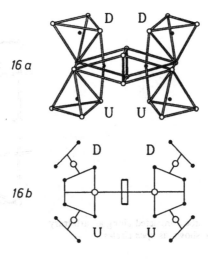

Structural Description of Complex Polyhedra Containing Polychalcogenide Anions 293

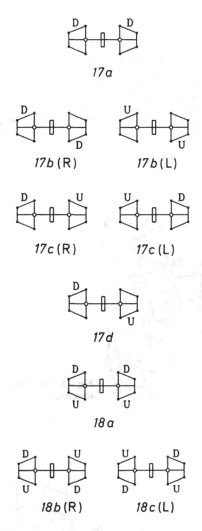

possibilities presented in **17**, and, for four connections by $(M_2(X_2)_2X_8)$ group (two for each H2 polyhedron), those shown in **18**.

This extra notation will allow us to better characterize certain structures and, consequently, to discover similarities between phases that apparently look different. Let us illustrate that notation with the following three compounds: MoS_2Cl_3 [5], NbS_2Cl_2 [12], and (2D) NbP_2S_8 [13]. Each one of them results from the **14a** H2[d2]H2 combination that allows a metal to metal bonding through the $(S_2)_2$ rectangular, common face. The $(M_2(X_2)_2X_8)$ clusters thus constituted are bonded through direct b links (**15**) for the first two phases and by $P_2S_6^{2-}$ polyanionic, bridging groups (D[b]D **19** units) for the last one. In MoS_2Cl_3, the **17d** type connections give rise to zigzagging infinite chains corresponding to a one-dimensional phase (Fig. 10), whereas for NbS_2Cl_2 a two-dimensional network results from the **18a** configuration (Fig. 11). Going from

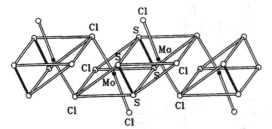

Fig. 10. Projection of a MoS_2Cl_3 chain that shows a **17a** type connection

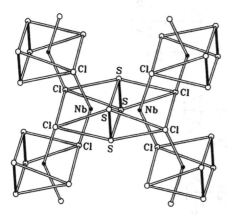

Fig. 11. 18a configuration in the (2D) NbS_2Cl_2 structure

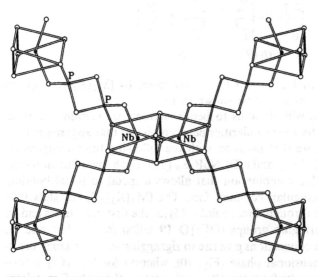

Fig. 12. The (2D) NbP_2S_8 structure where *open circles* stand for sulfur atoms

this last network with direct connection to polyanionic bridges yields the (2D) NbP$_2$S$_8$ structure (Fig. 12) in which chlorine atoms have been replaced by (P$_2$S$_6$) units **19**.

Both crystal networks of Ta$_4$P$_4$S$_{29}$ [14] and TaPS$_6$ [15] are based upon (M$_2$(X$_2$)$_2$X$_8$) clusters identical to those described above, except that no metal to metal bonds are found between d^0 tantalum atoms. The cluster to cluster interconnections are indirect through (PS$_4$) D polyanions and are of the **18b** type which presents two configurations: one right (R) and one left (L). These interconnections give rise to infinite chains (Fig. 13), right and left helices for the **18b(R)** and **18b(L)** configuration, respectively (here is found the origin of the labelling R and L of **17b**, **17c**, and **18b**). The juxtaposition of equivalent helices constitutes the tridimensional network **20** with alternatively wide and narrow tunnels (the wide tunnels correspond to the above defined helices and the narrow ones to inverse helices resulting from the interconnection of the first ones). The R and L label of a tridimensional lattice is taken from the corresponding R or L label of the encountered configuration and consequently gives the direction of rotation of the tunnels. If we reduce the representation of a (M$_2$(X$_2$)$_2$X$_8$) unit to a single segment and if we omit the polyanionic bridges [16], we obtain **21** for a counter clockwise (R) lattice.

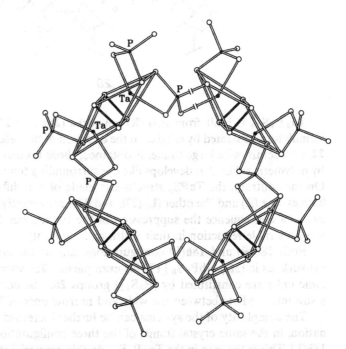

Fig. 13. Helix of Ta$_4$P$_4$S$_{29}$ that results from an .. H2[d2]H2[b]D[b]H2[d2] .. infinite condensation (projection along the c direction, the *broken lines* corresponding to a cell translation). *Open circles* stand for sulfur atoms

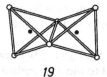

19

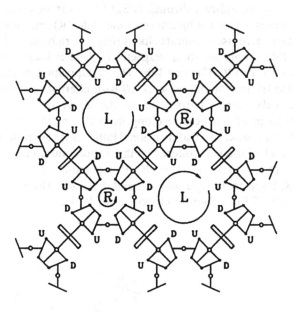

20

$Ta_4P_4S_{29}$ is built from two (R) networks (**21a** and **21b**), with the $TaPS_6$ formulation, separated by c/2 (i.e., in the direction of the helices). From that array **22**, a structure with large tunnel is obtained, those tunnels actually being filled by polymeric sulfur that develops like the surrounding tunnels, i.e., as (R) helices. On the contrary, the $TaPS_6$ structure is made of two different tridimensional lattices, one (R) and the other (L) (**23a** and **23b**, respectively). Their interlocking has as a consequence the suppression of the tunnels (see **24**) or more exactly a considerable reduction in their width and regularity.

Both **18b**(R) and **18b**(L) configurations can be encountered on the same network, as in (3D) NbP_2S_8 [17] for example (see **25**) where, this time, polyanionic links are constituted by (P_4S_{12}) groups **26**. The occurring tunnels have a size intermediate between the wide and narrow ones of $Ta_4P_4S_{29}$.

The complexity of the systems can be further increased through the combination, in the same crystal frame, of the three configurations **18a**, **18b**(R), and **18b**(L). This is the case in the $Ta_2P_2S_{11}$ double network introduced in Sect. 3.3 (in **27** is presented one of the two interlocked networks of the structure) for which we observe tunnels (made of five $(M_2(X_2)X_9)$ units) with clockwise and

Structural Description of Complex Polyhedra Containing Polychalcogenide Anions 297

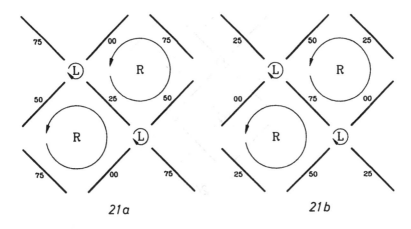

21a *21b*

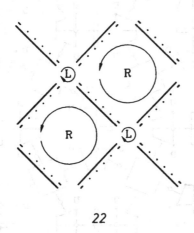

22

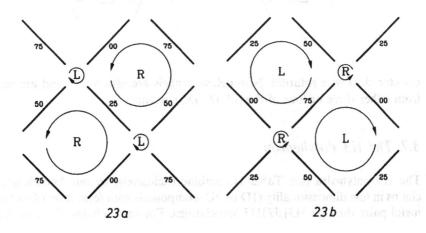

23a *23b*

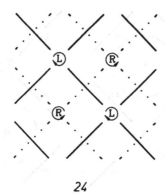

24

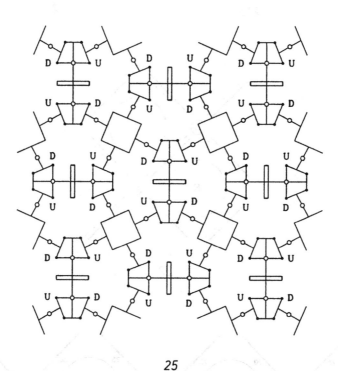

25

counter clockwise rotation. Non-helical tunnels are also found and are made from either three or four of the $(M_2(X_2)X_9)$ groups.

3.7 The H3 Polyhedron

The H3 polyhedra (see Table 1) combine exclusively, to our knowledge, in chains in low dimensionality (1D or 2D) compounds with formation of metal to metal pairs through H3[d2]H3 associations. For those chains, there exist six possible configurations of H3[d2]H3 combinations as shown in **28**. The H3[c]H3 combination **29** (**28d** configuration) leads to the (M_2X_9) rows of the (1D) Nb_2Se_9 [18] (Fig. 14) (to be compared with the (M_2X_9) chains already shown in Fig. 9) whereas the H3[c1]H3 combination **30** (**28b(R)** and **28b(L)**

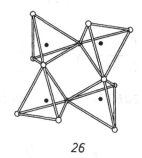

26

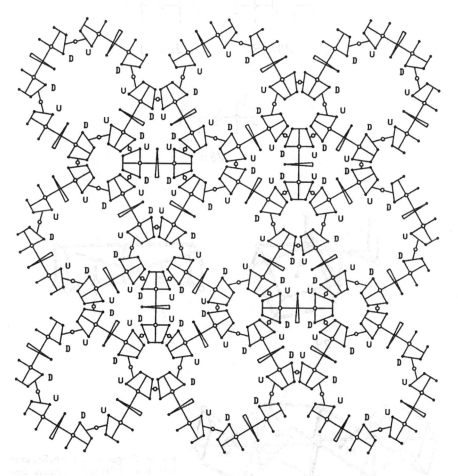

27

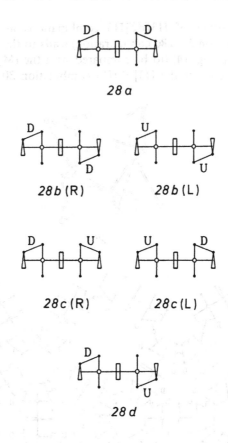

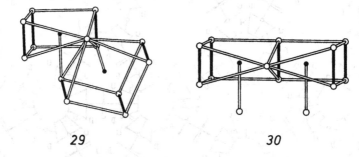

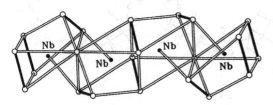

Fig. 14. Fragment of a Nb_2Se_9 chain that shows H3[c]H3 junctions. *Open circles* stand for selenium atoms

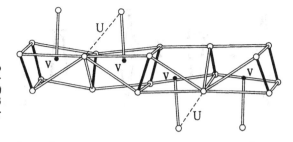

Fig. 15. V_2S_9 chain of $V_2P_2S_{10}$ built upon H3[c1]H3 combinations **30** in the **28b(R)** and **28b(L)** configurations and H3[d2]H3 links. *Open circles* stand for sulfur atoms

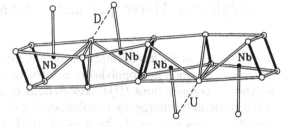

Fig. 16. Nb_2S_9 chain of Nb_2PS_{10} built upon H3[c1]H3 combinations **30** in the **28d** configuration and H3[d2]H3 links. *Open circles* stand for sulfur atoms

configurations forms the (M_2X_9) fibers of a third family (Fig. 15), basis for the V_2PS_{10} [19] and $Nb_4P_2S_{21}$ [20] structures and, finally, the H3[c1]H3 combination **30** (**28d** configuration) gives the (M_2X_9) chains (Fig. 16) of the Nb_2PS_{10} compound [21].

3.8 The H4 Polyhedron

This aptitude to constitute chains is still strong for the $(M(X_2)_4)$ H4 polyhedra. Given the restriction fixed in the development of our systematic classification, i.e., no occurrence of (X_3) groups, only one H3[i]H3 combination involving more than one apex remains possible: the sharing of a $(X_2)_2$ face.

The more simple form, but the occurrence of isolated H4 polyhedra, is encountered in $Mo_2S_{12}(NH_4)_2 \cdot 2H_2O$ [22] in which isolated $Mo_2(S_2)_6^{-2}$ anionic clusters with $Mo^V - Mo^V$ bond takes place (Fig. 17). In all other known examples, H4 polyhedra form infinite chains of the VS_4 structural type [23] (Fig. 18).

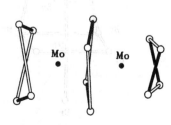

Fig. 17. Depiction of the (Mo_2S_{12}) cluster of $Mo_2S_{12}(NH_4)_2 \cdot 2H_2O$ that shows simple isolated H4[d2]H4 units. *Open circles* stand for sulfur atoms

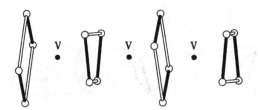

Fig. 18. Condensation of H4 polyhedra in VS$_4$ through H4[d2]H4 links. *Open circles* stand for sulfur atoms

4 Structural Arrangements Based on Different Polyhedra: Heteropolyhedral Construction I[i]I'

We have seen, in the few examples so far presented, the structural wealth of the systems based upon homopolyhedra constructions. The introduction of heteropolyhedral combinations I[i]I', associations of polyhedra of various natures, will considerably enlarge the possibilities. We do not intend to enumerate and describe in this section all the possible I[i]I' combinations (some of them: F[i]D, G1[i]D... have already been mentioned in Sect. 3); we will limit ourselves to three of them: F[b]G1, F[b]H2, and H3[d2]H4.

4.1 The F[b]G1 Hetero-Construction

The F[b]G1 heteropolyhedral association **31** is found in the Nb$_4$Se$_3$Br$_{10}$(NCMe)$_4$ molecular structure [24] along with the **8a** homopolyhedral combinations (Fig. 19). In the cluster, the metal atoms in a G1

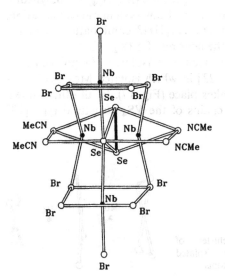

Fig. 19. The structure of Nb$_4$Br$_{10}$Se$_3$(NCMe)$_4$ in which is observed the heteropolyhedral association F[b]G1

31

environment present an oxidation state of IV and are engaged in a d^1–d^1 bond. The remnant metal atoms, in the center of the F units, are trivalent.

4.2 The F[b]H2 Hetero-Construction

It is in $Nb_3Se_5Cl_7$ phase [11] that we find an example of the F[b]H2 junction **32** (see Fig. 20). As in $Nb_4Se_3Br_{10}(NCMe)_4$, each group corresponds to a different oxidation state of the metal: IV for H2 and V for F. The H2 polyhedra are condensed in (M_2X_9) chains (see Sect. 3.6) and the F octahedra are grafted alternatively on each side of the chains. The F environment of Nb^V is insured by five chlorine atoms (three belonging to the (M_2X_9) frame) and one selenium (remarkably close to the metal, $d_{Nb-Se} = 225$ pm).

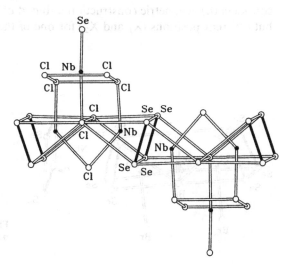

Fig. 20. Example of F[b]H2 heteropolyhedral association in the halogeno-chalcogenide of niobium $Nb_3Se_5Cl_7$

32

4.3 The H3[d2]H4 Hetero-Construction

The H2, H3 and H4 groups that are incorporated in the making of (M_2X_9) (for the first two) and (MX_4) (for the third) infinite rows present at least one rectangular $(X_2)_2$ d2 face. Therefore, it is not surprising to find various chains built from those polyhedra. This is the case in $Nb_6Se_{20}Br_6$ [25] where we observe the ... H3[c1]H3[d2]H4 ... succession (Fig. 21). Many one-dimensional structures based upon this principle, i.e., upon H2, H3, and H4 alternating groups, can be imagined.

This H3[d2]H4 association is also present in the $M_4Se_{16}Br_2$ phases (M = Nb, Ta) [26] along with the unexpected H3[c1]H4 one. In this last connection the triangular c1 face of H3 matches one part of the rectangular d2 face of H4 (see Fig. 22). Even more surprising is the statistical distribution of one anion, observed but not explained in the original structural description of $M_4Se_{16}Br_2$. This distribution is quite evident, in view of our analysis, if one considers the symmetric construction a–b/c–d of Fig. 22 that implies two close but different positions (X_1 and X_2) for one of the apex.

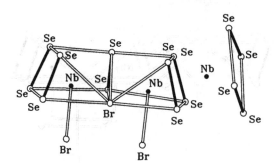

Fig. 21. Condensation of H3 and H4 polyhedra in $Nb_6Se_{20}Br_6$

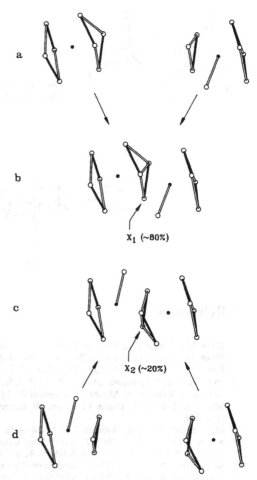

Fig. 22. Origin of the statistical distribution in Nb$_4$Se$_{16}$Br$_2$. In b is shown an Nb$_4$Se$_{16}$Br$_2$ chain fragment involving the X$_1$ position (80%) and resulting from the c1 face sharing of the H4 and H3 polyhedra presented in a. Symmetrically, in c is shown a fragment with the X$_2$ position (20%) issued from the H3 and H4 polyhedra presented in d. X1 and X2 are mutually exclusive (dX1-X2 = 71 pm)

5 Conclusion

In this article we have restricted ourselves to D, F, G, and H polyhedra. Other groups are of course possible and known. Let us just mention the Mo$_2$O$_2$S$_2$(S$_2$)$_2^{-2}$ E1 anion (Fig. 23) of [N(C$_2$H$_5$)$_4$]$_2$[Mo$_2$O$_2$S$_2$(S$_2$)$_2$] [27].

In addition to the already-put-forward advantages, i.e., better understanding of original coordination polyhedra, classification, and simplification of the structural descriptions with easier comparative studies, the developed method allows the conception of new structures. Except for the suggestion outlined in Sect. 4.3 to prepare new linear networks based upon H2, H3, and H4 groups, many new structural types can be imagined either by simple modification of existing networks or by creation of entirely new arrangements. For instance, one can envision the interesting low dimensional MX$_4$ 33 structure.

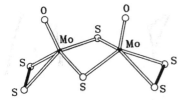

Fig. 23. E1 environment in the $Mo_2O_2S_2(S_2)_2^{-2}$ anion of $[N(C_2H_5)_4]_2[Mo_2O_2S_2(S_2)_2]$

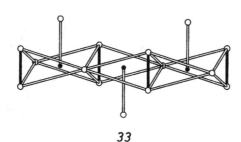

33

6 References

1. Drew MGB, Rice DA, Williams DM (1985) J. Chem. Soc. Dalton Trans. 1985: 417
2. Evain M, Brec R, Ouvrard G, Rouxel J (1985) J. Solid State Chem. 56: 12
3. Drew MGB, Rice DA, Williams DM (1985) J. Chem. Soc. Dalton Trans. 1985: 2251
4. Evain M, Lee S, Queignec M, Brec R (1987) J. Solid State Chem. 71: 139
5. Marcoll J, Rabenau A, Mootz D, Wunderlich H (1974) Rev. Chim. Min. 11: 607
6. Müller A, Pohl S, Dartmann M, Cohen JP, Bennett JM, Kirchner RM (1979) Z. Naturforsch, B34: 434
7. Drew MGB, Rice DA, Williams DM (1984) Acta Cryst. C40: 1547
8. Kronert VW, Plieth K (1965) Z. Anorg. Allg. Chem. 336: 207
9. Furuseth S, Brattas L, Kjekshus A (1975) Acta Chem. Scand. A29: 623
10. Rijnsdorp J, Jellinek F (1978) J. Solid State Chem. 25: 325
11. Rijnsdorp J, Jellinek F (1979) J. Solid State Chem. 28: 149
12. Rijnsdorp J, De Lange GJ, Wiegers GA (1979) J. Solid State Chem. 30: 365
13. Grenouilleau P, Brec R, Evain M, Rouxel J (1983) Rev. Chim. Min. 20: 628
14. Evain M, Queignec M, Brec R, Rouxel J (1985) J. Solid State Chem. 56: 148
15. Fiechter S, Kuhs WF, Nitsche R (1980) Acta Cryst. B36: 2217
16. Evain M, Brec R, Whangbo MH, (1987) J. Solid State Chem. 71: 244
17. Evain M, Brec R, Ouvrard G, Rouxel J (1984) Mat Res. Bull 19: 41
18. Meerschaut A, Guémas L, Berger R, Rouxel J (1979) Acta Cryst. B35: 1747
19. Brec R, Ouvrard G, Evain M, Grenouilleau P, Rouxel J (1983) J. Solid State Chem. 47: 174
20. Brec R, Evain M, Grenouilleau P, Rouxel J (1983) Rev. Chim. Min. 20: 283
21. Brec R, Grenouilleau P, Evain M, Rouxel J (1983) Rev. Chim. Min. 20: 295
22. Müller A, Nolte WO, Krebs B (1980) Inorg. Chem. 19: 2835
23. Allmann R, Baumann L, Kutoglu A, Rösch H, Hellner E (1964) Naturwiss. 51: 263
24. Benton AJ, Drew MGB, Rice DA (1981) J. Chem. Soc. Chem. Comm. 1241
25. Meerschaut A, Grenouilleau P, Rouxel J (1986) J. Solid State Chem. 61: 90
26. Grenouilleau P, Meerschaut A, Guémas L, Rouxel J (1987) J. Solid State Chem. 66: 293
27. Clegg W, Sheldrick GM, Garner CD, Christou G (1980) Acta Cryst. B36: 2784

Crystal Chemistry of Inorganic Nitrides

Nathaniel E. Brese and Michael O'Keeffe

Department of Chemistry, Arizona State University, Tempe, AZ 85287-1604, USA

The crystal chemistry of solid inorganic nitrides, azides, amides and imides is reviewed systematically. Many of the structures unique to nitrides are illustrated and analyzed. We give full crystallographic data for all compounds with well determined structures and list derived bond valences and Madelung potentials. Similarities to and differences from oxide chemistry are pointed out and general principles relating to structure and bonding in nitrides are elucidated.

1	Introduction	309
2	Methods	310
3	Systematic Review	311
	3.1 Alkali Metals	311
	3.1.1 H	311
	3.1.2 Li	312
	3.1.3 Na	319
	3.1.4 K	320
	3.1.5 Rb	321
	3.1.6 Cs	321
	3.2 Alkaline Earth Metals	321
	3.2.1 Be	321
	3.2.2 Mg	323
	3.2.3 Ca	324
	3.2.4 Sr	328
	3.2.5 Ba	329
	3.3 Group III	331
	3.3.1 B	331
	3.3.2 Al, Ga, In	331
	3.3.3 Tl	331
	3.4 Group IV	332
	3.4.1 Si	332
	3.4.2 Ge, Sn	333
	3.4.3 Pb	333
	3.5 Group V	334
	3.6 Lanthanides and Actinides	335
	3.7 Transition Metals	338
	3.7.1 Ti, Zr, Hf	339
	3.7.2 V, Nb, Ta	340

	3.7.3	Cr, Mo, W	343
	3.7.4	Mn, Re	344
	3.7.5	Fe	344
	3.7.6	Co	345
	3.7.7	Ni	345
	3.7.8	Cu, Ag, Au	345
	3.7.9	Zn, Cd, Hg	346
4	Discussion		346
	4.1	Stability	346
	4.2	Structure	347
	4.3	Bonding	348
	4.4	Physical Properties	349
5	Tables		350
6	References		371

1 Introduction

Nitrides are fascinating materials. Although the air we breathe is 78% nitrogen, essentially 100% of the land we walk upon is made of oxides. It would take an encyclopaedia to review oxide crystal chemistry, on the other hand we can present here a comprehensive review of solid state nitrides. Indeed until recently, solid state nitride chemistry was largely neglected, with the notable exception of the work of Juza and collaborators in the immediate post-war decades.

Why are nitrides not as numerous as oxides? The answer is that although generally nitrides are comparable in stability to oxides with respect to dissociation to atoms, the greater stability of N_2 as compared to O_2 means that nitrides are less stable with respect to dissociation to the combined elements. Thus compare the free energy changes [1] in kJ mol^{-1} for the following processes at 1500 K:

$$1/2\ Si_3N_4(s) \rightarrow 3/2\ Si(g) + 2\ N(g) \qquad 1228$$

$$SiO_2(s) \rightarrow Si(g) + 2\ O(g) \qquad 1183$$

$$1/2\ Si_3N_4(s) \rightarrow 3/2\ Si(s) + N_2(g) \qquad 125$$

$$SiO_2(s) \rightarrow Si(s) + O_2(g) \qquad 644$$

A result of such thermodynamic properties is that generally nitrides are refractory in the sense of being difficult to sinter and to crystallize, yet at the same time they have high dissociation pressures. Nitride chemistry is also made more difficult by the fact that many nitrides readily react with water (or moist air) ultimately to form hydroxides and ammonia. It is almost certainly true that many of the "nitrides" reported in the older literature contain significant amounts of O and/or H.

Despite the above remarks, it is our belief that solid state nitrides represent a rich lode of fascinating compounds that remains largely unexploited. Many known nitrides have novel compositions and coordination environments. The following review of their crystal chemistry emphasizes this point and suggests possible compositions for likely new nitrides.

Past reviews of nitride crystal chemistry [2, 3] described cation coordination and some aspects of crystal growth techniques. More recent developments suggest that it is now timely to take a more in-depth review of known structures and to examine their crystal chemistry.

We have included not only pure nitride materials but also many amides, azides, oxynitrides (but not nitrates and nitrites), nitride chalcogenides, and nitride halides. Specifically excluded are polymeric compounds (e.g. S_xN_y), cluster compounds (e.g. Sc_2Cl_2N), coordination compounds (e.g. [Pd(NS$_3$)$_2$]), ternary azides and amides, as well as ternary "interstitial" transition-metal nitrides. We have attempted to include all important structures and all type compounds. This review is also unique in that we attempt to evaluate the reliability of structure determinations using crystal–chemical indicators. For some compounds we also use some new descriptions to clarify their structures previously considered rather complex.

2 Methods

We prefer an empirical approach to bonding as opposed to any complex theories of binding [4] or bonding [5, 6, 7]. The bond valence method for determining the strength of bonds in crystals has developed from Pauling's concept of bond strength [8] and recently has been documented extensively [9, 10, 11].

An individual bond valence for bonds between atoms i and j is taken to be

$$v_{ij} = \exp\left\{\frac{R_{ij} - d}{b}\right\} \tag{1}$$

where v_{ij} is an individual bond valence, d is the bond length, b is a constant taken as 0.37 Å after Brown [12], and R_{ij} is the bond valence parameter for bonds between i and j atoms. The individual bond valences should sum to the actual atomic valence:

$$V_i = \Sigma_j v_{ij} \tag{2}$$

This method is preferred over any sum-of-radii method [13, 14], since the R_{ij} parameters for all atom pairs are either known or can be estimated [15, 16], and since radii do not allow for the variations in bond lengths within one coordination sphere which are often observed. Apparent atomic valences calculated from Eq. (2) that are very different from the expected valence are usually an indicator of an incorrect structure.

The Madelung potential at an atomic site in a crystal is approximately proportional to the formal charge q (in units of the electron charge e) of the ion in that site, that is

$$\phi = -\alpha q, \tag{3}$$

where α is roughly 12 V for oxides [17] and nitrides. More precisely α scales with the "size" of the atom (inversely with bond length). Even for materials for which the ionic model seems quite inappropriate, Madelung potentials can be a useful indicator of the validity of a structure [18]. For example, a warning flag for an incorrect structure would be like atoms having very different site potentials or, at an extreme, anions[1] having negative or cations having positive

---

[1] We use the terms "anion" and "cation" to mean the more electronegative and electropositive element, respectively and do not imply any specific applicability of the ionic model.

potentials. Our local program utilizes the Ewald method for spherical atoms [19].

3 Systematic Review

The bulk of structural information is tabulated; data for amides, azides, and transition metals have been tabulated separately in Tables 1, 2, and 9. Materials with common structure-types are listed separately for ease of comparison. The tables list composition, space group and relevant cell constants on the first line; the following entries are atomic positions, average bond length to atoms of different kind (e.g. cations to anions only), the sum of the individual bond valences, the charge used for Madelung potential calculations, and the actual site potential calculated in volts. For amides the N bond length average includes only bonds to cations other than H; the H bond length average includes only bonds to the nearest N atom. Similarly for azides, the central nitrogen is only bonded to its two end nitrogen atoms; only N–M bonds are averaged for the terminal N atoms.

The following discussion is broken down by chemical family, however, to display the lack of work on some systems. For reasons to become apparent later, it is appropriate to start with the more electropositive elements. Only those structures unique to nitrides are described and illustrated; the more common structure types can be found in standard texts [20, 21].

3.1 Alkali Metals

3.1.1 H

Because we consider amides and imides, it is appropriate to summarize structural data for some N–H binary compounds first.

Ammonia crystallizes with discrete molecules in approximate cubic eutaxy[2] as shown by a neutron diffraction study of ND_3 [22]. The N atoms have 6 N neighbors at 3.35 Å and 6 N neighbors at 3.88 Å which indicates distortion from true eutaxy. The near N–D distances are 1.01(2) Å, while the next D$\cdots$N distance is 2.37(3) Å; the D–N–D angles are 110(2)°.

Hydrogen azide (hydrazoic acid) has been studied in the gas phase by electron diffraction [23] and by rotational spectroscopy [24]. The molecule

---

[2] Eutaxy is a more appropriate term than "close-packing", since atoms in crystals are not actually hard spheres in contact. Eutaxy describes points in space that correspond to the centers of close-packed spheres.

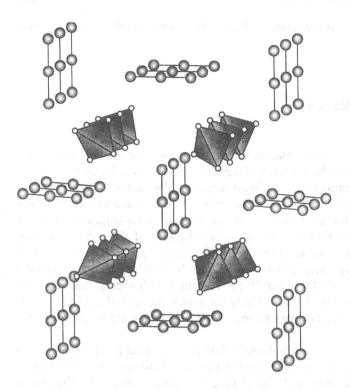

Fig. 1. Structure of NH$_4$N$_3$ in clinographic projection on (010). *Tetrahedra* represent NH$_4$ groups. The *linear chains* are azide ions

is planar but asymmetric (H–N–N–N = 0.98(2), 1.237(2), 1.133(2) Å) with a H–N–N angle of 114°. Molecular orbital calculations at the 631G** level [25] confirm these observations and suggest a N–N–N angle of 170° as opposed to the assumed 180° angle. It is suggested that the H–N–N bending removes the degeneracy of two nonbonding π molecular orbitals [26].

Ammonium azide contains isolated H$_4$N$^+$ ions in cubic eutaxy and linear N$_3^-$ ones as shown by neutron diffraction [27]. Strong N–H$\cdots$N hydrogen bonds (1.00–1.04, 1.94–1.98 Å) link these ions, and its structure is shown in Fig. 1. The H–N–H angles range from 106° to 110°. H$_5$N$_5$ contains staggered H$_5$N$_2^+$ ions and linear N$_3^-$ ones [28].

3.1.2 Li

Red lithium nitride, Li$_3$N, can be prepared from the elements at 400 °C and has a unique structure [29]. Li atoms link into six-membered rings in a planar fashion, as is in graphite and BN. The six-membered rings are centered by N atoms and the sheets are linked by Li atoms. Nitrogen is, therefore, 8-coordinated in a hexagonal bipyramidal arrangement as shown in Fig. 2.

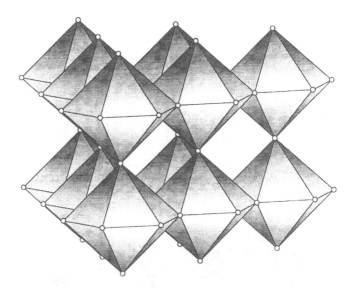

Fig. 2. Structure of α-Li$_3$N in clinographic projection on (100). The *circles* represent Li atoms, and N atoms center the hexagonal bipyramids

Li$_3$N is a very good ionic conductor. Li ion vacancies in the Li$_2$N planes have been suggested from careful X-ray diffraction experiments [30]; these vacancies undoubtedly provide a mechanism for conduction. A hopping model for conduction along [001] has been supported by ^{7}Li NMR data [31]. Large single-crystals can be pulled from melts using the Czochralski technique [32]. Nickel, cobalt, and copper can partially substitute for the lithium between the sheets at slightly higher temperatures (600 °C) [33].

In Li$_4$FeN$_2$, half the sites linking the Li$_2$N sheets of the Li$_3$N structure are filled by Fe atoms in two-coordination [34]. Li$_4$SrN$_2$ can be prepared at 700 °C, and its structure also resembles that of Li$_3$N; N atoms center pentagonal bipyramids in LiSrN$_2$ layers [35].

A high-pressure (β-) form of Li$_3$N is observed at 600 MPa in a diamond anvil cell and can be made in bulk at 4 GPa [36]. It crystallizes with the Na$_3$As structure and is, therefore, isotypic with Li$_3$P. Note that the substitution of larger atoms, either cations *or* anions, mimics the effect of pressure. In contrast to the Li$_2$N layers in the α-structure, the β-form contains LiN layers. The nitrogen atoms are shifted from simple hexagonal packing to hexagonal eutaxy ("close packing"), and the symmetry is changed (from *P6/mmm* to *P6$_3$/mmc*). N atoms center trigonal prisms which are capped on all 5 faces; alternatively, a NLi$_5$ trigonal bipyramid interpenetrates a NLi$_6$ trigonal prism. The low-pressure form is recovered at 200 °C and atm pressure. The β- structure is illustrated in Fig. 3.

Lithium azide, LiN$_3$, has the delafossite (CuFeO$_2$) structure (with N$_3$ replacing linear O–Cu–O groups); the metal atoms center octahedra [37]. The

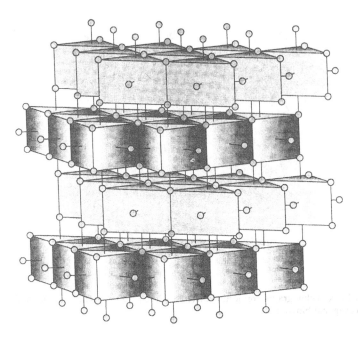

Fig. 3. Structure of β-Li$_3$N in clinographic projection on (100). N atoms center trigonal prisms which are capped on all faces. β-Li$_3$N is isostructural with Li$_3$P

azide ions center trigonal prisms and are directed towards the triangular-face centers. The Li-(N$_3$)$_6$ octahedra share edges to form CdCl$_2$-type layers which are linked by bridging azide ions. Figure 4 shows the structure in projection.

Lithium also forms an imide, Li$_2$(NH) as well as a beige nitride–hydride, The Li, N structure is that of sphalerite; the doubling of the c axis is undoubtedly a result of hydrogen bonding but the hydrogen atom positions were not located.

Lithium also forms an imide, Li$_2$(NH) as well as a beige nitride–hydride, Li$_4$NH [39]. It was assumed that the nitride–hydride formed with an antifluorite superstructure in a large tetragonal cell; however, our neutron diffraction studies [40] have not confirmed this behavior. It is formed from the nitride and hydride directly at 500 °C or the decomposition of the amide under vacuum. The imide also forms from decomposition of the amide and has the antifluorite structure [41] with a rotationally disordered NH^{2-} group.

By far the largest class of ternary lithium nitrides are those with the antifluorite structure, prepared largely by Juza, and reviewed by him [42]. The ternaries are more thermally stable than the binaries which would seem to indicate a relaxation of the internal strain (both cation–cation and anion–anion repulsions) of binary nitrides. Many of these compounds are in fact ordered superstructures of antifluorite and it is remarkable that most of the transition metals are observed in high oxidation states—much higher than those in the binaries (e.g. Li$_7$Mn^{5+}N$_4$ vs Mn$_3^{2+}$N$_2$). They are Li$^+$ conductors at elevated

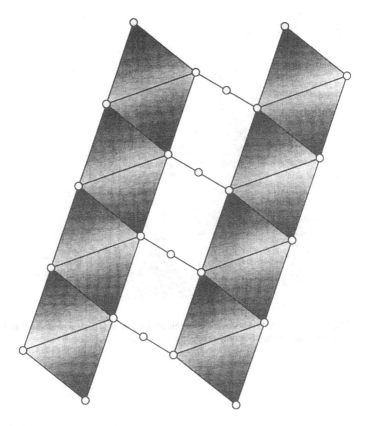

Fig. 4. Structure of LiN$_3$ projected on (010). *Small circles* represent N atoms. Slabs of edge-sharing, LiN$_6$ octahedra are linked by azide groups

temperatures, and many form solid solutions with lithium chalcogenides [43, 44]. The stoichiometric compounds include the following: LiMgN [45], LiZnN [45], Li$_3$AlN$_2$ [46], Li$_3$GaN$_2$ [46], Li$_3$FeN$_2$ [47], Li$_5$SiN$_3$ [48], Li$_5$GeN$_3$ [48], Li$_5$TiN$_3$ [48], Li$_7$MnN$_4$ [49], Li$_7$PN$_4$ [50], Li$_7$VN$_4$ [51], Li$_7$NbN$_4$ [51], Li$_7$TaN$_4$ [51], and Li$_9$CrN$_5$ [52]. The structures of the defective compounds Li$_6$MoN$_4$ [53], Li$_6$WN$_4$ [53], Li$_{14}$Cr$_2$N$_8$O [53], and Li$_{15}$Cr$_2$N$_9$ [53] also are based on the fluorite structure.

Another large class of ternary lithium nitrides contains those with the anti-La$_2$O$_3$ structure. The series Li$_2$MN$_2$ (M = Ce, Th, U, Zr, Hf) has been reported [54, 55]. The La$_2$O$_3$ structure has been the subject of much discussion because of the odd 7-coordination of the La atoms, but it is readily understood in terms of coordination of anions by cations [56]. In the antistructure nitrides, the regular octahedra and tetrahedra of *anions* are suitable sites for the M atoms and Li atoms, respectively. In contrast, the anion-centered description is more appropriate for U$_2$N$_2$X (X = As, S, Se) compounds with the Ce$_2$O$_2$S structure (Fig. 5) and with the U$_2$N$_2$Te structure (Fig. 6).

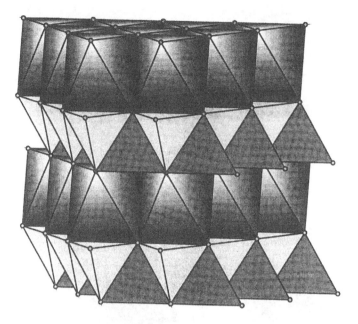

Fig. 5. Structure of U_2N_2S in clinographic projection on (100). *Circles* represent U atoms. N atoms center the tetrahedra. S atoms center the octahedra

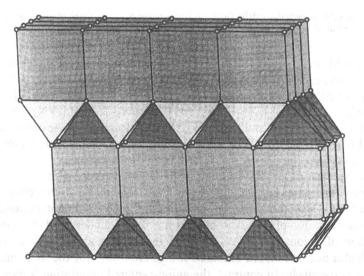

Fig. 6. Structure of U_2N_2Te in clinographic projection on (010). *Circles* represent U atoms. N atoms center the tetrahedra. Te atoms center the cubes

LiPN$_2$ forms with the filled cristobalite structure [57, 58]. The PN$_2^-$ array mimics that of the isoelectronic SiO$_2$, with the Li$^+$ ions in tetrahedral interstices. Alternatively, the structure may be considered a ternary derivative of the wurtzite structure (as exemplified by AlN).

Related wurtzite superstructures are found in LiSi$_2$N$_3$ [59] and LiGe$_2$N$_3$ [60]. Note that Li$_2$SiO$_3$ and Li$_2$SiO$_3$ are isostructural but that the structural rôles of Li and Si (Ge) are reversed.

Yet another wurtzite superstructure is found in gray α-LiSiNO whose structure has been determined using time-of-flight neutron diffraction on powdered samples [61]. There are two distinct ways of decorating the wurtzite structure to order the cations; LiSiXY may have either XLi$_3$Si and YSi$_3$Li tetrahedra or XSi$_2$Li$_2$ and YLi$_2$Si$_2$ tetrahedra available for the anions. The former (with $X = $ O and $Y = $ N) is known for the α-form, while the latter ordering is found in materials such as BeSiN$_2$ (see Fig. 7). α-LiSiNO is prepared by heating mixtures of Li$_4$SiO$_4$ and Si$_3$N$_4$ at 1150 °C in sealed tubes and 1050 °C under ammonia, while the β-form is prepared by quenching the α-form from 1450 °C [62]. The β-form crystallizes in a monoclinic system.

LiSrN [63] crystallizes with the tetragonal YCoC structure [64] which is displayed in Fig. 8. Strings of edge-sharing SrN$_4$ tetrahedra run parallel to [001]; the strings are bridged along [100] and [010] by linear NLiN groups. An equally pleasing description is afforded by considering anion-centered polyhedra. Chains of edge-sharing NLi$_4$Sr$_2$ octahedra run parallel to [100] and

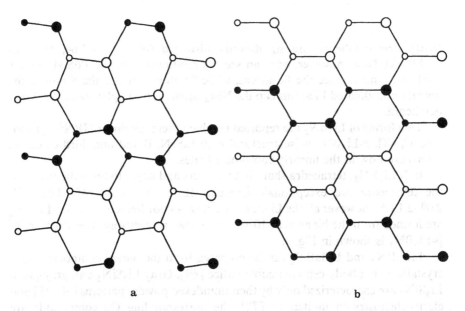

Fig. 7a. One layer of the α-LiSiON structure projected on (001). *Large, empty* and *filled* circles represent Li and Si atoms, respectively. *Small, empty* and *filled* circles represent O and N atoms, respectively. **b)** One layer of the BeSiN$_2$ structure projected on (001). Be atoms are now the *large empty circles*

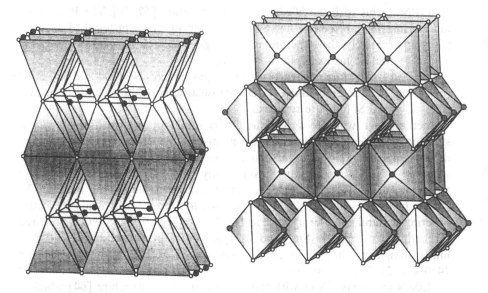

Fig. 8. Clinographic projection of the LiSrN structure on (100) emphasizing cation-centered polyhedra (**left**) and anion-centered polyhedra (**right**). Li atoms are represented by *large dark circles*. N or Sr atoms are represented by *small light circles*. Sr atoms center the tetrahedra (left), while N atoms center the octahedra (right)

[010]. The octahedra are significantly distorted (N − 4Sr = 2.64, N − 2Li = 1.96 Å). Now, however, one can see the relationship to the PdS structures [65]. N atoms replace the Pd atoms, while Sr atoms replace the S atoms; the linearly-coordinated Li atoms cap the NSr$_4$ squares to complete corner-sharing octahedra.

Two forms of Li$_3$BN$_2$ are reported in which there are linear NBN^{3-} groups [66, 67, 68]. α-Li$_3$BN$_2$ is isostructural with LiSrN; B and one third of the Li atoms are now in the linearly-coordinated sites.

In β-Li$_3$BN$_2$, tetrahedra share both corners and edges to form the network; the tetrahedra are exceptionally-irregular (Li − 4N = 1.98–2.45, 2.03–2.51, 2.09–2.19 Å), however all the Li atoms are now 4-coordinated. α- and β-Li$_3$BN$_2$ are formed from the binaries at 1070 and 1170 K, respectively. The structure of β-Li$_3$BN$_2$ is shown in Fig. 9.

LiCaBN$_2$ and LiBaBN$_2$ can be prepared from the binary components and crystallize on a body-centered cubic lattice [69]. Gray Li$_2$SiN$_2$ and gray-green Li$_8$SiN$_4$ are characterized only by their unindexed powder patterns [70, 71] and electrochemistry in molten Li [72]; the corresponding Ge compounds are similarly known [60]. Black LiBaN [73] and LiBeN [74] are prepared from the binaries and have been indexed with hexagonal and orthorhombic symmetry, respectively. LiCaN is also known with orthorhombic symmetry [35].

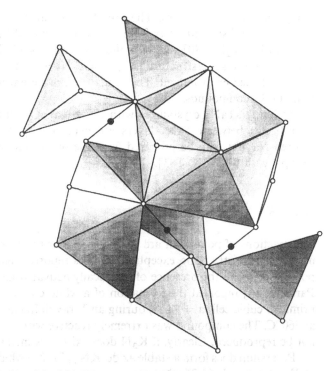

Fig. 9. Clinographic projection of the β-Li₃BN₂ structure on (100). Li atoms center the tetrahedra. *Small light circles* represent N atoms. *Large dark circles* represent B atoms in linear coordination

3.1.3 Na

Sodium metal reportedly reacts with electrically-activated nitrogen at low pressures to yield a nitride whose color is dependent upon temperature [75, 76] (orange at −180 °C vermillion at room temperature, black at 300 °C). Extended reaction times lead to the azide. Its density is reported as 1.7 g/cm³, while those for Na metal and NaN₃ are 1.0 g/cm³ and 1.9 g/cm³, respectively. We suspect that the composition Na₃N actually consists of a mixture of azide and unreacted metal or is some other intermediate along the path to azide formation, since no one has been able to isolate this "nitride".

Sodium does form two binary azides; α-NaN₃ transforms to β-NaN₃ at 18 °C [37]. The higher-temperature (β-) form has been studied by neutron diffraction [77]. The β-structure is essentially cubic eutaxy of both the Na ions and the centers of the azide ions; the Na atoms center octahedra. β-NaN₃, therefore, has the NaHF₂ structure. α-NaN₃ distorts from β-NaN₃ by a shear of the layers; the azide centers and the sodium ions remain in cubic eutaxy. α-NaN₃ is isostructural with LiN₃ and, therefore, has the delafossite (CuFeO₂) structure.

Yellow sodium amide is prepared by reacting molten sodium with ammonia at 300 °C; its structure consists of NH₂⁻ ions in cubic eutaxy with the Na⁺ ions filling the tetrahedral holes. The four tetrahedral Na⁺ ions surrounding the NH₂⁻ ion are displaced away from the positive side of the dipole [78, 79].

Gray NaGe₂N₃ [80] and the white oxynitrides NaGeON and NaSiON form with a wurtzite superstructure [81]. NaGe₂N₃ was the first example of Na in a fully nitrided environment. The structures are similar to those of the corresponding Li compounds.

Yellow NaTaN₂ crystallizes with the α-NaFeO₂ structure [82]; the octahedral sites between TaN₂ layers are occupied by Na atoms. Dark-honey colored Na₃BN₂ can be prepared from the binaries at 1000 °C at 4 GPa; it is isotypic with β-Li₃BN₂ [83].

3.1.4 K

The reactions of potassium are generally similar to those of sodium, and their nitrogen chemistry is no exception. K₃N is reported again as a fleeting compound formed in the presence of electrically activated nitrogen [75]. However, Parsons [84] reported the formation of a white compound which has a large primitive cubic cell (a = 9.6 Å) during an X-ray diffraction experiment on KN₃ at 360 °C. The compound was extremely reactive with air, and his results could not be reproduced. Clearly, if K₃N does exist, it is metastable.

Potassium does form a stable azide, KN₃ [77, 85] which crystallizes with the KHF₂ structure [86]. The N₃⁻ groups centering alternate cubes are skew, so that looking down the unique axis of the tetragonal cell, $\bar{4}$ symmetry is apparent and the KN₈ coordination polyhedron is a square antiprism. The structure is illustrated in Fig. 10.

The amide of potassium, KNH₂, has been known for considerable time [87], and its structure has been explored at temperatures from 31 K to 353 K [88, 89, 90, 91]. Above 347 K, KNH₂ has the rocksalt structure with a freely-

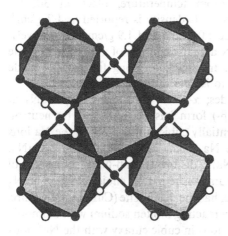

Fig. 10. Structure of KN₃ projected on (001). *Filled* and *open* circles represent N atoms at z = 0 and z = 1/2. K atoms at z = 1/4 and z = 3/4 center the square antiprisms

rotating amide group (γ), between 326 K and 347 K a tetragonal distortion of rocksalt (β), and below 326 K a further distortion to monoclinic symmetry (α).

The yellow oxynitride KGeON crystallizes with the wurtzite structure [92]. Unlike LiSiON, however, KGeON forms with only a single polytype; the tetrahedron around Ge containing 1 oxygen and 3 nitrogen atoms. The K atom actually forms 6 bonds (K–N]2.81, K–5O = 2.63–3.05 Å). This increasing distortion from LiSiON to KGeON suggests that Rb and Cs would not form oxynitrides with wurtzite superstructures.

$K_3Co_2N_3$ can be prepared from $Co_2(NH_2)_3(NHK)_3$ with loss of ammonia [93]. It is steel-gray to black, pyrophoric and crystalline. The X-ray diffraction pattern was not indexed. $KOsO_3N$ has the scheelite ($CaWO_4$) structure [94].

3.1.5 Rb

Rubidium forms no binary nitride. Like potassium, its amide forms with the cubic rocksalt structure (above 40 °C) which distorts to monoclinic symmetry as it is cooled [88, 95]. Its azide forms with the KHF_2 structure, analogous to NaN_3 and KN_3 [77, 85].

3.1.6 Cs

Cesium also forms no binary nitride. Its azide forms with the KHF_2 structure [85], and its low-temperature amide (below 50 °C) [95] forms with a tetragonal distortion of the CsCl structure. The high-temperature form (above 35 °C) of $CsNH_2$ has the cubic CsCl structure with a rotationally-disordered NH_2^- group. The hydrogen atoms have not been located for the amides. $CsTaN_2$ has been reported with a disordered structure resembling β-cristobalite [82]. Yellow $CsOsO_3N$ has the $BaSO_4$ structure [96].

3.2 Alkaline Earth Metals

3.2.1 Be

Be_3N_2 is polymorphic with a transition temperature around 1400 °C [97]. The structure of white α-Be_3N_2 is antibixbyite in which all the Be atoms are tetrahedrally-coordinated, and the nitrogen atoms center octahedra [98]. The high-temperature form is based on close-packed layers of Be in the sequence *cch*; meanwhile, the nitrogen atoms are packed *ch* and fill two-thirds of the octahedral holes [97, 99]. This high-temperature compound, as reported by Eckerlin, contained trace amounts of Si; whether the presence of Si is an integral part of the β form is unknown.

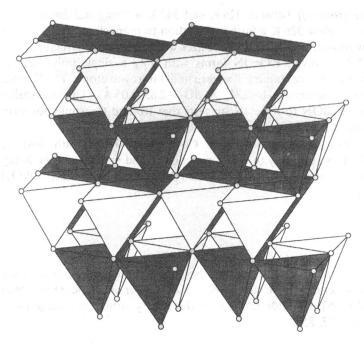

Fig. 11. Structure of BeSiN$_2$ in clinographic projection on (010). Be and Si center the light- and dark-colored tetrahedra, respectively

Colorless BeSiN$_2$ is a stoichiometric compound which crystallizes with a wurtzite superstructure [100]; each nitrogen atom is surrounded by two Be and two Si atoms as shown in Fig. 11; tetrahedra filled with like atoms form zigzag chains. Electron microscopy revealed coherent intergrowths of β-Be$_3$N$_2$ and BeSiN$_2$ [101].

BeThN$_2$ forms from reaction of the binary nitrides and crystallizes on an oddly-shaped hexagonal cell (10.501 × 3.955 Å) [54]. We [40] collected X-ray and neutron diffraction data from a powdered sample in hopes of locating the Th atoms using the X-ray data and the light atoms using the neutron data. The refinements did not proceed directly to a solution, although we believe there to be BeN$_4$ tetrahedra, ThN$_6$ trigonal prisms and ThN$_8$ cubes.

Beryllium amide is formed under supercritical ammonia [102], and its structure has been determined [103]. It crystallizes with the ZnI$_2$ structure. Topologically, the Be atoms define two unconnected but interpenetrating D4 nets. (The D4 net is the diamond net decorated with a tetrahedron at each lattice point.) Therefore, each diamond lattice point centers a group of four BeN$_4$ tetrahedra inside a larger tetrahedron. This is illustrated in Figs. 12 and 13. The hydrogen atoms were located from the X-ray diffraction data, and they are involved in hydrogen bonding between the nets. The H–N next-closest distances are 2.85–2.96 Å (compared with 2.94–3.15 Å in NH$_4$N$_3$ and 3.35 Å in H$_3$N).

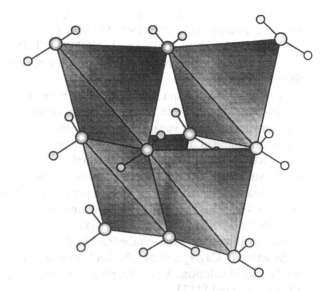

Fig. 12. Structure of Be(NH$_2$)$_2$. Be atoms center the tetrahedra. *Large circles* represent N atoms. *Small circles* represent H atoms. Notice the empty octahedron in the center

Fig. 13. Structure of Be(NH$_2$)$_2$. The large tetrahedra are made up of the groups shown in Fig. 12. The topology is that of two interpenetrating diamond nets

3.2.2 Mg

The nitride chemistry of magnesium is remarkably similar to that of beryllium. Olive Mg$_3$N$_2$ has the antibixbyite structure [104, 105], and the amide of Mg is formed at elevated pressures [102]. The amide is isostructural with that of Be discussed above [106].

Several hygroscopic compounds of magnesium form with the wurtzite structure; among these are violet MgSiN$_2$ [107], gray MgGeN$_2$ [108, 109], yellow-gray Mg$_2$PN$_3$ [110], and MgAlON [111]. The nitrides are ordered, but only unit cell and spacegroup information were discerned about MgAlON from electron diffraction data.

A series of magnesium nitride fluorides have been characterized [112]. The structure of Mg$_3$NF$_3$ can be described as rocksalt with every fourth cation position vacant. These vacancies are ordered, as are the anions, such that each Mg is surrounded by 4 fluorine and 2 nitrogen atoms. The structure of L-Mg$_2$NF (the lower-pressure form) is halfway between the rocksalt and zincblende structures. L-Mg$_2$NF crystallizes in the space group I4$_1$/amit. The 5-coordination around the Mg atom is capped by a third fluorine 0.7 Å further away (2.1 vs 2.8 Å). The yellow, high-pressure form, H-Mg$_2$NF, has the rocksalt structure (cubic eutaxy of the Mg atoms and equal amounts of N and F atoms disordered throughout the octahedral sites).

Reaction of Ca$_3$N$_2$ and Mg$_3$N$_2$ leads to a unique, violet compound and not merely a solid solution. A powder pattern of Mg$_3$Ca$_3$N$_4$ was collected but has not been indexed [113].

3.2.3 Ca

The normal, brown calcium nitride (α-Ca$_3$N$_2$) also forms with the antibixbyite structure [114]. However powder patterns for black hexagonal β-Ca$_3$N$_2$ and yellow orthorhombic γ-Ca$_3$N$_2$ have been reported [115].

A reddish-brown tetragonal Ca$_{11}$N$_8$ has also been claimed [116]; at first glance, one might guess that it is an oxynitride (e.g. Ca$_{11}$N$_6$O$_2$). The refinement of the X-ray data was poor, and since the compound is extremely air-sensitive, the truth is hard to ascertain.

Greenish-black hexagonal Ca$_2$N has been reported with the anti-CdCl$_2$ structure [117]. However, the crystal was lost during measurement, and the lack of an absorption correction renders the structure determination of this compound controversial. It is a poor conductor ($\sigma = 20$ Sm^{-1}). It has been suggested that there are hydrogen atoms between the Ca$_2$N layers [118], but compare Sr$_2$N discussed below.

Red Ca$_2$NCl and brown Ca$_2$NBr crystallize with stuffed CdCl$_2$-type layers (anti- α-NaFeO$_2$ structure, itself a relative of rocksalt) [119]. Bright-yellow Ca$_2$NF has the rocksalt structure (disordered anions) [120, 121], and no structural information is known about black Ca$_2$NI [122]. A nitride-hydride Ca$_2$NH is reportedly based on a disordered rocksalt structure as well, although the refinement of the neutron data required placement of H atoms on partially-occupied general positions [123].

Black "CaN" (which is likely the imide) has been reported with the rocksalt structure [124]. Accordingly, Ca(NH) has the rocksalt structure [125].

Calcium forms numerous ternary nitrides. CaGeN$_2$ crystallizes with the stuffed β-cristobalite structure [126]. Gray CaSiN$_2$ has been synthesized but not analyzed with diffraction techniques [127], although it is likely isostructural.

CaGaN has the NbCrN structure [128] which is unique to nitrides. The structure contains Ga–N–Ca–N–Ga layers; CaN$_5$ and inverted CaN$_5$ square pyramids share edges to build up the layers (see Fig. 14). Each N atom is bonded to a Ga atom which points into the interlayer gap. This structure is reminiscent of the layered structure of LiOH in which OH groups point into the gap, but Li atoms center tetrahedral sites instead of the square pyramidal ones (see Fig. 15). Metal–metal bonding (akin to hydrogen bonding in the hydroxides) is evident from the four close Ga–Ga contacts of 2.83 Å. This metal–metal bonding is also suggested by the low Ga valence sum (0.9 using only Ga–N, 1.9 using Ga–N and Ga–Ga interactions).

Brown Ca$_2$ZnN$_2$ is insulating and diamagnetic [129]. The structure of Ca$_2$ZnN$_2$ is similar to that of CaGaN. Instead of terminal metal atoms in the interlayer gap, however, the layers are shifted by $\frac{1}{2}$ (a + b), so the metal atoms join two layers as shown in Fig. 16. Calcium atoms again center square pyramids, and the Zn atoms join the layers. A pleasing description is obtained

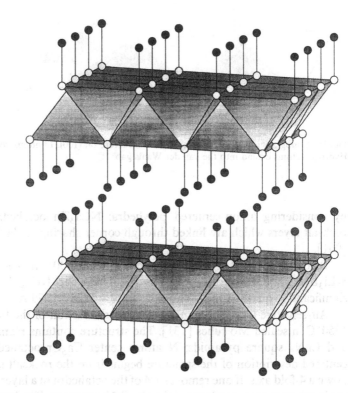

Fig. 14. Clinographic projection of the CaGaN structure on (100). Ca atoms center the square pyramids. Ga atoms are the tails extending into the interlayer gap. *Light circles* represent N atoms

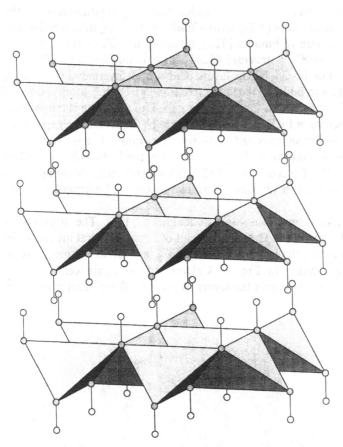

Fig. 15. Clinographic projection of the LiOH structure on (100). Li atoms center the tetrahedra. Hydrogen atoms extend into the van der Waals gap

by considering anion-centered polyhedra: NCa_5Zn octahedra form double rocksalt layers which are linked through corner-sharing of the Zn atoms along [001].

Black CaNiN is metallic and paramagnetic [130]. It is isostructural with α-Li_3BN_2 (and therefore YCoC and LiSrN). The NCa_4Ni_2 octahedra are significantly squashed (N − 4Ca = 2.50, N − 2Ga = 1.79 Å).

Air-sensitive Ca_3CrN_3 can be made from reaction of the binary nitrides at 1350 °C in sealed Mo tubes [131]. The structure contains planar CrN_3 groups and CaN_5 square pyramids; N atoms center Ca_5Cr octahedra. The anion-centered description of the structure begins from the rocksalt structure viewed down a 4-fold axis. If one removes 1/4 of the octahedra in a layer and then shears the layers, one obtains a layer of the Ca_3CrN_3 structure. The layers are joined by edge-sharing as shown in Fig. 17.

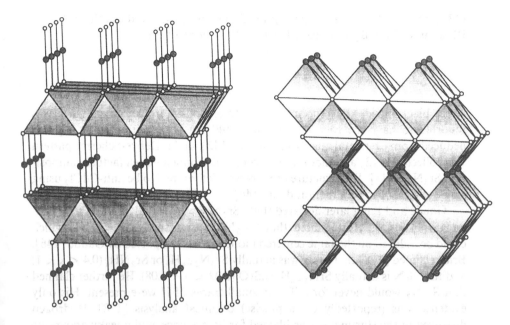

Fig. 16. Clinographic projection of the Ca_2ZnN_2 structure on (100) emphasizing cation-centered polyhedra (**left**) and anion-centered polyhedra (**right**). Zn atoms are represented by *large dark circles*. N or Ca atoms are represented by *small light circles*. Ca atoms center the square pyramids (left), while N atoms center the octahedra (right). Notice the similarity to the CaGaN structure and to rocksalt

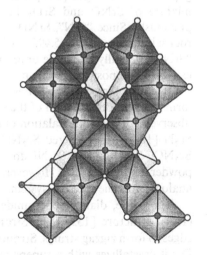

Fig. 17. Structure of Ca_3CrN_3 projected on (001). N atoms center the Ca_5Cr octahedra. The *larger circles* represent Ca atoms, while the *smaller circles* represent Cr atoms. *Shading* indicates displacement by c/2

Dark-green $CaTaO_2N$ has a rhombohedral structure based on perovskite [132]. Little structural data are available for the following compounds: red-brown, monoclinic $Ca_{21}Fe_3N_{17}$ and $Ca_{21}Ga_3N_{17}$ [133], black Ca_3N_4 [134, 135], green-brown Ca_4SiN_4 [136], brown, hexagonal Ca_5MoN_5 and Ca_5WN_5

[137], rose $Ca_5Si_2N_6$ [136], dark-gray $Ca_6Ga_2N_5$ [133], and $Ca(NH_2)_2$ [138]. IR studies of $Ca_3B_2N_4$ suggest linear NBN^{3-} ions [139].

3.2.4 Sr

Three black nitrides have been reported to crystallize from Sr metal. SrN was reported to have the rocksalt structure [140]. A composition of Sr_3N_2 was deduced from elemental analysis [138, 141, 142], and a Debye-Scherrer pattern was collected [122] and indexed on a centered monoclinic cell (actual composition Sr_5N_8) [143]. The structure of Sr_2N was determined to be anti-$CdCl_2$ using single-crystal X-ray diffraction data [144].

Gaudé and Lang later declared that "Sr_3N_2" was actually a mixture of SrN and Sr_2N [145]. Motte stated that "Sr_3N_2" was actually $Sr_3N_{2-x}H_y$ which could be indexed on the same centered monoclinic cell reported by Gaudé [146]. Brice claimed [118] that Sr_2N was actually $Sr_3N_{2-x}H_y$ or Sr_2NH_x ($0.4 < x < 1$) and that SrN is actually $SrN_{1-x}H_x \cdot nSrO$ ($0.02 < n < 0.08$). He further claimed that Sr_3N_2 would never form if even small traces of H were present. His only evidence was (reportedly quite precise) chemical analysis [118]. Hydrogen dissolved in the starting Sr was blamed for its presence, and a leaky apparatus was blamed for the oxygen contamination. Hulliger also questions the existence of the alkaline earth nitrides M_2N since they are reportedly not metallic conductors [147].

Neutron diffraction data on a pure sample of $Sr_{2.01(2)}N$ found the interlayer gap void of hydrogen [148]. Further its reaction with D_2 leads smoothly to a mixture of SrND and SrD_2 [149, 150] and not to Sr_2ND_x as suggested previously. Since "SrN", SrNH, and SrO_xN_y are all black materials with the rocksalt structure (a = 5.495, 5.46, and 5.473 Å, respectively), it is likely that "SrN" is actually the imide or an oxynitride solid solution. We believe that the expected composition Sr_3N_2 has yet to be observed as a pure phase.

By filling the octahedral sites between the layers of the $CdCl_2$ structure, one forms a slight deformation of the rocksalt structure (α-$NaFeO_2$). This has been observed with the intercalation of fluorine and hydrogen between the layers of Sr_2N [121, 151]. Ordered Sr_2NH is yellow, while the parent Sr_2N is black. Sr_2NCl, Sr_2NBr, and Sr_2NI do not appear to be cubic from their observed powder patterns [122]. However, similarity to stuffed $CdCl_2$ is expected by analogy with the calcium system.

The white, diamagnetic amide of strontium crystallizes with the anatase (TiO_2) structure [152, 153]. Strontium atoms center octahedra which share edges to form zigzag strings. Strontium imide forms from a reaction of Sr_2N and D_2; it crystallizes with a superstructure of rocksalt in which the ND group is rotationally disordered [149].

Strontium azide can be prepared from the carbonate in aqueous hydrazoic acid; its structure contains square antiprisms centered by Sr atoms [37, 152]. The structure is shown in Fig. 18 and is reminiscent of the structure of the alkali

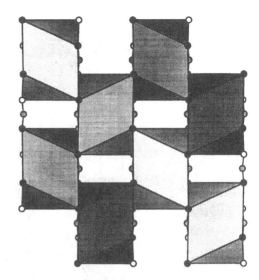

Fig. 18. Structure of Sr(N$_3$)$_2$ projected on (001). *Circles* represent N atoms. *Shading* of the faces of the square antiprisms indicates displacement by c/4

azides (see Fig. 10). However, the square faces of the antiprisms are now distorted to parallelograms (angle = 67°). In addition, the cations now lie on a diamond net.

SrTaO$_2$N crystallizes with a tetragonal distortion of the perovskite structure [154]. No atomic coordinates are known for black, orthorhombic Sr$_9$Re$_3$N$_{10}$ [155], monoclinic Sr$_{21}$Ga$_3$N$_{17}$ [133], or the cubic series of brown nitrides Sr$_{27}$M$_5$N$_{28}$ (M = Mo, W, Re) with a = 5.25 Å [137]. Only unindexed powder patterns are known for yellow Sr$_3$B$_2$N$_4$ [156] and Sr$_6$NI$_9$ [122]. Olive Sr$_3$N$_4$ [125, 157], yellow-green SrSiN$_2$ and reddish-brown Sr$_4$SiN$_4$ have not been characterized by diffraction techniques at all [158].

3.2.5 Ba

Just as the nitride chemistries of beryllium and magnesium are similar, so too are those of strontium and barium.

The black, stoichiometric nitride Ba$_3$N$_2$ is prepared directly from the elements or decomposition of the azide. It is poorly crystalline [159, 160]. We have been unable to successfully index a neutron diffraction pattern, since it exhibits broad diffraction peaks. Ba$_2$N with the anti-CdCl$_2$ structure is reported [161]. Since Ba$_2$NF [121], Ba$_2$NH [162], and Ba(NH) [125, 138, 163] are known to have the rocksalt structure, we again suggest that the other nitride halides Ba$_2$NX (X = Cl, Br, I) are also based on a stuffed anti-CdCl$_2$ structure [122, 164, 165].

The white amide of barium has been prepared in liquid ammonia solution over the period of 2 months. Its monoclinic structure assigns Ba to irregular 7 and 8-coordination. The hydrogen atoms were not found [166].

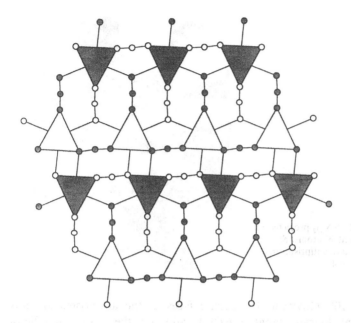

Fig. 19. Projection of the Ba(N$_3$)$_2$ structure on (010). Ba atoms center the tricapped trigonal prisms. *Shading* indicates displacement by b/2. Condensing the azide groups to a single site yields the PbFCl structure

Barium azide has a structure related to that of PbFCl [167]; the azide ions afford regular 9-coordination for the barium ions (see Fig. 19). The decomposition of Ba(N$_3$)$_2$ results in mixtures of Ba$_3$N$_2$ and Ba$_3$N$_4$ [168], however, the reported tetragonal cell of Ba$_3$N$_4$ does not match the observed powder pattern [134, 135]. The supernitride BaN$_2$ reported to be formed at high pressure is amorphous [169].

BaNiN is prepared by heating Ba$_3$N$_2$ in a Ni crucible under dry nitrogen at 1000 °C [170]. The structure contains N atoms coordinated by four Ba and two Ni atoms in distorted octahedra. Zigzag Ni-N chains run throughout the structure such that N(2) atoms have Ni atoms in *cis* positions (Ni-N-Ni = 83°) while N(1) atoms have Ni atoms in a *trans* position (Ni-N-Ni = 176°). Each "zig" or "zag" on the chain consists of a nearly linear N$_4$Ni$_3$ segment. Ba atoms center distorted N$_4$ tetrahedra. We have not found any relationship between this structure and any other simple structure. BaNiN shows Curie-Weiss behavior ($T_c = 310$ K, $\mu = 1$ μ_B) down to about 40 K. Bond valence analysis suggests a valence of less than 2 for Ba and more than 1 for Ni as assigned by the authors (recall similar behavior for CaNiN).

The structure of reddish-brown Ba$_8$Ni$_6$N$_7$ has recently been reported [171]. Again the nitrogen atoms center octahedra, and Ni atoms are linearly-coordinated. Ba atoms center tetrahedra and trigonal planes. Bond valence sums at the Ba positions average 1.4 while those at the Ni sites average 1.8 suggesting that the Ba atoms are not fully oxidized, similar to BaNiN.

The cubic perovskites BaNbO$_2$N and BaTaO$_2$N are prepared from binary oxides or carbonates, and they have been characterized by neutron diffraction [172]. Yellow, orthorhombic BaSiN$_2$ (spacegroup $Cmc*$) possibly has a superstructure of wurtzite [173]. The structures of tetragonal Ba$_3$Ge$_3$N$_6$ [174] and orthorhombic Ba$_9$M$_3$N$_{10}$ (M = Os, Re) [155] are not known. The diffraction pattern of Ba$_6$NI$_9$ is known but has not been indexed [165, 175]. Ba$_3$B$_2$N$_4$ is known by elemental analysis; IR measurements suggest linear NBN^{3-} ions [139].

3.3 Group III

3.3.1 B

White BN is isoelectronic with C, so it is not surprising that the most stable form of BN is that with a graphite-like structure [176]. As opposed to the graphite structure, however, the vertices of the hexagons in the layers of BN are not staggered along [001] but aligned. At extreme temperatures and pressures (1800 °C and 8.5 GPa), BN turns black and transforms to the sphalerite structure [177, 178, 179]. BN shocked to 55 GPa is light-gray and is reported with the wurtzite structure [180]. Careful studies of the phase transitions have been conducted [181]. Cubic BN is a hard material, just as carbon with the diamond structure, and has been useful in applications requiring hardness at elevated temperatures.

3.3.2 Al, Ga, In

Light-brown AlN, light-gray GaN, and dark-brown InN form with the wurtzite structure [182, 183]. The electrical conductivity increases from AlN to GaN to InN [184], in addition GaN and InN both show diamagnetic susceptibility (χ_M = −2.8 and −4.1 × 10^{-5}, respectively) [185]. In$_{32}$ON$_{17}$F$_{43}$ is known with an ordered fluorite superstructure as determined by neutron diffraction [186].

An important class of ceramic materials is the "sialons" of variable composition (e.g. Si$_{6-z}$Al$_z$O$_z$N$_{8-z}$ z = 0–4). To a first approximation, these can be considered as solid solutions of Al$_3$O$_3$N and Si$_3$N$_4$. The more common structures are variants of Si$_3$N$_4$ or AlN, and a common feature is the SiN$_4$ tetrahedron. Lanthanides are often incorporated to modify their structures and enhance their ceramic properties. These materials have been extensively reviewed elsewhere [187, 188, 189].

3.3.3 Tl

The lone pair elements (Tl and Pb) generally combine to form many unusual structures. Little nitrogen chemistry has been explored with Tl. TlN is unknown, and TlN$_3$ has the KHF$_2$ structure [77, 85].

3.4 Group IV

3.4.1 Si

β-Si$_3$N$_4$ has the phenacite (Be$_2$SiO$_4$) structure, although the c axis in phenacite is tripled, since the cations are ordered [190]. The α- and β- structures are shown in Figs. 20 and 21 for comparison. Hyde and Andersson describe this structure as triply-twinned *hcp* of the nitrogen atoms with the Si atoms filling the tetrahedral holes [21]. The choice of spacegroup $P6_3/m$ was found to be superior to the choice of $P6_3$ on the basis of convergent-beam electron diffraction experiments [191].

The other polymorph, α-Si$_3$N$_4$, contains the same local geometries as the β form [192]. However, alternate SiN$_4$ tetrahedra now cover the Si$_6$N$_6$ and Si$_4$N$_4$ channels from the β-structure. α-Si$_3$N$_4$ is a good insulator (conductivity $< 10^{-10}$ Sm^{-1}) [193]. ^{15}N MAS NMR can easily distinguish between the two polymorphs [194]. The α-form may be an oxynitride with up to 3% oxygen substituting for nitrogen [195, 196, 197].

Si$_2$N$_2$O [198] has the high pressure B$_2$O$_3$ structure. At high pressures ($P \leq 2.3$ GPa), the SiN$_3$O tetrahedra in Si$_2$N$_2$O rotate slightly, and the Si–O–Si angles decrease [199]. Similar behavior occurs at high temperature [200]. Si$_2$ON is reported with an orthorhombic unit cell [201].

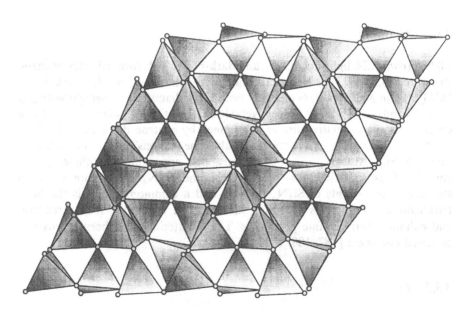

Fig. 20. Structure of α-Si$_3$N$_4$ tipped 10 degrees from [001]. Si atoms center the tetrahedra

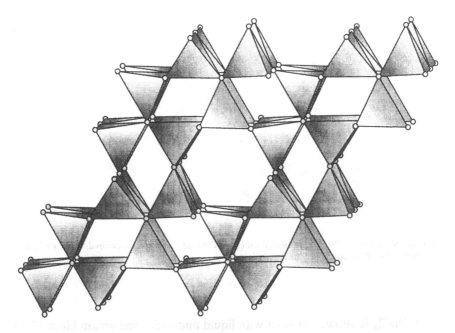

Fig. 21. Structure of β-Si₃N₄ tipped 10 degrees from [001]. Si atoms center the tetrahedra

3.4.2 Ge, Sn

α- and β-Ge₃N₄ are isostructural with their silicon analogues [202]. β-Ge₃N₄ is a poor conductor (10^{-6} Sm^{-1}) and is diamagnetic ($\chi_M = -9.1 \times 10^{-5}$) [203]. The oxynitride Ge₂N₂O is isostructural with Si₂N₂O as shown by neutron diffraction [204].

The structure of GeCr₃N is related to the perovskite structure; the octahedra are rotated as in the U₃Si-type structure [205].

Brown, amorphous SnNH was obtained from a reaction of SnCl₂ and KNH₂ in liquid ammonia. When heated above 340 °C in vacuum, Sn₃N₂ is obtained [206]. If SnCl₄ is reacted in liquid ammonia and the product is heated to 100 °C, white Sn(NH₂)₃Cl is produced. Further heating results in brown SnNCl (above 270 °C) and finally white Sn₃N₄ [207, 208].

3.4.3 Pb

Pb(N₃)₂ is explosive, and its structure contains 8-coordinated Pb atoms [209]. Figure 22 displays the structure in which Pb₂N₁₂ "dimers" share edges. The dimers are linked by edge-sharing and by bridging azide groups. Lead does not form a stable nitride.

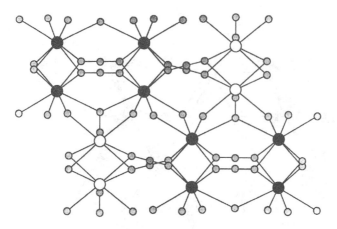

Fig. 22. Structure of Pb(N$_3$)$_2$ projected on (100). Pb(1) atoms are *light-colored*, while Pb(2) atoms are *dark-colored*. N atoms are *medium-colored*

If PbCl$_4$ is allowed to react with liquid ammonia, one obtain black PbNCl or Pb$_5$N$_6$Cl$_4$ depending upon the reaction time and ammonia excess. Both are explosive, the latter extremely so [208]. PbCl$_2$ and N$_2$ are the products of these explosions.

3.5 Group V

P$_3$N$_5$ can be prepared by simple reaction of the elements and is commercially available in high purity. It crystallizes with extremely fine particles which are amorphous to diffraction techniques. However, powder diffraction peaks were reported from a poorly-characterized sample of crystalline "P$_3$N$_5$" [210]. It is an ideal insulator for use in MIS devices in which InP is the semiconductor and can be applied by CVD techniques [211].

PON formed under high pressure is crystalline and has the α-quartz structure [212]. PON is isoelectronic with SiO$_2$.

Phospham or phosphonitrilic amide, PNNH, is a fascinating material also isoelectronic with SiO$_2$ [213]. Its formula suggests a relationship to LiPN$_2$ with the filled high cristobalite structure, but it is amorphous to X-ray analysis. Phospham has been used in the synthesis of phosphorus-containing nitrides [111]. There is an extensive literature on phosphonitrile polymers (for a review, see [214]).

Antimony and bismuth form the explosive nitrides SbN and BiN. These are prepared from reactions of their halides with KNH$_2$ in liquid ammonia [215, 216]. SbN forms via the intermediate Sb(NH)Cl [207].

3.6 Lanthanides and Actinides

Most of the lanthanide and actinide elements, as well as the group III transition metals, combine with nitrogen to form hard materials with the rocksalt structure, although some of them are slightly nitrogen deficient [217, 218, 219, 220]. Their magnetic behavior has been studied extensively, for example by neutron diffraction [221]. Brooks has calculated the electronic structures of the actinides with the rocksalt structure [222]. In addition, most lanthanides and actinides form oxynitrides [223, 224], and their ferromagnetic transition temperatures can thus be tailored [225, 226].

Lanthanides and actinides generally do not form azides.

Antiperovskites are known to form with rare earth elements, most notably the series M_3AlN (M = Ce, La, Nd, Pr, Sm [227]. Those with the actual perovskite structure are oxynitrides LaWO$_{0.6}$N$_{2.4}$ and NdWO$_{0.6}$N$_{2.4}$ [228] and have been the subject of some theoretical discussion [229]. The magnetic phase transitions can be explained by assuming that the Fermi level lies near a singularity in the electronic density of states [230]. XANES and heat capacity measurements confirm these magnetic transitions [231, 232].

Several other series are known with mineral structures, since isomorphous substitution with rare earth elements is common. Materials known include those with the apatite $\{Ca_5F(PO_4)_3\}$ structure (Ln, M)$_{10}$Si$_6$(O, N)$_{26}$ (Ln = La, Nd, Sm, Gd, Yb, Y; M = Ti, Ge, V) [233, 234], melilite $\{(Ca, Na)_2(Mg, Al, Si)_3O_7\}$ structure Ln$_2$Si$_3$O$_3$N$_4$ (Ln = Dy, Er, Gd, Ho, La, Nd, Sm, Y, Yb) [235], cuspidine $\{Ca_4Si_2O_7(F, OH)_2\}$ structure Ln$_4$Si$_2$O$_7$N$_2$ (Ln = Dy, Er, Gd, Ho, La, Nd, Sm, Y, Yb) [235], and scheelite $\{CaWO_4\}$ structure LnWO$_3$N (Ln = Nd, Sm, Gd, Dy) [236]. Also known are compounds with the CaFe$_2$O$_4$ structure BaCeLn(O, N)$_4$ (Ln = La, Ce) [237] and the K$_2$NiF$_4$ structure Ln$_2$AlO$_3$N (Ln = Ce, Eu, La, Nd, Pr, Sm) [238, 239].

Under high temperatures and pressures (3 MPa), some of the rare earths take up nitrogen to form dinitrides, LnN$_2$ (Ln = Ce, Nd, Pr), which have lattice constants similar to La$_2$O$_3$ (no true structural work has been done) [240]. Although the stoichiometry is 1:2 and thus isomorphism is impossible, hexagonal eutaxy of Ln is probable.

Uranium and thorium form a series of compounds with structures related to La$_2$O$_3$ (*hcp* metal atoms with alternate layers of tetrahedral and octahedral holes filled). The compounds M_2N$_2$X (M = U, Th; X = As, P, S, Se) have the Ce$_2$O$_2$S structure (La$_2$O$_3$ with S replacing O in octahedral sites) [241]; there are NM_4 tetrahedra and XM_6 octahedra. The compounds M_2N$_2$X (M = U, Th; X = Bi, Sb, Te) have the (Na, Bi)$_2$O$_2$Cl structure, which differs from that of La$_2$O$_3$ by a shift to tetragonal symmetry [242]. The metal atoms are now 8- instead of 7-coordinated (and the N and X atoms are now 4- and 8-coordinated). Th$_2$N$_2$O and Th$_2$N$_2$(NH) have the Ce$_2$O$_2$S structure [243, 244]; green Ce$_2$N$_2$O is also isostructural [55].

Despotoví'c and Ban reported the formation of U$_2$S$_2$N during a TGA experiment, and matched the calculated and observed powder diffraction pattern to arrive at a structural model [245]. The model uses the same spacegroup as that known for U_2SN_2 and uses partial occupancy of sites for nitrogen atoms which are the least sensitive to X-rays. We believe their stoichiometry and structure to be unreliable.

Uranium and thorium nitride–halides are related to the above. The series MNX (M = U, Th; X = Cl, Br, I) have the PbFCl structure (see Fig. 23) [246, 247, 248]. The corresponding fluorides have the LaOF structure [249, 250] which is an ordered superstructure of fluorite with a rhombohedral or tetragonal distortion [251, 252]. UNF and ThNF, therefore, crystallize with MN_4F_4 cubes while the anions center the tetrahedral sites (see Fig. 24). Yellow ThNH crystallizes on a face-centered cubic lattice [253].

The amides and imides of thorium can be formed at high pressures (400–600 MPa), although $Th(NH_2)_4$ also can be formed from $K_2Th(NO_3)_6$ in liquid ammonia [254]. $Th(NH_2)_4$ loses ammonia to form $Th_2(NH)_3(NH_2)_2$, bright-yellow $Th(NH_2)$, and Th_3N_4 at 50, 100, and 130 °C, successively [253]. $K_2Th_2(NH)_5$ is also formed which decomposes at 270 °C to form explosive, dark-yellow, amorphous $K_3Th_3N_5$. $Th(NH)_2$ crystallizes with the Th_3N_4 structure in which some of the Th sites are vacant. $Th_3N_2(NH)_3$ can be described as $Th_3N_4 + 2Th(NH)_2$. No structural data are known for brown $ThN(NH)_2$.

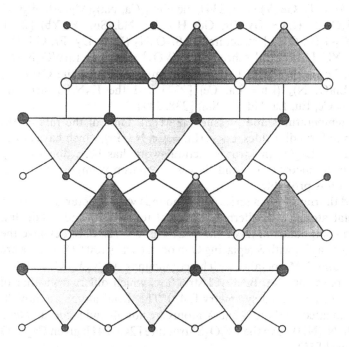

Fig. 23. Structure of UNCl with the PbFCl structure projected on (100). U atoms center tricapped trigonal prisms. *Large circles* represent Cl atoms. *Small circles* represent N atoms. *Light* and *dark shading* indicates displacement by a/2

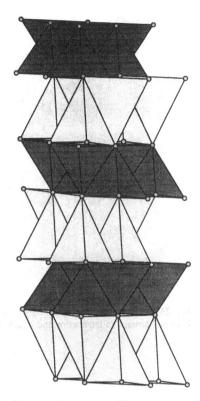

Fig. 24. Structure of ThNF in clinographic projection on (100). Th atoms are represented by *circles*. N atoms center the short, dark tetrahedra. F atoms center the stretched, light tetrahedra

The structure of brown Th_3N_4 is best described as *hhc* packing of Th atoms with nitrogen atoms in two layers of octahedral holes followed by a layer in the tetrahedral holes [255]. Figure 25 displays the structure.

At high pressures (34 GPa), UN transforms from cubic (rocksalt structure [256]) to hexagonal symmetry. At 10 MPa, UN_2 is known to form with the fluorite structure ($a = 5.31$ Å) [257]. U_2N_3 has the bixbyite structure [257]. UN is paramagnetic with a susceptibility maximum at 53 K. The higher nitrides are nonstoichiometric and have antiferromagnetic ordering temperatures from 94 K to 8 K ($UN_{1.55}$ to $UN_{1.80}$) [258].

Gray CeN_xF_{3-3x} and yellow-green PrN_xF_{3-3x} have the fluorite structure [259]. Orange $Eu(NH_2)_2$ and dark-brown $Yb(NH_2)_2$ have the anatase structure [260].

$Ce_{15}B_8N_{25}$ has a large, complicated structure [261]. Planar BN_3 groups are evident. Ce atoms are 8- and 9-coordinated by nitrogen, while N atoms are 5-, 6-, and 7-coordinated by Ce. The N atoms also form one strong bond to B atoms. Although $Ce_{15}B_8N_{25}$ has a rather high linear absorption coefficient (μ (MoKα) = 264 cm^{-1}), no absorption correction was made due to the irregularity of the faces.

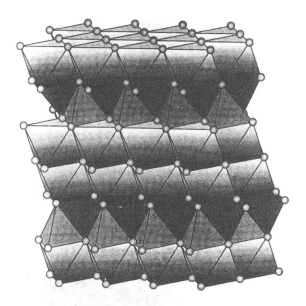

Fig. 25. Structure of Th$_3$N$_4$ in clinographic projection on (010). *Circles* represent Th atoms. N atoms center the tetrahedra and octahedra

La$_6$Cr$_{21}$N$_{23}$ crystallizes with a superstructure of rocksalt and is a superconductor below 2.7 K [262]. The partial occupancy of the nitrogen positions was fixed at the experimental stoichiometry. La$_2$U$_2$N$_5$ formed at high pressure reportedly has a structure similar to CsCl [263]. La(NH$_2$)$_3$ is formed under ammonia at 400 MPa and crystallizes with amide groups in approximate cubic eutaxy [264].

Sm$_3$Si$_6$N$_{11}$ is reported with a large tetragonal cell [265].

3.7 Transition Metals

The chemistry of the transition metal nitrides is made uncertain by their nonstoichiometry and affinity for incorporating oxygen. For example, eight crystallographically distinct forms of TaN$_x$ are reported which vary primarily in their oxygen and carbon content. As a class, the transition metal nitrides have been termed interstitial in nature, since their structures are generally based on eutaxy of metal atoms with nitrogen atoms partially filling the available sites. We, therefore, mainly discuss those stoichiometric nitrides with interesting structures.

Some of the transition metal nitrides with the rocksalt structure, especially those in groups V and VI, are well known for their "high temperature" superconducting transition temperatures [266, 267, 268, 269]. Band structure calculations have been used to predict transition temperatures [270, 271, 272]. For example, stoichiometric MoN was predicted to have a critical temperature of 30 K. (An unimpressive T_c since the developments in oxide super-conductors but still interesting).

These materials are generally black and hard; since thin films are easy to lay down [273], some are used as tool coatings [274]. These desirable mechanical properties have made transition metal nitrides the subject of several reviews [275, 276, 277] and books [278, 279].

3.7.1 Ti, Zr, Hf

ε-Ti$_2$N is known with the antirutile (anti-TiO$_2$) structure [280]. Neutron diffraction studies have also shown δ-Ti$_2$N to have an ordered structure with a unit cell twice the volume of the original rocksalt cell [281]. The ε-form seems to have a wider range of stoichiometry and is generally more deficient in nitrogen than the δ-form [282]. The expected nitride Ti$_3$N$_4$, is known as the product of Ti(NH$_2$)$_4 \cdot$ TiBr$_4 \cdot$ 8NH$_3$ + KNH$_2$ but no structural work was done [283]. Ti$_3$Al$_2$N$_2$ forms with partial filling of metals atoms in an $hhhcc$ packing sequence of N atoms [284]. Ti$_3$AlN has the antiperovskite structure [284]. The nitride halides of titanium have the FeOCl structure [285].

The decomposition products of zirconium halides heated under ammonia are a blue nitride with the rocksalt structure δ-Zr$_x$N (yellow at perfect stoichiometry) and a brown compound of composition Zr$_3$N$_4$; both are diamagnetic, but δ-ZrN is a metallic conductor (superconducting T_c = 10 K), whereas Zr$_3$N$_4$ is an insulator [286, 287]. Band structure calculations suggest ZrN–Zr$_3$N$_4$ would form a good Josephson junction [288].

The nitride halides of zirconium are polymorphic. The low-temperature forms of ZrNCl and ZrNBr, as well as ZrNI and ZrN(NH$_2$) have the FeOCl structure-type [285, 289, 290]. Above a few hundred degrees, ZrNCl and ZrNBr transform to an ordered CdCl$_2$-type structure [291, 292].

ZrN$_x$F$_{4-3x}$ (x = 0.91–0.94) crystallizes with a fluorite superstructure [293]. Only the subcell is given in Table 11, since there are 59 independent positions of variable occupation reported for the true cell.

Zr$_3$AlN is reported with the filled Re$_3$B structure [294]. AlZr$_8$ bicapped-trigonal prisms form columns along [100]. Squashed octahedra are filled with N atoms (Fig. 26). Since the N–M distances seem prohibitively short [N–Zr = 1.52 (2$\times$), 2.84 (4$\times$)], we find this structure implausible.

The nitrides Hf$_3$N$_4$, ε-Hf$_3$N$_2$, and ζ-Hf$_4$N$_3$ are known [295]. The latter two are characterized by eutaxy of metal atoms with the nitrogen atoms partially filling the octahedral sites; electron microscopy revealed stacking faults along {111} of the ζ-phase [296]. ε-Hf$_3$N$_2$ has hhc packing of the cations and ζ-Hf$_4$N$_3$ has $hhcc$ packing; for a further discussion of these structures see [56].

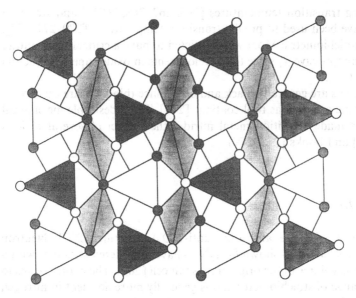

Fig. 26. Structure of Zr$_3$AlN projected on (100). *Large circles* represent Zr atoms. Al atoms center the bicapped trigonal prisms. N atoms center the squashed octahedra. *Shading* indicates displacement by a/2

3.7.2 V, Nb, Ta

β-V$_2$N is hexagonal, and its structure has been refined from neutron diffraction data [297]. It is isostructural with ε-Fe$_2$N, β-Nb$_2$N and β-Ta$_2$N. The structure consists of metal atoms in hexagonal eutaxy and N atoms in octahedral holes. Its formula is more appropriately written as N(1)$_2$N(2)V$_6$, since it is isostructural with Li$_2$ZrF$_6$. Although V is not a good neutron scatterer, the nitrogen content was unambiguously quantified.

V$_3$AsN and V$_3$PN have the filled Re$_3$B structure [298], while V$_5$P$_3$N has the filled Mn$_5$Si$_3$ structure [299].

The plethora of structures known in the niobium-nitrogen system appears to result from varying amounts of oxygen contamination [300, 301]. The phase diagram has been established [302, 303, 304, 305]. NbN, β-Nb$_2$N and γ-Nb$_4$N$_3$ are known from neutron diffraction experiments [306, 307], but the fully nitrided Nb$_3$N$_5$ has not been observed [308].

NbCrN and NbMoN are isostructural and have the NbCrN structure which was discussed under CaGaN. NbON has the baddelyite (ZrO$_2$) structure [309] in which monocapped, twisted NbN$_4$O$_3$ trigonal prisms are staggered to form zigzag layers parallel to (010). The simplicity of the structure is best appreciated by noting the regular ONb$_3$ and NNb$_4$ groups as shown in Figs. 27 and 28.

The nitrogen chemistry of Ta is similar to that of Nb; however, Ta seems to more readily incorporate nitrogen [310]. Several binary compounds have

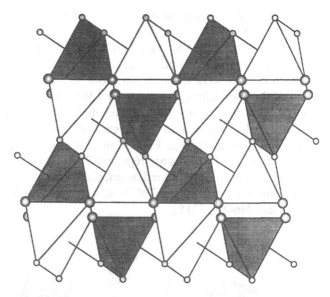

Fig. 27. Projection of the NbON structure on (010). *Dark circles* represent N atoms and *light circles* represent O atoms. Light and dark, Nb-centered monocapped trigonal prisms are displaced by b/2

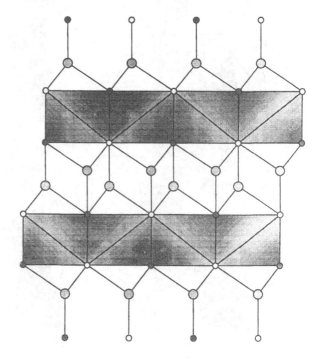

Fig. 28. Projection of NbON structure on (010) emphasizing the anion coordination. *Small circles* represent Nb atoms; light and dark circles are displaced by b/2. N atoms center the tetrahedra. O atoms center the Nb_3 triangles

a Ta:N ratio of greater than one, although many of these can only be prepared as thin films [311].

The most interesting tantalum nitride in the present context is the fully nitrided vermillion compound Ta_3N_5 which has the pseudobrookite (Fe_2TiO_5) structure. If it were truly isomorphous with high-temperature Ti_3O_5, Ta_3N_5 would be monoclinic as suggested [312]. Its structure has been refined as monoclinic and orthorhombic, and metrically it is nearly tetragonal [313]. Neutron diffraction data [314] confirm the spacegroup to be *Cmcm*. The structure is displayed in Fig. 29 and contains Ta atoms in highly irregular octahedra. The N atoms are regularly 3- and 4-coordinated as seen in Fig. 30. Notice that an anion-centered description of the Ta_3N_5 structure shows remarkable similarity to that of TaON which forms with the baddelyite structure (compare NbON) [315].

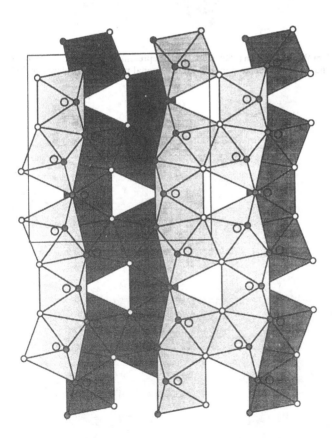

Fig. 29. Ta_3N_5 structure projected on (100). Ta atoms center the octahedra. *Small, light* and *dark circles* represent the 3- and 4- coordinated N atoms respectively

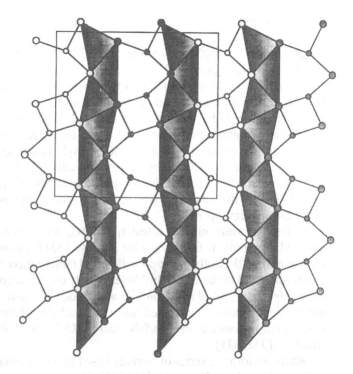

Fig. 30. Projection of the Ta$_3$N$_5$ structure on (100). *Large circles* represent N atoms, while *small circles* represent Ta atoms. *Dark* and *light circles* are displaced by a/2. N atoms center the shaded tetrahedra

3.7.3 Cr, Mo, W

Both CrN and Cr$_2$N can be formed from CrB heated under ammonia [316] as well as from the elements.

Both the Mo and W nitride systems exhibit tremendous polymorphism, although most are contaminated with oxygen. A partial Mo-N phase diagram has been constructed [317].

The W system is quite similar to the Mo one, although an even larger number of hexagonal supercells have been observed [318, 319]. Many of these can only be prepared as thin films and seen in the electron microscope.

The expected composition, WN$_2$, was observed as a brown coating on W filaments [320, 321, 322]. A composition near WN$_2$ (trigonal unit cell) has been observed in the electron microscope in the thinnest regions of nitrided films, but the derived structure contains three notably short N-N distances of 1.71 Å with eight corresponding W-N ones of 2.91-3.03 Å [323]. The suggested structure is reminiscent of that of fluorite, but we do not feel that this compound is very well established.

3.7.4 Mn, Re

Hägg mapped out a phase diagram for the Mn–N system [324]. Juza adjusted the phase limits slightly [325].

ε-Mn$_4$N crystallizes with the antiperovskite structure. It is ferrimagnetic (1.2 μ$_B$) and forms solid solutions with Cu and Zn [326, 327]. The magnetic structure of Mn$_4$N has been studied with polarized neutrons; the moments of the corner Mn atoms are antiparallel to the three face-center moments [328].

The antiferromagnetic structure of ξ-Mn$_2$N (isostructural with Mo$_2$C or ζ-Fe$_2$N) has also been studied by neutron diffraction [329]. Subsequent neutron diffraction experiments fixed the composition at Mn$_2$N$_{0.86}$ and confirmed the partial filling of the octahedral sites in the hexagonal eutactic array of Mn atoms using the spacegroup $P6_322$ [330].

A high pressure nitride, called η-Mn$_3$N$_2$, was formed from reactions of Mn and ammonia at 600 MPa after 30 days [331]. From an X-ray structure determination, the authors propose distorted cubic eutaxy of Mn atoms with N atoms filling 2/3 of the octahedral holes. Vastly disparate coordination environments for the 2 types of Mn atoms (2- and 5-coordination), the small number of independent reflections collected and the small size of the crystal render this determination questionable. MnSiN$_2$ and MnGeN$_2$ crystallize with the wurtzite structure [332, 333].

Rhenium forms a series of nitrides based on cubic eutaxy of the Re atoms with the formula ReN$_x$ ($x \leq 0.43$) [334]. N atoms center the octahedra. The small number of polymorphs, however, may only indicate the expense of obtaining Re metal.

3.7.5 Fe

Iron nitrides have been studied extensively, since they are found in many hardened steels [335]. It has been suggested that stoichiometric nitrides do not form, rather solid solutions of N in Fe are the norm [336]. Several intermediate phases have been observed during the martensite to austenite transformation, and many iron "nitrides" are actually carbonitrides [337]. The phase diagram was worked out by Hägg [338] and by Jack [339]. The iron nitrides are analogous to the Mn nitrides in that the general theme is hexagonal eutaxy of metal atoms with N atoms in the octahedral sites. Several of these nitrides (ζ, ε, γ') can also be formed by reacting iron borides with ammonia [316].

Nitrogen martensite (α'-Fe$_4$N) undergoes a structural change to a tetragonal superstructure (α'') upon quenching [335]. This material further decomposes to cubic γ'-Fe$_4$N. Carbon readily substitutes for N in Fe$_4$N [340]. Electron diffraction on γ'-Fe$_4$N suggested antiperovskite to be its structure [341]. Solid solutions, Fe$_{4-x}$Ni$_x$N ($0 \leq x \leq 3$), allow tailoring of ferromagnetic moments [342].

ε-Fe$_x$N ($2 < x < 4$) is known with hexagonal eutaxy of the Fe atoms (spacegroup $P312$) in positions $6l$ and the N atoms successively fill $1d$, $1e$, and $1a$ [339]; it is striking that the N atoms do actually order. FeNiN crystallizes in a primitive tetragonal cell with an ordered arrangement of the metal atoms [343]; the structure approximates NaCl with alternate (001) layers of Na missing. Black amorphous Fe$_3$N$_2$ reportedly forms from reactions of ferrous thiocyanate and potassium amide in liquid ammonia [344].

3.7.6 Co

Juza and Sachsze indexed γ-Co$_3$N on a hexagonal cell (2.66×4.35 Å), although faint lines suggest long-range N order [345]. Terao confirmed several phases in the electron microscope and identified Co$_4$N with the antiperovskite structure ($a = 3.74$ Å) [346]. CoN has only been observed as the decomposition product of Co(NH$_2$)$_3$ and is amorphous to X-rays [347]. Black Co$_3$N$_2$ was formed from blue cobalt amide in liquid ammonia; it does not react with water [206, 344].

3.7.7 Ni

Only two nickel nitrides have been studied with diffraction techniques. Ni$_3$N crystallizes with hexagonal eutaxy of the Ni atoms; the N atoms fill 1/3 of the octahedral holes [348] and is isostructural with ε-Fe$_3$N. Ni$_4$N has the antiperovskite structure [349].

Ni$_3$N$_2$ has been observed as the thermal decomposition product of nickel amide between 120 and 360 °C before it decomposes to Ni$_3$N (analogous to cobalt). Ni$_3$N$_2$ is amorphous to X-ray diffraction [87, 350]. It has also been prepared from a reaction of NiCl$_2$ and ammonia at 500 °C as shown by IR [351].

3.7.8 Cu, Ag, Au

The coinage metals are rich in azide chemistry. White CuN$_3$ crystallizes with Cu$^+$ in 2-coordination, while explosive greenish-black Cu(N$_3$)$_2$ crystallizes with Cu^{2+} in 4-coordination [352, 353]. The structure reported for CuN$_3$ is unlikely as the N–N distances in the azide group are 1.64 Å (normally 1.17 Å). The copper(II) azide contains nearly planar CuN$_{12}$ groups, and its structure is displayed in Fig. 31. AgN$_3$ is obtained as a fine white powder from an aqueous preparation and is isomorphous to KN$_3$ [354, 355]. Dry AgN$_3$ has been used as a primary explosive [356].

On the other hand, the nitride chemistry of the coinage metals is poor. Cu$_3$N is well known with the anti-ReO$_3$ structure [357]; however, Ag$_3$N and Au$_3$N are elusive. Hahn and Gilbert describe a method to make black, face-centered-cubic

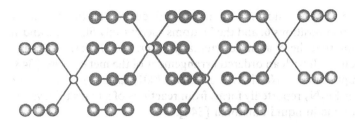

Fig. 31. Structure of Cu(N$_3$)$_2$ projected on (001). *Small light circles* represent Cu atoms. The azide groups above or below yield approximate 5-coordination for Cu

Ag$_3$N (a = 4.369 Å, density = 8.91 g cm^{-3}) [358]; however, their mild preparation using only aqueous AgCl, NH$_4$OH, and KOH, results in powerful explosions when the product is dry. We believe this explosive to be Ag$_3$N; possibly isostructural with Cu$_3$N. Since small explosions cannot easily propagate between isolated particles, its stability under water is understood. Au$_3$N has not been observed.

3.7.9 Zn, Cd, Hg

Zn$_3$N$_2$ crystallizes with the antibixbyite structure [359, 360]. It is black and diamagnetic. Cd$_3$N$_2$ and Hg$_3$N$_2$ can be prepared from their iodides by reaction with KNH$_2$ under liquid ammonia [87, 215]. The resultant black and chocolate-brown products, however, are extremely explosive when dry. For example, the decomposition of Cd$_3$N$_2$ into the elements releases 160 kJ mol^{-1} [360]. Cd$_3$N$_2$ has the antibixbyite structure but the structure of Hg$_3$N$_2$ is unknown [185, 360]. Their compositions have been confirmed from elemental analysis.

Yellow Zn(NH$_2$)$_2$ and Cd (NH$_2$)$_2$ are known [361]. Mercury azide crystallizes with metal atoms centering capped trigonal prisms [362]; a reported short N–N distance of 0.9 Å influences the Madelung potentials, particularly of N(6). ZnGeN$_2$ has the wurtzite structure as does ZnAlON [111, 363].

4 Discussion

The chemistry of nitrides provides some useful lessons in solid state inorganic chemistry. We focus here mainly on the "normal" (as opposed to metal-rich "interstitial") compounds.

4.1 Stability

A most striking observation (at first sight) is the instability of binary alkali metal nitrides other than Li$_3$N. However, this is a reflection of the importance of stoichiometry in influencing the structure and stability of solids [364]. Thus

Na$_3$N with Na in (say) four coordination would require N in twelve coordination by Na—an impossibly crowded situation for a large atom like Na. Similar considerations apply to the nitrides of the heavier alkaline earth nitrides which decompose at rather low temperatures. In contrast, by the time we reach groups III and IV, the metal:non-metal ratio is closer to one and stable ceramics such as AlN and Si$_3$N$_4$ are found.

The instability of the later transition element nitrides such as Co$_3$N$_2$ and Ni$_3$N$_2$ (compared to say the readily formed Mg$_3$N$_2$) is more readily understood in terms of the greater cohesive energy of the metallic elements. (The heat of sublimation of Mg at 0 K is 145 kJ mol^{-1}, while for Ni it is 428 kJ mol^{-1}).

We suspect that the lack of alkali nitrides on the one hand and of transitional metal nitrides on the other hand led to the neglect, until recently, of a very rich ternary nitride chemistry. We have recorded here some remarkable nitrides (mostly reported rather recently) containing transition metal and alkali metals. We suspect that some of them were made serendipitously, as crucibles of elements like iron are often used to contain alkali-rich materials (such as Li$_3$N). We observe that many ternary *etc.* oxides of elements such as Pt were similarly discovered as products of reactions in precious metal crucibles.

Equally striking, but understandable in the above terms, is the mutually exclusive relationship between nitrides and azides; only a few metals form both stable azides and nitrides (e.g. Li). This parallels the relationship between oxides and peroxides quite well: if the oxide is quite stable, the peroxide is generally unstable. The nitrides and azides of some elements are, of course, both unstable (e.g. Ag$_3$N and AgN$_3$).

The problem of synthesizing ternary nitrides is exacerbated by the requirement for high cation valences. Stable materials generally have more anions than cations [364], so cations with high charges are required to balance the relatively high -3 charge on nitrogen. While sufficient A^{+3} ions are available for a simple AN compound, materials such as ABN$_3$ require e.g. $(+4, +5)$ or $(+3, +6)$ charge pairs. It is notable that many recently-reported nitrides do not contain transition metals in their highest oxidation states (e.g. CaNiN, Ba$_8$Ni$_6$N$_7$) and one might expect that high pressure (high N$_2$ activity) syntheses will play an important rôle in the future.

4.2 Structure

Although there are similarities between nitride and oxide structures (e.g. Ta$_3$N$_5$ and Ti$_3$O$_5$ are isostructural as are Si$_3$N$_4$ and Be$_2$SiO$_4$) there are some important differences and we have given examples of structures that are (so far) unique to nitrides. Perhaps most striking are the low coordination numbers sometimes found. Examples include Fe(II), Zn, Ni and B in two coordination in Li$_4$FeN$_2$, Ca$_2$ZnN$_2$, CaNiN, BaNiN and Li$_3$BN$_2$; and Cr(III) in three coordination in Ca$_3$CrN$_3$. These can be understood in general terms as follows: The coordination number of N by metal atoms only exceeds six in Li-rich compounds. Nitrides with metal/N ratios greater than one therefore have metal coordination numbers less than six. A further generalization is that a given compound often forms with each (chemical) kind of atom in the same coordination environment when possible. Thus with coordination numbers as prefixed roman superscripts we have $^{iv}Ca_3{}^{vi}N_2$, $^{iii}Sr_2{}^{vi}N$, $^{ii}Li^{iv}Sr^{vi}N$, $^{iv}Ba^{ii}Ni^{vi}N$, $^{iii}Ba_2{}^{iv}Ba_6{}^{ii}Ni_6{}^{vi}N_7$, $^{v}Ca_2{}^{ii}Zn_2{}^{vi}N_2$ and $^{v}Ca_3{}^{iii}Cr^{vi}N_3$. The last four compounds illustrate the importance of stoichiometry (and the irrelevance of "radius ratio" rules) for determining the coordination numbers (compare ^{iii}Ba and ^{iv}Ba with ^{v}Ca). In contrast anion-rich compounds such as $^{vi}Ta_3{}^{iii}N_2{}^{iv}N_3$ and $^{vii}Nb^{iv}N^{iii}O$ have relatively high cation coordination numbers and low anion coordination numbers.

The generally larger metal:non-metal ratio in nitrides results in descriptions of structure in terms of anion-centered coordination polyhedra (e.g. NM_6 octahedra) being often more simple and useful than the more conventional cation-centered approach. We have given a number of examples (see Figs. 5, 6, 16, 17, 24, 28, 30).

Azides also present interesting analogies to other compounds with linear groups, and we have pointed out the structural similarities of some azides with oxides with O–Cu–O groups (delafossites) and with linear F–H–F groups.

The positions of H atoms in amides and imides have generally not been determined; as there are now both NH and NH_2 groups, the structural chemistry of these compounds should be at least as rich as the corresponding oxides (with only OH groups).

4.3 Bonding

Some nitride compositions provide interesting challenges to those who would attempt to understand (or to "explain") bonding in solids.

The reaction of the elements yields only Sr_2N rather than the expected Sr_3N_2, a compound which has not yet been convincingly demonstrated. The assignment of valences in compounds such as CaNiN and BaNiN is baffling, and prompts the remark that Ni is equally enigmatic in chalcogenides such as Ni_3S_2. On most electronegativity scales N and S have rather similar electronegativities. Certainly metal–metal bonding is more generally evident than in oxides of non-transition elements (see also e.g. CaGaN).

It is important to note that the concept of valence in solid materials is based on electroneutrality. The *valences* for azides found using bond valence parameters [365] generally are close to 5 for the central nitrogen and 3 for the end ones consistent with expected valences for nitrogen using the 8-n rule [366]. The actual charges on the nitrogen atoms are not presumed to be so large, indeed it has been suggested that charges of -0.8 and $+0.6$ for the azide nitrogen atoms are more appropriate [37, 367].

The crystal chemistry of azides also demonstrates the inadequacies of using ionic radii. The N–N bond length in azides is generally close to 1.17 Å; although the assigned radius of N^{3-} assumes a rather broad range (1.42–1.54 Å) [14], negative radii for the central N^{5+} "ions" result. Clearly, one should not regard eutactic arrays of nitrogen atoms as close-packed spheres of radius 1.5 Å. N ... N distances are rather large in most nitrides with metal/N ratios greater than one, but values as small as 2.65 Å are found in compounds such as Ta_3N_5 and UN_2.

We have listed expected bond lengths for cations in regular nitrogen environments in Table 12. These were derived from bond valence parameters [15] and should be of use for attempting isomorphic substitutions or for computer modelling of structures.

4.4 Physical Properties

Practical interest in nitrides is mainly focussed in their use as tough ceramics such as sialons and interstitial nitride coatings on metals [368, 369]. Silicon oxynitride glasses are of importance in fiber optics as their refractive indices depend on the nitrogen content. There are surely other interesting properties awaiting exploration.

One feature of nitrides is that they are frequently brightly colored (brilliant red Ta_3N_5 is particularly striking). This is presumably because "charge transfer" bands are in the visible region of the spectrum. Translated into the language of solid-state physics, this means that they have smaller band gaps than oxides. One might expect that useful semiconducting materials may well be found among them. Presumably their reactivity to atmospheric conditions will preclude their use as pigments. Other examples of colored non-transition metal nitrides are: Li_3N (red), LiCaN (orange), Mg_3N_2 (green), Ca_3N_2 (brown), $Ca_3Mg_3N_4$ (violet), $MgSiN_2$ (violet), AlN (brown). In fact, pure Si_3N_4 is a rare example of a colorless nitride.

Many new metallic compounds are being prepared. Their magnetic and electrical properties should provide a fruitful area of research.

Note added in proof

Recent papers attest to the current vitality of nitride chemistry. A theoretical study of the electronic properties of CaNiN has been published [370]. The "compounds" $Ca_{21}M_3N_{17}$ (M = Fe, Ga) of Table 11 have been shown [371] to be a mixture of Ca_3N_2 and Ca_6MN_5. In the latter there are planar MN_3 groups. The same planar group has been found [372] in Ba_3FeN_3 and Sr_3FeN_3. $CuTaN_2$ with the delafossite ($CuFeO_2$) structure has been made [373] by the exchange reaction of $NaTaN_2$ with CuI. A new lithium phosphonitride $Li_5P_2N_5$ has been characterized [374] with $[P_4N_{10}]^{10-}$ anions isostructural to P_4O_{10}. The structure of diamagnetic $Ni(NH_2)_2$ has been determined [375] to contain Ni_6 octahedral clusters with N bonded to two Ni atoms on each edge of the octahedron (thus the cluster is isostructural to that in β-$PtCl_2$). In this work the existence of (amorphous) Ni_3N_2 as a decomposition product of the amide was confirmed. The crystal structure of Mg_3BN_3 (containing linear BN_2 groups) has been determined [376]. Refinement [377] of the structure of Si_2N_2O results in bond lengths that give bond valence sums closer to the ideal values than those recorded in Table 10.

Acknowledgements. This material is based upon work supported under a National Science Foundation Graduate Fellowship to NEB and is part of a continuing program in crystal chemistry supported by the National Science Foundation (DMR 8813524).

5 Tables

In the following tables we give following the compound name, the space group and unique lattice parameters in the order a, b, c, α, β, γ (Å and deg.). x, y, and z are the coordinates of the atoms (with origin at a center in centrosymmetric crystals). d refers to average bond lengths as indicated in each table. Σv_i is the bond valence sum at each atom, V is the expected valence and the charge (in units of e) used to calculate the Madelung potential ϕ_i at each atom site.

Table 1. Structural data for selected amides. N–M and H–N distances are the only ones included in the N and H average bond lengths, respectively

Atom	x	y	z	d	Σv_i	V	ϕ_i
$Ba(NH_2)_2$	Cc	8.951	12.67 7.037	123.5			
Ba(1)	0	0.68670	0	8⟨2.98⟩	2.2	2.0	−14.67
Ba(2)	0.25735	0.10778	0.3484	7⟨2.93⟩	2.1	2.0	−14.55
N(1)	0.021	0.4943	0.245	4⟨2.97⟩	1.1	−1.0	9.03
N(2)	0.377	0.3039	0.603	4⟨2.95⟩	1.1	−1.0	−9.02
N(3)	0.114	0.0850	0.624	4⟨2.99⟩	1.1	−1.0	9.02
N(4)	0.217	0.2618	0.013	3⟨2.92⟩	0.9	−1.0	7.85
$Be(NH_2)_2$	$I4_1/acd$	10.17	16.137				
Be	0.3762	0.3218	0.0609	4⟨1.75⟩	2.1	2.0	−31.51
H(1)	0.008	0.007	0.410	1⟨0.86⟩	1.6	1.0	−28.83
H(2)	0.098	0.022	0.346	1⟨0.83⟩	1.7	1.0	−30.33
H(3)	0.055	0.212	0.219	1⟨0.86⟩	1.6	1.0	−28.58
H(4)	0.308	0.040	0.285	1⟨0.86⟩	1.6	1.0	−28.24
N(1)	0.0458	0.0616	0.3770	2⟨1.74⟩	4.3	−3.0	47.44
N(2)	0	1/4	0.2518	2⟨1.74⟩	4.2	−3.0	46.84
N(3)	0.3573	0	1/4	2⟨1.77⟩	4.1	−3.0	48.23
α-$CsNH_2$	$P4/nmm$	5.641	4.194				
Cs	1/4	1/4	1/2	8⟨3.51⟩	0.6	1.0	−7.21
N	3/4	1/4	0	8⟨3.51⟩	0.6	−1.0	7.21
β-$CsNH_2$	$Pm\bar{3}m$	4.063					
Cs	1/2	1/2	1/2	8⟨3.52⟩	0.6	1.0	−7.21
N	0	0	0	8⟨3.52⟩	0.6	−1.0	7.21
$Eu(NH_2)_2$	$I4_1/amd$	5.450	10.861				
Eu	0	3/4	1/8	6⟨2.71⟩	1.7	2.0	−15.46
N	0	3/4	0.371	3⟨2.71⟩	0.8	−1.0	9.34

Crystal Chemistry of Inorganic Nitrides

Table 1. (*Contd.*)

α-KNH$_2$	$P2_1/m$	4.586	3.904	6.223	95.80			
K	0.228	1/4	0.295		6⟨3.07⟩	0.7	1.0	− 8.32
H	0.29	0.04	0.88		1⟨1.04⟩	1.0	1.0	− 23.10
N	0.289	1/4	0.778		6⟨3.07⟩	2.7	− 3.0	37.62
β-KNH$_2$	$P\bar{4}$	4.282	6.182					
K	0	1/2	0.202		6⟨3.06⟩	0.7	1.0	− 8.02
N	0	1/2	0.737		6⟨3.06⟩	0.7	− 1.0	− 8.20
γ-KNH$_2$	$Fm\bar{3}m$	6.109						
K	0	0	0		6⟨3.05⟩	0.7	1.0	− 8.24
N	0	0	1/2		6⟨3.05⟩	0.7	− 1.0	− 8.24
La(NH$_2$)$_3$	$P2_1/c$	6.794	11.55	10.41	106.5			
La(1)	0.61302	0.36288	0.13283		8⟨2.74⟩	2.9	3.0	− 22.63
La(2)	0.27189	0.56931	0.49182		8⟨2.69⟩	3.3	3.0	− 20.68
N(1)	0.0730	0.9582	0.1599		3⟨2.68⟩	1.3	− 1.0	11.46
N(2)	0.4076	0.4575	0.2924		3⟨2.72⟩	1.1	− 1.0	11.02
N(3)	0.2749	0.0510	0.4642		2⟨2.63⟩	0.9	− 1.0	7.83
N(4)	0.4014	0.1615	0.0632		3⟨2.74⟩	1.0	− 1.0	11.01
N(5)	0.0743	0.7277	0.3349		2⟨2.57⟩	1.1	− 1.0	8.96
N(6)	0.3456	0.7249	0.1151		3⟨2.85⟩	0.8	− 1.0	9.50
LiNH$_2$	$I\bar{4}$	5.016	10.22					
Li(1)	0	0	1/4		4⟨2.24⟩	0.8	1.0	− 10.34
Li(2)	0	1/2	0		4⟨2.14⟩	1.0	1.0	− 11.28
N	0.232	0.232	0.116		4⟨2.19⟩	0.9	− 1.0	10.74
Mg(NH$_2$)$_2$	$I4_1/acd$	10.37	20.15					
Mg	0.3734	0.3612	0.06307		4⟨2.08⟩	2.1	2.0	− 24.41
H(1)	0.236	0.061	0.271		1⟨0.92⟩	1.3	1.0	− 26.97
H(2)	0.058	0.201	0.239		1⟨0.87⟩	1.5	1.0	− 30.12
H(3)	0.229	0.177	0.104		1⟨0.89⟩	1.5	1.0	− 28.52
H(4)	0.209	0.285	0.148		1⟨0.94⟩	1.3	1.0	− 26.10
N(1)	0.2867	0	1/4		2⟨2.09⟩	3.7	− 3.0	44.17
N(2)	0	1/4	0.2574		2⟨2.08⟩	4.2	− 3.0	45.13
N(3)	0.0128	0.0229	0.3757		2⟨2.08⟩	3.8	− 3.0	43.74
NaND$_2$	$Fddd$	8.949	10.456	8.061				
Na	1/8	0.2702	1/8		4⟨2.47⟩	0.9	1.0	− 10.07
D	0.165	0.068	0.440		1⟨0.94⟩	1.3	1.0	− 26.63
N	1/8	1/8	0.3615		4⟨2.47⟩	3.5	− 3.0	41.26
α-RbNH$_2$	$P2_1/m$	4.850	4.148	6.402	97.80			
Rb	0.203	1/4	0.295		6⟨3.23⟩	0.6	1.0	− 7.47
N	0.28	1/4	0.79		6⟨3.23⟩	0.6	− 1.0	7.47
RbNH$_2$	$Fm\bar{3}m$	6.39500						
Rb	0	0	0		6⟨3.20⟩	0.6	1.0	− 7.87
N	0	0	1/2		6⟨3.20⟩	0.6	− 1.0	7.87
Sr(NH$_2$)$_2$	$I4_1/amd$	5.451	10.91					
Sr	0	3/4	1/8		6⟨2.69⟩	1.7	2.0	− 11.65
H	0	0.16	0.70		1⟨0.86⟩	1.6	1.0	− 24.43
N	0	1/4	0.115		3⟨2.69⟩	4.0	− 3.0	44.08
Yb(NH$_2$)$_2$	$I4_1/amd$	5.193	10.419					
Yb	0	3/4	1/8		6⟨2.60⟩	1.6	2.0	− 16.10
N	0	3/4	3/8		3⟨2.60⟩	0.8	− 1.0	9.68
ZrN(NH$_2$)	$Pmmn$	4.21	3.43	7.52				
Zr	1/4	3/4	0.119		6⟨2.33⟩	3.7	4.0	− 35.59
N(1)	1/4	1/4	0.901		4⟨2.24⟩	3.0	− 3.0	37.84
N(2)	1/4	1/4	0.361		2⟨2.50⟩	0.7	− 1.0	7.25

Table 2. (Contd.)

Table 2. Structural data for selected azides. The average bond length for the center N in each azide group is only that to the two nearest N atoms

Atom	x	y	z		d	Σv_i	V	ϕ_i
AgN$_3$	Ibam	5.6170	5.9146	6.0057				
Ag	1/2	0	1/4		8⟨2.67⟩	0.9	1.0	− 12.56
N(1)	0	0	0		2⟨1.18⟩	4.2	5.0	− 68.87
N(2)	0.145	0.145	0		4⟨2.67⟩	2.6	− 3.0	48.31
Ba(N$_3$)$_2$	P2$_1$/m	9.63	4.41	5.43	99.57			
Ba	0.2181	1/4	0.1715		9⟨3.21⟩	2.7	2.0	− 21.15
N(1)	0.460	1/4	3/4		2⟨1.19⟩	4.2	5.0	− 62.46
N(2)	0.108	3/4	0.646		2⟨1.17⟩	4.6	5.0	− 74.33
N(3)	0.093	3/4	0.860		3⟨2.93⟩	3.0	− 3.0	40.79
N(4)	0.112	3/4	0.437		2⟨2.91⟩	3.1	− 3.0	45.36
N(5)	0.584	1/4	0.779		2⟨2.90⟩	2.9	− 3.0	52.89
N(6)	0.333	1/4	0.702		2⟨2.93⟩	2.6	− 3.0	50.20
CsN$_3$	I4/mcm		6.5412	8.0908				
Cs	0	0	1/4		8⟨3.28⟩	1.1	1.0	− 9.27
N(1)	1/2	0	0		2⟨1.17⟩	4.4	− 5.0	70.12
N(2)	0.374	0.126	0		4⟨3.28⟩	2.8	− 3.0	48.11
CuN$_3$	I4$_1$/a	8.653	5.594					
Cu	0	0	1/2		4⟨2.48⟩	0.5	1.0	− 10.34
N(1)	0	0	0		2⟨1.64⟩	1.6	5.0	− 58.20
N(2)	0.077	0.173	1/4		1⟨2.15⟩	1.9	− 3.0	26.73
Cu(N$_3$)$_2$	Pnma	13.454	3.079	9.084				
Cu	0.3967	1/4	0.5803		4⟨2.00⟩	1.4	2.0	− 27.80
N(1)	0.2541	3/4	0.4099		2⟨1.17⟩	4.4	5.0	− 67.92
N(2)	0.5307	3/4	0.7619		2⟨1.16⟩	4.7	5.0	− 69.46
N(3)	0.1879	3/4	0.3305		1⟨2.54⟩	2.5	− 3.0	48.21
N(4)	0.3208	3/4	0.4973		2⟨2.00⟩	2.9	− 3.0	50.68
N(5)	0.4782	3/4	0.6533		3⟨2.24⟩	2.8	− 3.0	48.51
N(6)	0.5818	3/4	0.8561		no N–Cu	2.8	− 3.0	51.20
NH$_4$N$_3$	Pmna	8.948	3.808	8.659				
H(1)	0.2916	0.7082	0.3355		1⟨1.04⟩	1.1	1.0	− 24.68
H(2)	0.3317	0.3927	0.2076		1⟨1.00⟩	1.2	1.0	− 26.33
N(1)	0	0	0		2⟨1.17⟩	4.4	5.0	− 67.92
N(2)	1/2	0	0		2⟨1.17⟩	4.4	5.0	− 68.44
N(3)	0.1305	0	0		2⟨1.94⟩	2.5	− 3.0	50.56
N(4)	1/2	0.1098	0.1263		2⟨1.98⟩	2.4	− 3.0	50.17
N(5)	1/4	0.5443	1/4		4⟨1.02⟩	4.1	− 3.0	48.13
N$_2$H$_5$N$_3$	P2$_1$/c	5.663	5.506	12.436	114.0			
H(1)	0.2749	0.1803	0.4335		3⟨1.63⟩	1.0	1.0	− 20.07
H(2)	0.6455	0.4696	0.1380		3⟨1.59⟩	1.4	1.0	− 25.69
H(3)	0.1245	0.5054	0.1549		3⟨1.66⟩	1.4	1.0	− 26.07
H(4)	0.0913	0.7990	0.4114		2⟨1.44⟩	1.4	1.0	− 22.91
H(5)	0.8507	0.7033	0.2119		3⟨1.67⟩	1.3	1.0	− 22.42
N(1)	0.3431	0.0238	0.0990		2⟨1.17⟩	4.4	5.0	− 66.76
N(2)	0.4052	0.2125	0.0750		2⟨1.81⟩	2.5	− 3.0	51.79
N(3)	0.2845	0.8359	0.1235		1⟨2.14⟩	2.4	− 3.0	50.33
N(4)	0.2104	0.0759	0.3486		5⟨1.39⟩	4.3	− 2.0	29.33
N(5)	0.9848	0.4084	0.1501		6⟨1.64⟩	3.9	− 2.0	27.52

Table 2. (*Contd.*)

α-Hg(N$_3$)$_2$	Pca2$_1$	10.632	6.264	6.323				
Hg	0.03183	0.23877	1/4		10⟨2.80⟩	2.3	2.0	− 23.40
N(1)	0.188	0.893	0.121		2⟨1.16⟩	4.6	5.0	− 69.73
N(2)	0.823	0.609	0.280		2⟨1.31⟩	5.1	5.0	− 69.21
N(3)	0.082	0.957	0.106		2⟨2.45⟩	3.1	− 3.0	46.97
N(4)	0.284	0.820	0.112		1⟨2.80⟩	2.7	− 3.0	51.78
N(5)	0.962	0.529	0.390		2⟨2.43⟩	1.4	− 3.0	31.43
N(6)	0.807	0.700	0.172		2⟨3.09⟩	4.7	− 3.0	67.41
KN$_3$	I4/mcm	6.1129	7.0943					
K	0	0	1/4		16⟨3.25⟩	1.4	1.0	− 10.73
N(1)	1/2	0	0		2⟨1.17⟩	4.6	5.0	− 69.62
N(2)	0.36424	0.13576	0		4⟨2.97⟩	2.9	− 3.0	48.17
LiN$_3$	C2/m	5.627	3.319	4.979	107.4			
Li	0	0	0		6⟨2.26⟩	1.0	1.0	− 18.97
N(1)	0	1/2	1/2		2⟨1.16⟩	4.5	5.0	− 67.03
N(2)	0.1048	1/2	0.7397		3⟨2.26⟩	2.8	− 3.0	49.22
α-NaN$_3$	C2/m	6.211	3.658	5.323	108.43			
Na	0	0	0		6⟨2.51⟩	1.3	1.0	− 15.76
N(1)	0	1/2	1/2		2⟨1.17⟩	4.4	5.0	− 67.48
N(2)	0.1016	1/2	0.7258		3⟨2.51⟩	2.9	− 3.0	48.97
β-NaN$_3$	R$\bar{3}$m	3.646	15.213	(hexagonal axes)				
Na	0	0	0		6⟨2.51⟩	1.2	1.0	− 15.86
N(1)	0	0	1/2		2⟨1.16⟩	4.5	5.0	− 67.64
N(2)	0	0	0.5764		3⟨2.51⟩	3.0	− 3.0	49.08
Pb(N$_3$)$_2$	Pnma	6.63	16.25	11.31				
Pb(1)	0.0605	1/4	0.8644		14⟨3.00⟩	2.4	2.0	− 23.07
Pb(2)	0.3387	0.0890	0.1231		13⟨2.98⟩	2.4	2.0	− 21.68
N(1)	0.1526	1/4	0.2785		2⟨1.17⟩	4.5	5.0	− 69.25
N(2)	0.3909	0.0666	0.7625		2⟨1.17⟩	4.5	5.0	− 68.36
N(3)	0.8426	0.0905	0.0014		2⟨1.17⟩	4.5	5.0	− 69.17
N(4)	0.5165	1/4	0.9609		2⟨1.17⟩	4.5	5.0	− 70.03
N(5)	0.1518	0.1783	0.2804		2⟨2.68⟩	2.9	− 3.0	48.48
N(6)	0.4340	0.9985	0.7359		2⟨2.62⟩	2.9	− 3.0	48.66
N(7)	0.3514	0.1341	0.7860		2⟨2.86⟩	2.7	− 3.0	49.12
N(8)	0.9929	0.1283	0.0055		2⟨2.65⟩	2.9	− 3.0	47.76
N(9)	0.6918	0.0536	0.9968		2⟨2.75⟩	2.8	− 3.0	49.27
N(10)	0.3612	1/4	0.0138		3⟨2.80⟩	2.8	− 3.0	46.64
N(11)	0.6638	1/4	0.9074		2⟨2.91⟩	2.8	− 3.0	48.90
RbN$_3$	I4/mcm	6.3098	7.5188					
Rb	0	0	1/4		8⟨3.10⟩	1.1	1.0	− 10.06
N(1)	1/2	0	0		2⟨1.17⟩	4.3	5.0	− 69.61
N(2)	0.36846	0.13154	0		4⟨3.10⟩	2.8	− 3.0	47.99
Sr(N$_3$)$_2$	Fddd	11.82	11.47	6.08				
Sr	1/8	1/8	1/8		12⟨2.92⟩	2.5	2.0	− 22.72
N(1)	0.99611	1/8	5/8		2⟨1.17⟩	4.5	5.0	− 68.65
N(2)	0.00220	0.05937	0.77150		2⟨2.69⟩	2.9	− 3.0	48.72
or N(3)	0.00220	0.19063	0.4785		2⟨2.69⟩	2.9	− 3.0	48.72
TlN$_3$	I4/mcm	6.208	7.355					
Tl	0	0	1/4		8⟨3.04⟩	1.0	1.0	− 10.28
N(1)	1/2	0	0		2⟨1.16⟩	4.5	5.0	− 70.50
N(2)	0.3679	0.1321	0		4⟨3.04⟩	2.9	− 3.0	48.64

Table 3. Structural data for nitrides with the bixbyite structure

Atom	x	y	z	d	Σv_i	V	ϕ_i
α-Be$_3$N$_2$	$Ia\bar{3}$	8.13					
Be	3/8	1/8	3/8	4⟨1.74⟩	2.1	2.0	−27.27
N(1)	1/4	1/4	1/4	6⟨1.76⟩	3.0	−3.0	40.08
N(2)	0.985	0	1/4	6⟨1.74⟩	3.2	−3.0	41.04
α-Ca$_3$N$_2$	$Ia\bar{3}$	11.473					
Ca	0.389	0.153	0.382	4⟨2.47⟩	1.7	2.0	−20.33
N(1)	1/4	1/4	1/4	6⟨2.46⟩	2.5	−3.0	28.45
N(2)	0.960	0	1/4	6⟨2.47⟩	2.5	−3.0	28.49
Mg$_3$N$_2$	$Ia\bar{3}$	9.964					
Mg	0.387	0.152	0.382	4⟨2.14⟩	1.8	2.0	−23.39
N(1)	1/4	1/4	1/4	6⟨2.1⟩	2.8	−3.0	33.09
N(2)	0.963	0	1/4	6⟨2.14⟩	2.7	3.0	32.83
U$_2$N$_3$	$Ia\bar{3}$	10.678					
U(1)	1/4	1/4	1/4	6⟨2.29⟩	5.3	6.0	−62.37
U(2)	0.982	0	1/4	6⟨2.30⟩	5.1	6.0	−61.51
N	0.383	0.145	0.381	4⟨2.30⟩	3.4	−4.0	42.95

Table 4. Structural data for materials with structures derived from that of fluorite

Atom	x	y	z	d	Σv_i	V	ϕ_i
In$_{32}$ON$_{17}$F$_{43}$	$Ia\bar{3}$	10.536					
In(1)	1/4	1/4	1/4	8⟨2.36⟩			
In(2)	0.5355	0	1/4	8⟨2.32⟩			
N, F	0.1326	0.1326	0.1326	4⟨2.13⟩			
F, N	0.3598	0.0797	0.3604	4⟨2.39⟩			
LiMgN	$Fm\bar{3}m$	4.97					
Li, Mg	1/4	1/4	1/4	4⟨2.15⟩	1.4	1.5	−17.69
N	0	0	0	8⟨2.15⟩	2.8	−3.0	32.88
LiZnN	$F\bar{4}3m$	4.877					
Li	3/4	3/4	3/4	4⟨2.11⟩	1.0	1.0	−12.87
Zn	1/4	1/4	1/4	4⟨2.11⟩	1.6	2.0	−23.19
N	0	0	0	8⟨2.11⟩	2.6	−3.0	33.51
Li$_2$(NH)	$Fm\bar{3}m$	5.047					
Li	1/4	1/4	1/4	4⟨2.19⟩	0.8	1.0	−11.61
N	0	0	0	8⟨2.19⟩	1.7	−2.0	21.59
Li$_3$AlN$_2$	$Ia\bar{3}$	9.461					
Li	0.160	0.382	0.110	4⟨2.15⟩	0.9	1.0	−12.16
Al	0.115	0.115	0.115	4⟨1.88⟩	3.1	3.0	−37.46
N(1)	0	0	0	8⟨2.08⟩	2.9	−3.0	35.20
N(2)	0.205	0	1/4	8⟨2.08⟩	2.9	−3.0	35.32
Li$_3$FeN$_2$	$Ibam$	4.872 9.641	4.792				
Li(1)	0	0.7411	1/4	4⟨2.16⟩	0.9	1.0	−14.18
Li(2)	1/2	0	1/4	4⟨2.11⟩	1.0	1.0	−13.15
Fe	0	0	1/4	4⟨1.96⟩	3.1	3.0	−31.77
N	0.2237	0.8860	0	8⟨2.10⟩	3.0	−3.0	34.97
Li$_3$GaN$_2$	$Ia\bar{3}$	9.592					
Li	0.152	0.381	0.114	4⟨2.15⟩	0.9	1.0	−12.59
Ga	0.117	0.117	0.117	4⟨1.94⟩	3.0	3.0	−35.65

Table 4. (*Contd.*)

N(1)	0	0	0	8⟨2.10⟩	2.9	−3.0	34.77
N(2)	0.215	0	1/4	8⟨2.10⟩	2.9	−3.0	34.78
Li$_5$GeN$_3$	$Ia\bar{3}$	9.61					
Li	1/8	3/8	1/8	4⟨2.16⟩	1.0	1.0	−15.09
Ge	0.115	0.115	0.115	4⟨1.91⟩	3.7	4.0	−46.30
N(1)	0	0	0	8⟨2.04⟩	3.5	−3.0	35.69
N(2)	0.205	0	1/4	8⟨2.12⟩	3.4	−3.0	34.75
Li$_5$SiN$_3$	$Ia\bar{3}$	9.44					
Li	1/8	3/8	1/8	4⟨2.10⟩	1.2	1.0	−15.55
Si	0.117	0.117	0.117	4⟨1.91⟩	2.7	4.0	−46.03
N(1)	0	0	0	8⟨2.01⟩	3.2	−3.0	35.85
N(2)	0.215	0	1/4	8⟨2.07⟩	3.1	−3.0	34.94
Li$_5$TiN$_3$	$Ia\bar{3}$	9.70					
Li	1/8	3/8	1/8	4⟨2.13⟩	1.0	1.0	−15.49
Ti	0.120	0.120	0.120	4⟨2.02⟩	3.1	4.0	−43.17
N(1)	0	0	0	8⟨2.08⟩	3.2	−3.0	34.10
N(2)	0.230	0	1/4	8⟨2.11⟩	3.0	−3.0	33.32
Li$_{15}$Cr$_2$N$_9$	$P4/ncc$	10.233	9.389				
Li(1)	0.9150	0.7301	0.2349	4⟨2.10⟩	1.1	1.0	−15.49
Li(2)	0.2150	0.4123	0.0066	4⟨2.17⟩	1.0	1.0	−17.34
Li(3)	0.4000	0.5655	0.9779	4⟨2.13⟩	1.0	1.0	−13.15
Li(4)	3/4	1/4	1/4	4⟨2.04⟩	1.3	1.0	−10.28
Li(5)	3/4	1/4	0	4⟨2.12⟩	1.0	1.0	−14.54
Li(6)	0.9031	0.0969	1/4	4⟨2.17⟩	0.9	1.0	−10.51
Cr	0.4195	0.5805	1/4	4⟨1.77⟩	5.0	6.0	−62.84
N(1)	0.0617	0.4156	0.1447	8⟨2.11⟩	2.9	−3.0	38.13
N(2)	0.0855	0.7830	0.1326	8⟨2.09⟩	2.9	−3.0	38.87
N(3)	1/4	1/4	0.3911	8⟨2.04⟩	2.5	−3.0	25.89
Li$_{14}$Cr$_2$N$_8$O	$P\bar{3}$	5.799	8.263				
Li(1)	0.9283	0.3028	0.0923	4⟨2.10⟩	1.1	1.0	−14.78
Li(2)	0.3555	0.3181	0.4101	4⟨2.09⟩	1.0	1.0	−13.22
Li(3)	0	0	0.2501	4⟨2.10⟩	1.0	1.0	−15.03
Cr	1/3	2/3	0.2382	4⟨1.75⟩	5.2	6.0	−63.47
N(1)	0.0608	0.6889	0.1676	7⟨2.05⟩	2.9	−3.0	38.32
N(2)	1/3	2/3	0.4528	7⟨2.06⟩	2.8	−3.0	38.64
O	0	0	1/2	8⟨2.09⟩	1.5	−2.0	21.33
Li$_6$MoN$_4$	$P4_2/nmc$	6.673	4.925				
Li(1)	3/4	3/4	0.8358	4⟨2.13⟩	1.0	1.0	−16.81
Li(2)	0.0368	0.5368	3/4	4⟨2.12⟩	1.0	1.0	−14.38
Mo	3/4	1/4	3/4	4⟨1.88⟩	6.2	6.0	−59.78
N	3/4	0.4905	0.5517	7⟨2.09⟩	3.1	−3.0	36.26
Li$_6$WN$_4$	$P4_2/nmc$	6.679	4.927				
Li(1)	3/4	3/4	0.804	4⟨2.09⟩	1.1	1.0	−17.71
Li(2)	0.039	0.539	3/4	4⟨2.12⟩	1.0	1.0	−14.32
W	3/4	1/4	3/4	4⟨1.91⟩	5.9	6.0	−58.44
N	3/4	0.497	0.553	7⟨2.08⟩	3.0	−3.0	35.77
Li$_7$MnN$_4$	$P\bar{4}3n$	9.571					
Li(1)	0	1/2	1/2	4⟨2.07⟩	1.1	1.0	−10.49
Li(2)	1/4	0	1/2	4⟨2.07⟩	1.1	1.0	−16.21
Li(3)	1/4	1/4	1/4	4⟨2.11⟩	1.0	1.0	−16.36
Li(4)	1/4	0	0	4⟨2.05⟩	1.2	1.0	−15.01
Li(5)	1/4	1/4	0	4⟨2.09⟩	1.1	1.0	−14.14
Mn(1)	0	0	0	4⟨1.91⟩	3.6	5.0	−51.19

Table 4. (*Contd.*)

Mn(2)	1/4	1/2	0	4⟨2.07⟩	2.3	5.0	−44.07
N(1)	0.115	0.115	0.115	8⟨2.08⟩	2.8	−3.0	34.67
N(2)	3/8	3/8	1/8	8⟨2.07⟩	2.6	−3.0	34.47
Li$_7$PN$_4$	P$\bar{4}$3n	9.3648					
Li(1)	0	1/2	1/2	4⟨1.99⟩	1.4	1.0	−12.91
Li(2)	1/4	0	1/2	4⟨2.09⟩	1.1	1.0	−15.52
Li(3)	0.2245	0.2245	0.2245	4⟨2.17⟩	0.9	1.0	−14.17
Li(4)	0.2493	0.2386	0.9742	4⟨2.10⟩	1.1	1.0	−13.69
Li(5)	0.2576	0	0	4⟨2.09⟩	1.1	1.0	−12.69
P(1)	1/2	0	1/4	4⟨1.73⟩	4.0	5.0	−57.11
P(2)	0	0	0	4⟨1.69⟩	4.5	5.0	−58.22
N(1)	0.3532	0.3844	0.1012	8⟨2.05⟩	2.9	−3.0	37.40
N(2)	0.1040	0.1040	0.1040	8⟨2.03⟩	3.3	−3.0	38.80
Li$_7$VN$_4$	P$\bar{4}$3n	9.604					
Li(1)	0	1/2	1/2	4⟨2.08⟩	1.1	1.0	−10.62
Li(2)	1/4	0	1/2	4⟨2.08⟩	1.1	1.0	−16.38
Li(3)	1/4	1/4	1/4	4⟨2.10⟩	1.1	1.0	−16.84
Li(4)	1/4	0	0	4⟨2.07⟩	1.2	1.0	−14.34
Li(5)	1/4	1/4	0	4⟨2.09⟩	1.1	1.0	−14.31
V(1)	0	0	0	4⟨2.00⟩	2.8	5.0	−47.53
V(2)	1/4	1/2	0	4⟨2.08⟩	2.2	5.0	−44.14
N(1)	0.120	0.120	0.120	8⟨2.08⟩	2.6	−3.0	34.30
N(2)	3/8	3/8	1/8	8⟨2.08⟩	2.5	−3.0	34.19

Table 5. Structural data for materials with perovskite-related structures

Atom	x	y	z	d	Σv_i	V	ϕ_i
BaNbO$_2$N	Pm$\bar{3}$m	4.1283					
Ba	0	0	0	12⟨2.92⟩	2.2	2.0	−21.59
Nb	1/2	1/2	1/2	6⟨2.06⟩	4.0	5.0	−53.07
O	0	1/2	1/2	6⟨2.63⟩	2.1	−2.0	22.18
BaTaO$_2$N	Pm$\bar{3}$m	4.1128					
Ba	0	0	0	12⟨2.91⟩	2.3	2.0	−21.67
Ta	1/2	1/2	1/2	6⟨2.06⟩	4.2	5.0	−53.27
O	0	1/2	1/2	6⟨2.62⟩	2.1	−2.0	22.27
LaWO$_{0.6}$N$_{2.4}$	I$\bar{4}$	5.6523	8.0084				
La(1)	0	0	0	12⟨2.83⟩	3.3	3.0	−29.27
La(2)	0	0	1/2	12⟨2.83⟩	3.6	3.0	−29.79
W(1)	0	1/2	1/4	6⟨1.99⟩	7.2	6.0	−67.27
W(2)	0	1/2	3/4	6⟨2.03⟩	6.5	6.0	−65.36
N(1)	0	1/2	0.0023	6⟨2.55⟩	3.4	−3.0	35.03
N(2)	0.2815	0.2865	0.2439	6⟨2.56⟩	3.4	−3.0	34.74
SrTaO$_2$N	I$\bar{4}$2m	5.6919	8.0905				
Sr(1)	0	0	0	12⟨2.85⟩	1.8	2.0	−19.98
Sr(2)	0	0	1/2	12⟨2.85⟩	2.0	2.0	−20.36
Ta	0	1/2	1/4	6⟨2.02⟩	4.9	5.0	−53.80
O	0.2703	0.2703	0.2387	6⟨2.58⟩	2.1	−2.0	24.79
N	0	1/2	0	6⟨2.57⟩	2.6	−3.0	32.75

Table 6. Structural data for materials with wurtzite-related structures

Atom		x	y	z	d	Σv_i	V	ϕ_i
AlN	$P6_3mc$	3.110	4.980					
Al		1/3	2/3	0	4⟨1.89⟩	3.0	3.0	−37.48
N		1/3	2/3	0.3821	4⟨1.89⟩	3.0	−3.0	37.48
BeSiN$_2$	$Pna2_1$	4.977	5.747	4.674				
Be		1/12	5/8	0	4⟨1.76⟩	2.0	2.0	−32.16
Si		1/12	1/8	0	4⟨1.76⟩	4.1	4.0	−48.54
N(1)		1/12	1/8	3/8	4⟨1.76⟩	3.1	−3.0	40.56
N(2)		1/12	5/8	3/8	4⟨1.76⟩	3.1	−3.0	40.14
BN	$P6_3mc$	2.553	4.228					
B		1/3	2/3	0.371	4⟨1.57⟩	3.0	3.0	−45.12
N		1/3	2/3	0	4⟨1.57⟩	3.0	−3.0	45.12
GaN	$P6_3mc$	3.190	5.189					
Ga		1/3	2/3	0	4⟨1.95⟩	3.0	3.0	−36.35
N		1/3	2/3	0.377	4⟨1.95⟩	3.0	−3.0	36.35
InN	$P6_3mc$	3.533	5.692					
In		1/3	2/3	3/8	4⟨2.15⟩	2.9	3.0	−32.93
N		1/3	2/3	0	4⟨2.15⟩	2.9	−3.0	32.93
KGeON	$Pca2_1$	5.7376	8.0535	5.2173				
K		0.556	0.355	0.017	6⟨2.78⟩	1.2	1.0	−11.92
Ge		0.093	0.096	0	4⟨1.84⟩	4.1	4.0	−44.43
O		0.471	0.310	0.526	6⟨2.60⟩	1.9	−2.0	21.72
N		0.127	0.043	0.649	4⟨2.10⟩	3.4	−3.0	41.57
LiGe$_2$N$_3$	$Cmc2_1$	9.523	5.497	5.041				
Li		0	1/3	0	4⟨1.93⟩	1.7	1.0	−22.96
Ge		1/6	5/6	0	4⟨1.93⟩	3.5	4.0	−43.77
N(1)		1/6	5/6	3/8	4⟨1.93⟩	3.1	−3.0	39.26
N(2)		0	1/3	3/8	4⟨1.93⟩	2.6	−3.0	31.98
LiSi$_2$N$_3$	$Cmc2_1$	9.186	5.302	4.776				
Li		0	0.333	0	4⟨2.10⟩	1.1	1.0	−17.01
Si		0.168	0.834	0.984	4⟨1.75⟩	4.3	4.0	−49.51
N(1)		0.192	0.860	0.348	4⟨1.86⟩	3.3	−3.0	42.04
N(2)		0	0.290	0.424	4⟨1.87⟩	3.0	−3.0	37.41
LiSiON	$Pca2_1$	5.1986	6.3893	4.7398				
Li		0.5989	0.3797	0.0019	4⟨2.04⟩	1.0	1.0	−17.19
Si		0.0817	0.1167	0	4⟨1.72⟩	4.2	4.0	−47.10
O		0.5403	0.3460	0.5727	4⟨1.89⟩	1.8	−2.0	24.25
N		0.1080	0.0844	0.6278	4⟨1.87⟩	3.4	−3.0	44.45
Mg$_2$PN$_3$	$Cmc2_1$	9.759	5.635	4.743				
Mg		0.167	0.819	0.938	5⟨2.18⟩	2.8	2.0	−27.32
P		0	0.366	0	4⟨1.75⟩	3.9	5.0	−56.33
N(1)		0.150	0.753	0.406	5⟨2.15⟩	2.9	−3.0	37.04
N(2)		0	0.324	0.384	4⟨1.84⟩	3.6	−3.0	43.26
MgGeN$_2$	$Pna2_1$	5.494	6.611	5.166				
Mg		1/12	5/8	0	4⟨1.99⟩	2.8	2.0	−28.31
Ge		1/12	1/8	0	4⟨1.95⟩	3.3	4.0	−43.79
N(1)		1/12	1/8	0.380	4⟨1.97⟩	3.0	−3.0	36.12
N(2)		1/12	5/8	0.400	4⟨1.97⟩	3.1	−3.0	36.23
MgSiN$_2$	$Pna2_1$	5.279	6.476	4.992				
Mg		0.076	5/8	0.995	4⟨2.09⟩	2.1	2.0	−24.85
Si		0.072	0.131	0	4⟨1.76⟩	4.1	4.0	−50.10
N(1)		0.049	0.095	0.356	4⟨1.93⟩	3.1	−3.0	38.07
N(2)		0.110	0.652	0.414	4⟨1.92⟩	3.2	−3.0	38.96

Table 6. (*Contd.*)

Atom	x	y	z	d	Σv_i	V	ϕ_i
MgGeN$_2$	Pna2$_1$	5.486	6.675 5.246				
Mn	0.076	0.615	0	4⟨2.10⟩	2.2	2.0	−25.55
Ge	0.076	0.117	0.992	4⟨1.89⟩	3.9	4.0	−46.10
N(1)	0.063	0.113	0.356	4⟨2.00⟩	3.0	−3.0	36.53
N(2)	0.098	0.642	0.405	4⟨1.98⟩	3.1	−3.0	36.67
MnSiN$_2$	Pna2$_1$	5.248	6.511 5.070				
Mn	0.072	0.628	0.995	4⟨2.12⟩	2.1	2.0	−24.29
Si	0.071	0.130	0	4⟨1.76⟩	4.2	4.0	−50.40
N(1)	0.055	0.082	0.351	4⟨1.94⟩	3.1	−3.0	38.28
N(2)	0.102	0.655	0.410	4⟨1.93⟩	3.2	−3.0	38.51
NaGe$_2$N$_3$	Cmc2$_1$	9.8662	5.7820 5.1221				
Na	0	0.345	0	4⟨2.43⟩	1.1	1.0	−13.30
Ge	0.169	0.841	0.024	4⟨1.85⟩	4.4	4.0	−47.37
N(1)	0.211	0.870	0.377	4⟨2.03⟩	3.3	−3.0	39.12
N(2)	0	0.255	0.456	4⟨2.07⟩	3.2	−3.0	35.82
ZnGeN$_2$	Pna2$_1$	5.454	6.441 5.194				
Zn	0.083	0.620	0	4⟨2.02⟩	2.0	2.0	−26.98
Ge	0.083	1/8	0	4⟨1.89⟩	3.9	4.0	−45.85
N(1)	0.070	0.115	0.365	4⟨1.96⟩	2.9	−3.0	36.79
N(2)	0.095	0.640	0.385	4⟨1.95⟩	3.1	−3.0	37.19

Table 7. Lattice parameters for materials with structures based on rocksalt

Compound	a	d	Σv_i	V	ϕ_i
AmN	4.995	2.4975	3.0	3	− 30.2
Ba$_2$NF	5.69	2.845			
Ba$_2$NH	5.86	2.93			
BaNH	5.84	2.92	1.4	2	− 17.2
BaND	5.8613	2.9307	1.7	2	− 17.2
Ca$_2$NF	4.937	2.4685			
Ca$_2$NH	10.13	5.065			
CaNH	5.16	2.58	1.9	2	− 19.5
CeN	5.022	2.511	4.0	4	− 43.2
CfN	4.95	2.475	3.0	3	− 30.5
CmN	5.041	2.5205	3.0	3	− 29.9
DyN	4.905	2.4525	3.0	3	− 30.8
ErN	4.842	2.421	3.0	3	− 31.3
EuN	5.014	2.507	3.1	3	− 30.2
GdN	4.999	2.4995	3.0	3	− 30.3
HoN	4.874	2.437	3.0	3	− 31.0
LaN	5.305	2.6525	2.7	3	− 28.6
LuN	4.765	2.3825	3.0	3	− 28.5
NdN	5.151	2.5755	3.0	3	− 29.3
NpN	4.887	2.4435	3.0	3	− 30.8
PaN	5.047	2.5235	3.0	3	− 29.9
PrN	5.167	2.5835	2.8	3	− 29.2
PuN	4.9060	2.453	3.0	3	− 36.9
ReN$_{0.43}$	3.92	1.96			
ScN	4.502	2.251	4.0	3	− 33.9
SmN	5.0481	2.5241	3.0	3	− 29.9
SrN	5.495	2.7475			
Sr$_2$NH	10.90	2.73			
SrNH	5.45	2.725	1.7	2	− 18.5
SrO$_x$N$_y$	5.473	2.7365			
Sr$_2$NF	5.38	2.69			
TbN	4.933	2.4665	3.0	3	− 30.6
TcN$_{0.75}$	3.985	1.9925			
ThN	5.1666	2.5833	3.1	3	− 29.2
TmN	4.809	2.4045	3.0	3	− 31.4
UN	4.880	2.44	4.1	3	− 30.9
YbN	4.7852	2.3926	2.9	3	− 31.6
YN	4.877	2.44	3.0	3	− 31.0

Table 8. Structural data for U_2N_2X materials. Those crystallizing in spacegroup $P\bar{3}m1$ have the Ce_2O_2S structure and those in $I4/mmm$ have the $(Na, Bi)_2O_2Cl$ structure

Atom	x	y	z	d	Σv_i	V	ϕ_i
$Th_2(N,O)_2As$	$P\bar{3}m1$	4.041	6.979				
Th	1/3	2/3	0.293	7⟨2.69⟩	4.9	4.0	− 35.70
As	0	0	0	6⟨3.10⟩	2.7	− 3.0	24.31
N	1/3	2/3	0.635	4⟨2.39⟩	3.5	− 3.0	33.89
$Th_2(N,O)_2As$	$P\bar{3}m1$	4.0285	6.835				
Th	1/3	2/3	0.278	7⟨2.66⟩	4.8	4.0	− 36.23
P	0	0	0	6⟨3.00⟩	2.9	− 3.0	26.82
N	1/3	2/3	0.630	4⟨2.41⟩	3.3	− 3.0	32.15
Th_2N_2Bi	$I4/mmm$	4.075	13.62				
Th	0	0	0.344	8⟨2.99⟩	4.8	4.0	− 37.20
Bi	0	0	0	8⟨3.58⟩	2.9	− 3.0	16.18
N	1/2	0	1/4	4⟨2.41⟩	3.3	− 3.0	27.94
Th_2N_2Sb	$I4/mmm$	4.049	13.57				
Th	0	0	0.344	8⟨2.98⟩	4.5	4.0	− 37.42
Sb	0	0	0	8⟨3.56⟩	2.1	− 2.0	16.29
N	1/2	0	1/4	4⟨2.39⟩	3.5	− 3.0	28.08
Th_2N_2S	$P\bar{3}m1$	4.008	6.920				
Th	1/3	2/3	0.278	7⟨2.67⟩	4.4	4.0	− 38.08
S	0	0	0	6⟨3.01⟩	2.2	− 2.0	19.37
N	1/3	2/3	0.626	4⟨2.41⟩	3.3	− 3.0	28.82
Th_2N_2Se	$P\bar{3}m1$	4.0287	7.156				
Th	1/3	2/3	0.293	7⟨2.71⟩	4.6	4.0	− 37.68
Se	0	0	0	6⟨3.13⟩	2.2	− 2.0	17.03
N	1/3	2/3	0.628	4⟨2.39⟩	3.5	− 3.0	30.38
Th_2N_2Te	$I4/mmm$	4.094	13.014				
Th	0	0	0.344	8⟨2.96⟩	4.3	4.0	− 37.43
Te	0	0	0	8⟨3.54⟩	1.6	− 2.0	16.03
N	1/2	0	1/4	4⟨2.38⟩	3.5	− 3.0	28.65
U_2N_2As	$P\bar{3}m1$	3.833	6.737				
U	1/3	2/3	0.2789	7⟨2.56⟩	5.1	4.0	− 37.76
As	0	0	0	6⟨2.90⟩	3.7	− 3.0	28.36
N	1/3	2/3	0.6190	4⟨2.31⟩	3.3	− 3.0	32.93
U_2N_2Bi	$I4/mmm$	3.9292	12.548				
U	0	0	0.344	8⟨2.85⟩	5.3	4.0	− 36.70
Bi	0	0	0	8⟨3.40⟩	3.6	− 3.0	24.88
N	1/2	0	1/4	4⟨2.29⟩	3.5	− 3.0	32.85
U_2N_2Sb	$I4/mmm$	3.8937	12.3371				
U	0	0	0.344	8⟨2.81⟩	5.2	4.0	− 39.38
Sb	0	0	0	8⟨3.36⟩	2.9	− 2.0	16.85
N	1/2	0	1/4	4⟨2.27⟩	3.7	− 3.0	30.17
U_2N_2S	$P\bar{3}m1$	3.818	6.610				
U	1/3	2/3	0.2798	7⟨2.54⟩	4.8	4.0	− 39.98
S	0	0	0	6⟨2.88⟩	2.5	− 2.0	20.05
N	1/3	2/3	0.6264	4⟨2.29⟩	3.5	− 3.0	30.48
U_2N_2Se	$P\bar{3}m1$	3.863	6.867				
U	1/3	2/3	0.3054	7⟨2.61⟩	4.9	4.0	− 39.14
Se	0	0	0	6⟨3.06⟩	2.3	− 2.0	15.93
N	1/3	2/3	0.6306	4⟨2.26⟩	3.8	− 3.0	33.42
U_2N_2Te	$I4/mmm$	3.9631	12.561				
U	0	0	0.344	8⟨2.86⟩	4.2	4.0	− 38.69
Te	0	0	0	8⟨3.42⟩	1.8	− 2.0	16.55
N	1/2	0	1/4	4⟨2.31⟩	3.3	− 3.0	29.64

Table 9. Structural data for some binary transition metal nitrides

Atom	x		y	z	d(X − X)	d(X − Y)
Co_3N	$P6_322$	2.659	4.346			
Co	2/3		0	0	6⟨1.54⟩	4⟨1.40⟩
N(1)	1/3		2/3	1/4	3⟨1.54⟩	6⟨1.40⟩
N(2)	1/3		2/3	3/4	3⟨1.54⟩	6⟨1.40⟩
Co_4N	$Pm\bar{3}m$	3.738				
Co(1)	0		0	0	12⟨2.64⟩	8⟨3.24⟩
Co(2)	1/2		1/2	0	12⟨2.64⟩	2⟨1.87⟩
N	1/2		1/2	1/2	6⟨3.74⟩	6⟨1.87⟩
$\varepsilon\text{-}Fe_2N\text{-}Fe_3N$	P312	2.7	4.4			
Fe	0		2/3	1/4	6⟨1.56⟩	3⟨1.42⟩
N(1)	1/3		2/3	1/2	18⟨2.70⟩	6⟨1.42⟩
N(2)	2/3		1/3	0	3⟨1.56⟩	6⟨1.42⟩
N(3)	0		0	0	3⟨1.56⟩	6⟨1.42⟩
Fe_4N	$Pm\bar{3}m$	3.797				
Fe(1)	0		0	0	12⟨2.68⟩	8⟨3.29⟩
Fe(2)	1/2		1/2	0	12⟨2.68⟩	2⟨1.90⟩
N	1/2		1/2	1/2	6⟨3.80⟩	6⟨1.90⟩
$\varepsilon\text{-}Hf_3N_2$	$R\bar{3}m$	3.206	23.26	(hexagonal axes)		
Hf(1)	0		0	2/9	12⟨3.19⟩	6⟨2.26⟩
Hf(2)	0		0	0	12⟨3.19⟩	6⟨2.26⟩
N(1)	0		0	7/18	1⟨2.58⟩	6⟨2.26⟩
N(2)	0		0	1/2	2⟨2.58⟩	6⟨2.26⟩
$\zeta\text{-}Hf_4N_3$	$R\bar{3}m$	3.214	31.12	(hexagonal axes)		
Hf(1)	0		0	1/8	12⟨3.20⟩	6⟨2.06⟩
Hf(2)	0		0	7/24	12⟨3.20⟩	3⟨2.26⟩
N(1)	0		0	0	6⟨3.21⟩	6⟨2.26⟩
N(2)	0		0	1/2	2⟨1.19⟩	6⟨2.26⟩
N(3)	0		0	5/12	2⟨1.79⟩	3⟨1.86⟩
$\varepsilon\text{-}Mn_4N$	$Pm\bar{3}m$	3.86				
Mn(1)	0		0	0	12⟨2.73⟩	6⟨3.34⟩
Mn(2)	0		1/2	1/2	12⟨2.73⟩	2⟨1.93⟩
N	1/2		1/2	1/2	6⟨3.86⟩	6⟨1.93⟩
Mn_3N_2	$I4/mmm$	2.974	12.126			
Mn(1)	0		0	0	12⟨2.93⟩	2⟨1.93⟩
Mn(2)	0		0	0.33341	12⟨2.93⟩	5⟨2.11⟩
N	0		0	0.1593	8⟨3.01⟩	6⟨2.08⟩
$\beta\text{-}Nb_2N$	$P\bar{3}1m$	5.267	4.988			
Nb	1/3		0	1/4	11⟨3.05⟩	3⟨2.15⟩
N(1)	1/3		2/3	1/2	3⟨3.04⟩	6⟨2.15⟩
N(2)	0		0	0	12⟨3.93⟩	6⟨2.15⟩
$\gamma\text{-}Nb_4N_3$	$I4/mmm$	4.386	8.683			
Nb(1)	0		1/2	0	12⟨3.09⟩	6⟨2.19⟩
Nb(2)	0		0	0.2453	12⟨3.09⟩	6⟨2.19⟩
N(1)	0		0	0	12⟨3.09⟩	6⟨2.18⟩
N(2)	0		0	1/2	12⟨3.09⟩	6⟨2.19⟩
N(3)	0		1/2	1/4	12⟨3.09⟩	6⟨2.19⟩
Ni_3N	$P6_322$	2.664	4.298			
Ni	2/3		0	0	6⟨1.54⟩	4⟨1.39⟩
N(1)	1/3		2/3	1/4	3⟨1.54⟩	6⟨1.39⟩
N(2)	1/3		2/3	3/4	3⟨1.54⟩	6⟨1.39⟩

Table 9. (*Contd.*)

Atom	x	y	z	$d(X-X)$	$d(X-Y)$
Ni$_4$N	$Pm\bar{3}m$ 3.72				
Ni(1)	0	0	0	12⟨2.63⟩	8⟨3.22⟩
Ni(2)	1/2	1/2	0	12⟨2.63⟩	2⟨3.22⟩
N	1/2	1/2	1/2	6⟨2.63⟩	8⟨3.22⟩
β-Ta$_2$N	$P\bar{3}1m$ 5.285 4.919				
Ta	1/3	0	1/4	11⟨3.04⟩	3⟨2.15⟩
N(1)	1/3	2/3	1/2	3⟨3.05⟩	6⟨2.15⟩
N(2)	0	0	0	12⟨3.92⟩	6⟨2.15⟩
ε-Ti$_2$N	$P4_2/mnm$ 4.9452 3.0342				
Ti	0.296	0.296	0	11⟨3.04⟩	3⟨2.08⟩
N	0	0	0	2⟨3.03⟩	6⟨2.08⟩
δ-Ti$_2$N	$I4_1/amd$ 4.1493 8.7858				
Ti	0	3/4	0.8890	12⟨3.00⟩	6⟨2.12⟩
N(1)	0	3/4	1/8	12⟨2.99⟩	6⟨2.08⟩
N(2)	0	3/4	5/8	12⟨2.99⟩	6⟨2.16⟩
β-V$_2$N	$P\bar{3}1m$ 4.917 4.568				
V	0.323	0	0.272	11⟨2.82⟩	3⟨1.98⟩
N(1)	1/3	2/3	1/2	3⟨2.84⟩	6⟨1.96⟩
N(2)	0	0	0	12⟨2.64⟩	6⟨2.02⟩
WN$_2$	$R\bar{3}m$ 2.89 16.40 (hexagonal axes)				
W	0	0	0	6⟨2.89⟩	8⟨3.01⟩
N	0	0	0.1785	3⟨1.71⟩	4⟨3.01⟩

Crystal Chemistry of Inorganic Nitrides

Table 10. Structural data for materials with various structures

Atom		x	y	z	d	Σv_i	V	ϕ_i
BN	$F\bar{4}3m$	3.6157						
B		0	0	0	4⟨1.57⟩	3.1	3.0	−45.20
N		1/4	1/4	1/4	4⟨1.57⟩	3.1	−3.0	45.20
BN	$P6_3mc$	2.5040	6.6612					
B		1/3	2/3	0	3⟨1.45⟩	3.2	3.0	−46.08
N		0	0	0	3⟨1.45⟩	3.2	−3.0	46.06
BaNiN	Pnma	9.639	13.674	5.432				
Ba(1)		0.3281	0.0880	0.9737	4⟨2.86⟩	1.4	2.0	−16.57
Ba(2)		0.4974	1/4	0.5175	4⟨2.79⟩	1.7	2.0	−16.81
Ni(1)		0.1479	0.1613	0.4460	2⟨1.81⟩	1.7	1.0	−17.64
Ni(2)		0	0	0	2⟨1.78⟩	1.8	1.0	−19.00
N(1)		0.0759	0.0773	0.227	6⟨2.48⟩	3.4	−3.0	27.22
N(2)		0.224	1/4	0.658	6⟨2.53⟩	3.0	−3.0	27.22
$Ba_8Ni_6N_7$	C2/c	9.487	16.578	12.137	107.05			
Ba(1)		0.41431	0.03196	0.85639	4⟨2.77⟩	1.9	2.0	−19.31
Ba(2)		0.6535	0.29711	0.60337	4⟨2.93⟩	1.3	2.0	−17.46
Ba(3)		0.20468	0.9037	0.9828	4⟨2.97⟩	1.0	2.0	−16.41
Ba(4)		0	0.24783	1/4	3⟨2.75⟩	1.4	2.0	−15.87
Ba(5)		0	0.91744	1/4	3⟨2.77⟩	1.3	2.0	−15.73
Ni(1)		0.2466	0.8820	0.0994	2⟨1.81⟩	1.7	1.0	−18.30
Ni(2)		0.2511	0.2923	0.1329	2⟨1.76⟩	1.9	1.0	−19.64
Ni(3)		0.8591	0.0790	0.1139	2⟨1.80⟩	1.7	1.0	−18.33
N(1)		0.239	0.8309	0.229	6⟨2.49⟩	3.3	−3.0	27.11
N(2)		0.308	0.9439	0.999	6⟨2.53⟩	3.2	−3.0	25.55
N(3)		0	0.0854	1/4	6⟨2.53⟩	3.2	−3.0	27.27
N(4)		0	0.5931	1/4	6⟨2.77⟩	2.7	−3.0	23.32
N(5)		1/4	1/4	0	6⟨2.58⟩	3.0	−3.0	25.74
β-Be_3N_2	$P6_3/mmc$	2.8413	9.693					
Be(1)		1/3	2/3	0.071	4⟨1.77⟩	1.9	2.0	−26.96
Be(2)		0	0	1/4	5⟨1.95⟩	2.2	2.0	−30.98
N(1)		0	0	0	8⟨1.94⟩	3.0	−3.0	41.34
N(2)		1/3	2/3	1/4	5⟨1.68⟩	3.1	−3.0	39.46
CaGaN	P4/nmm	3.570	7.558					
Ca		1/4	1/4	0.6503	5⟨2.51⟩	1.9		
Ga		1/4	1/4	0.0839	1⟨1.86⟩	0.9		
N		1/4	1/4	0.330	6⟨2.40⟩	2.8		
$CaGeN_2$	$I\bar{4}2d$	5.426	7.154					
Ca		0	0	1/2	8⟨2.79⟩	2.0	2.0	−20.58
Ge		0	0	0	4⟨1.85⟩	4.3	4.0	−48.28
N		0.164	1/4	1/8	6⟨2.48⟩	3.2	−3.0	35.49
CaNiN	$P4_2/mmc$	3.5809	7.0096					
Ca		0	0	1/4	4⟨2.51⟩	1.5		
Ni		1/2	1/2	0	2⟨1.79⟩	1.8		
N		0	1/2	0	6⟨2.27⟩	3.3		
Ca_2N	$R\bar{3}m$	3.6379	18.7802	(hexagonal axes)				
Ca		0	0	0.2680	3⟨2.43⟩	1.4		
N		0	0	0	6⟨2.43⟩	2.7		
Ca_2NBr	$R\bar{3}m$	3.7166	21.558	(hexagonal axes)				
Ca		0	0	0.2258	6⟨2.83⟩	1.6	2.0	−19.40
N		0	0	1/2	6⟨2.50⟩	2.3	−3.0	27.53
Br		0	0	0	6⟨3.16⟩	1.0	−1.0	9.16

Table 10. (*Contd.*)

Atom	x	y	z	d	Σv_i	V	ϕ_i
Ca$_2$NCl	$R\bar{3}m$	3.6661	19.711	(hexagonal axes)			
Ca	0	0	0.2290	6⟨2.70⟩	1.9	2.0	−20.07
N	0	0	1/2	6⟨2.45⟩	2.6	−3.0	27.84
Cl	0	0	0	6⟨2.95⟩	1.2	−1.0	10.28
Ca$_2$ZnN$_2$	$I4/mmm$	3.5835	12.6583				
Ca	0	0	0.3360	5⟨2.52⟩	1.8	2.0	−21.88
Zn	0	0	0	2⟨1.84⟩	1.6	2.0	−21.48
N	0	0	0.1455	6⟨2.41⟩	2.6	−3.0	29.37
Ca$_3$CrN$_3$	$Cmcm$	8.503	10.284	5.032			
Ca(1)	0	0.1079	1/4	5⟨2.54⟩	1.8	2.0	−21.48
Ca(2)	0.2843	0.8826	1/4	5⟨2.55⟩	1.7	2.0	−20.02
Cr	1/2	0.1945	1/4	3⟨1.80⟩	3.5	3.0	−36.66
N(1)	1/2	0.3757	1/4	6⟨2.36⟩	3.1	−3.0	30.00
N(2)	0.6918	0.1286	1/4	6⟨2.45⟩	2.8	−3.0	31.42
Ca$_{11}$N$_8$	$P4_2nm$	14.45	3.60				
Ca(1)	0.2058	0.9807	1/2	4⟨2.41⟩	2.0		
Ca(2)	0.3821	0.1046	1/2	3⟨2.42⟩	1.4		
Ca(3)	0	0	0	4⟨2.49⟩	1.6		
Ca(4)	0.1661	0.1661	0	5⟨2.69⟩	1.3		
N(1)	0.3050	0.0290	0	6⟨2.48⟩	2.6		
N(2)	0.084	0.084	1/2	6⟨2.42⟩	2.9		
N(3)	0.308	0.308	0	3⟨2.64⟩	0.9		
Ce$_{15}$B$_8$N$_{25}$	$R\bar{3}c$	10.946	82.96°				
Ce(1)	0.28623	0.58605	0.41194	9⟨2.71⟩	3.7		
Ce(2)	0.16099	0.42249	0.90475	8⟨2.65⟩	3.9		
Ce(3)	0.10064	0.39936	1/4	8⟨2.62⟩	3.9		
B(1)	0.165	0.165	0.165	3⟨1.45⟩	3.2		
B(2)	0.157	0.066	0.655	3⟨1.46⟩	3.1		
N(1)	0	0	0	6⟨2.66⟩	2.5		
N(2)	0.096	0.750	0.161	6⟨2.48⟩	3.1		
N(3)	0.072	0.604	0.563	6⟨2.46⟩	3.4		
N(4)	0.075	0.152	0.273	6⟨2.42⟩	3.8		
N(5)	0.635	0.270	0.132	6⟨2.52⟩	3.2		
Cr$_3$GeN	$P\bar{4}2_1m$	5.375	4.012				
Cr(1)	0.2041	0.7041	0.0655	2⟨1.95⟩	1.5		
Cr(2)	0	0	1/2	2⟨2.01⟩	1.3		
Ge	0	1/2	0.5480	10⟨2.64⟩		(Ge–Cr	only)
N	0	0	0	6⟨1.97⟩	4.4		
CsOsO$_3$N	$Pnma$	8.409	7.242	8.089			
Cs	0.1309	1/4	0.3486	10⟨3.24⟩	1.2	1.0	−12.73
Os	0.12004	1/4	0.84355	4⟨1.72⟩	9.7	8.0	−74.88
O(1)	0.122	0.05400	0.720	4⟨2.84⟩	2.4	−2.0	30.03
O(2)	0.958	1/4	0.977	3⟨2.72⟩	2.2	−2.0	32.30
N	0.287	1/4	0.956	3⟨2.80⟩	3.8	−3.0	38.91
CsTaN$_2$	$Fd\bar{3}m$	8.7431					
Cs	3/8	3/8	3/8	8⟨3.47⟩	0.7		
Ta	1/8	1/8	1/8	4⟨1.93⟩	5.2		
1/6N	0	0.0311	0.9689	6⟨2.96⟩	3.0		
Cu$_3$N	$Pm\bar{3}m$	3.813					
Cu	1/2	0	0	2⟨1.91⟩	0.9	1.0	−14.03
N	0	0	0	6⟨1.91⟩	2.7	−3.0	31.06

Table 10. (*Contd.*)

Atom	x	y	z	d	Σv_i	V	ϕ_i
FeNiN	P4/mmm	2.830	3.713				
Fe	0	0	0	4⟨2.00⟩	2.7		
Ni	1/2	1/2	1/2	2⟨1.86⟩	1.5		
N	1/2	1/2	0	6⟨1.95⟩	4.2		
Ge$_2$N$_2$O	Cmc2$_1$	9.312	5.755	5.105			
Ge	0.173	0.149	0.321	4⟨1.81⟩	4.4	4.0	−46.41
N	0.205	0.136	0.666	3⟨1.83⟩	3.4	−3.0	40.14
O	0	0.246	0.230	2⟨1.77⟩	1.9	−2.0	26.81
α-Ge$_3$N$_4$	P3̄1c	8.1960	5.9301				
Ge(1)	0.080	0.508	0.643	4⟨1.82⟩	5.1	4.0	−49.95
Ge(2)	0.256	0.169	0.446	4⟨1.89⟩	4.0	4.0	−45.31
N(1)	0.643	0.597	0.381	3⟨1.83⟩	3.8	−3.0	39.89
N(2)	0.313	0.329	0.703	3⟨1.83⟩	3.4	−3.0	39.29
N(3)	1/3	2/3	0.55	3⟨1.90⟩	2.8	−3.0	34.40
N(4)	0	0	0.35	3⟨1.93⟩	2.6	−3.0	38.70
β-Ge$_3$N$_4$	P6$_3$/m	8.0276	3.0774				
Ge	0.1712	0.7658	1/4	4⟨1.83⟩	4.5	4.0	−48.06
N(1)	1/3	2/3	1/4	3⟨1.83⟩	3.4	−3.0	38.01
N(2)	0.3335	0.0295	1/4	3⟨1.83⟩	3.4	−3.0	39.29
La$_6$Cr$_{21}$N$_{23}$	Fm3̄m	12.98					
La	1/3	0	0	6⟨2.16⟩	9.7	3.0	−34.89
Cr(1)	0	1/3	1/3	6⟨2.16⟩	2.6	3.0	−34.90
Cr(2)	1/3	1/3	1/3	6⟨2.16⟩	2.6	3.0	−34.90
Cr(3)	0	0	0	6⟨2.16⟩	2.6	3.0	−34.89
N(1)	1/2	1/6	1/6	6⟨2.16⟩	4.9	−3.0	34.90
N(2)	1/6	0	0	6⟨2.16⟩	3.8	−3.0	34.90
N(3)	1/6	1/6	1/6	6⟨2.16⟩	2.6	−3.0	34.90
N(4)	1/2	1/2	1/2	6⟨2.16⟩	9.7	−3.0	34.89
LiPN$_2$	I4̄2d	4.575	7.118				
Li	0	0	1/2	4⟨2.09⟩	1.1	1.0	−16.45
P	0	0	0	4⟨1.64⟩	5.0	5.0	−61.77
N	0.170	1/4	1/8	4⟨1.87⟩	3.1	−3.0	42.83
LiSrN	P4$_2$/mmc	3.924	7.085				
Li	1/2	1/2	0	2⟨1.96⟩	0.8	1.0	−14.70
Sr	0	0	1/4	4⟨2.64⟩	1.3	2.0	−18.37
N	0	1/2	0	6⟨2.42⟩	2.1	−3.0	27.96
Li$_2$CeN$_2$	P3̄m1	3.557	5.496				
Li	1/3	2/3	0.65	4⟨2.10⟩	1.1	1.0	−18.32
Ce	0	0	0	6⟨2.53⟩	3.6	4.0	−32.47
N	1/3	2/3	0.27	7⟨2.28⟩	2.8	−3.0	30.40
Li$_2$ThN$_2$	P3̄	6.398	5.547				
Li	0.35	0.03	0.63	4⟨2.16⟩	1.0	1.0	−17.94
Th(1)	0	0	0	6⟨2.62⟩	2.9	4.0	−32.20
Th(2)	1/3	2/3	0.02	6⟨2.60⟩	2.9	4.0	−32.34
N	0.34	0.01	0.27	7⟨2.35⟩	2.4	−3.0	29.64
Li$_2$ZrN$_2$	P3̄m1	3.282	5.460				
Li	1/3	2/3	0.61	4⟨2.08⟩	1.1	1.0	−16.73
Zr	0	0	0	6⟨2.27⟩	3.9	4.0	−38.76
N	1/3	2/3	0.23	7⟨2.16⟩	3.0	−3.0	33.86
α-Li$_3$N	P6/mmm	3.648	3.875				
Li(1)	0	0	1/2	2⟨1.94⟩	0.8	1.0	−13.11
Li(2)	1/3	2/3	0	3⟨2.11⟩	0.8	1.0	−12.49
N	0	0	0	8⟨2.06⟩	2.4	−3.0	31.08

Table 10. (*Contd.*)

Atom	x	y	z	d	Σv_i	V	ϕ_i
β-Li₃N	P6₃/mmc	3.552	6.311				
Li(1)	0	0	1/4	3⟨2.05⟩	0.9	1.0	−12.43
Li(2)	1/3	2/3	0.583	4⟨2.25⟩	0.7	1.0	−10.99
N	1/3	2/3	1/4	11⟨2.20⟩	2.4	−3.0	31.79
α-Li₃BN₂	P4₂2₁2	4.6435	5.2592				
Li(1)	0	0	1/2	2⟨1.95⟩	0.8	1.0	−12.61
Li(2)	0	1/2	1/4	4⟨2.13⟩	1.0	1.0	−13.44
B	0	0	0	2⟨1.34⟩	2.9	3.0	−44.00
N	0.2962	0.2962	1/2	6⟨1.96⟩	2.9	−3.0	38.94
β-Li₃BN₂	P2₁/c	5.1502	7.0824 6.7908	112.956			
Li(1)	0.24995	0.48518	0.49811	4⟨2.20⟩	0.9	1.0	−13.26
Li(2)	0.25228	0.01212	0.37502	4⟨2.21⟩	0.9	1.0	−11.87
Li(3)	0.74360	0.20714	0.31402	4⟨2.13⟩	1.0	1.0	−14.19
B	0.21509	0.31982	0.17648	2⟨1.34⟩	2.9	3.0	−43.10
N(1)	0.43403	0.43735	0.21904	7⟨2.05⟩	2.9	−3.0	38.81
N(2)	0.99439	0.20472	0.13461	7⟨2.08⟩	2.8	−3.0	38.79
L-Mg₂NF	I4₁/amd	4.186	10.042				
Mg	0	1/4	0.1595	5⟨2.13⟩	1.9	2.0	−23.43
N	0	1/4	3/8	6⟨2.14⟩	2.8	−3.0	32.31
F	0	3/4	1/8	4⟨2.12⟩	0.9	−1.0	13.71
Mg₃NF₃	Pm$\bar{3}$m	4.216					
Mg	0	1/2	1/2	6⟨2.11⟩	2.0	2.0	−22.96
N	1/2	1/2	1/2	6⟨2.11⟩	3.0	−3.0	33.08
F	1/2	0	0	4⟨2.11⟩	1.0	−1.0	12.85
NaTaN₂	R$\bar{3}$m	3.139	16.925	(hexagonal axes)			
Na	0	0	1/2	6⟨2.51⟩	1.3	1.0	−16.53
Ta	0	0	0	6⟨2.11⟩	4.4	5.0	−49.98
N	0	0	0.2690	6⟨2.31⟩	2.8	−3.0	36.45
Na₃BN₂	P2₁/c	5.717	7.931 7.883	111.32			
Na(1)	0.2511	0.4687	0.4921	4⟨2.53⟩	0.8	1.0	−10.67
Na(2)	0.2413	0.0218	0.3778	4⟨2.58⟩	0.7	1.0	−9.24
Na(3)	0.7424	0.2080	0.3131	4⟨2.45⟩	1.0	1.0	−12.03
B	0.2133	0.3226	0.1760	2⟨1.34⟩	2.8	3.0	−44.23
N(1)	0.4158	0.4237	0.2241	7⟨2.34⟩	2.7	−3.0	36.44
N(2)	0.0120	0.2206	0.1303	7⟨2.37⟩	2.7	−3.0	36.45
NbMoN	P4/nmm	3.095	7.799				
Nb	3/4	3/4	0.335	5⟨2.22⟩	3.3		
Mo	3/4	3/4	0.900	1⟨2.21⟩	0.6		
N	3/4	3/4	0.617	6⟨2.21⟩	3.9		
NbON	P2₁/c	4.970	5.033 5.193	100.23			
Nb	0.2911	0.0472	0.2151	7⟨2.09⟩	5.5	5.0	−54.54
O	0.0636	0.3244	0.3476	3⟨2.07⟩	2.0	−2.0	24.84
N	0.4402	0.7546	0.4782	4⟨2.11⟩	3.5	−3.0	35.91
Nd₂AlO₃N	I4mm	3.7046	12.5301				
Al	0	0	0	6⟨1.94⟩	3.0	3.0	−38.10
Nd(1)	0	0	0.3649	9⟨2.60⟩	2.8	3.0	−26.61
Nd(2)	0	0	0.6517	9⟨2.57⟩	3.4	3.0	−32.08
O(1)	0	1/2	0.0086	6⟨2.34⟩	2.3	−2.0	24.23
O(2)	0	0	0.8335	6⟨2.49⟩	1.9	−2.0	23.61
N	0	0	0.1700	6⟨2.52⟩	2.7	−3.0	26.77

Table 10. (*Contd.*)

Atom	x	y	z	d	Σv_i	V	ϕ_i
Si$_2$N$_2$O	Cmc2$_1$	8.843	5.473	4.835			
Si	0.1763	0.1509	0.2898	4⟨1.69⟩	4.5	4.0	−49.95
N	0.218	0.121	0.642	3⟨1.72⟩	3.5	−3.0	42.55
O	0	0.214	0.230	2⟨1.62⟩	2.0	−2.0	29.74
α-Si$_3$N$_4$	P31c	7.818	5.591				
Si(1)	0.829	0.5135	0.6558	4⟨1.74⟩	4.3	4.0	−50.20
Si(2)	0.2555	0.1682	0.4509	4⟨1.74⟩	4.3	4.0	−50.25
N(1)	0.6558	0.6075	0.4320	3⟨1.75⟩	3.2	−3.0	41.23
N(2)	0.3154	0.3192	0.6962	5⟨2.16⟩	3.4	−3.0	41.70
N(3)	1/3	2/3	0.5926	6⟨2.23⟩	3.3	−3.0	40.56
N(4)	0	0	0.4503	3⟨1.76⟩	3.1	−3.0	40.81
β-Si$_3$N$_4$	P6$_3$/m	7.606	2.909				
Si	0.174	0.766	1/4	4⟨1.74⟩	4.4	4.0	−50.46
N(1)	1/3	2/3	1/4	3⟨1.72⟩	3.4	−3.0	42.93
N(2)	0.321	0.025	1/4	3⟨1.74⟩	3.2	−3.0	40.90
Sr$_2$N	R$\bar{3}$m	3.8570	20.6979	(hexagonal axes)			
Sr	0	0	0.26736	3⟨2.61⟩	1.1		
N	0	0	0	6⟨2.61⟩	2.1		
TaON	P2$_1$/c	4.9581	5.0267	5.1752 99.640			
Ta	0.292	0.046	0.213	7⟨2.09⟩	5.4	5.0	−54.70
O	0.064	0.324	0.345	3⟨2.07⟩	2.0	−2.0	24.81
N	0.4449	0.7566	0.4810	4⟨2.10⟩	3.3	−3.0	36.13
Ta$_3$N$_5$	Cmcm	3.88618	10.2119	10.2624			
Ta(1)	0	0.1971	1/4	6⟨2.08⟩	4.9	5.0	−54.69
Ta(2)	0	0.13455	0.55905	6⟨2.09⟩	4.9	5.0	−54.00
N(1)	0	0.76322	1/4	4⟨2.14⟩	2.8	−3.0	37.27
N(2)	0	0.04700	0.11949	3⟨2.01⟩	2.9	−3.0	35.50
N(3)	0	0.30861	0.7378	4⟨2.11⟩	3.0	−3.0	38.48
ThNF	R$\bar{3}$m	7.13	32.66°				
Th	0.239	0.239	0.239	8⟨2.49⟩	4.5	4.0	−39.08
N	0.122	0.122	0.122	4⟨2.38⟩	3.6	−3.0	29.33
F	0.368	0.368	0.368	4⟨2.61⟩	0.9	−1.0	11.97
ThNBr	P4/nmm	4.110	7.468				
Th	3/4	3/4	0.151	9⟨2.92⟩	4.8	4.0	−37.78
N	3/4	1/4	0	4⟨2.34⟩	4.0	−3.0	30.39
Br	3/4	3/4	0.637	5⟨3.37⟩	0.9	−1.0	7.80
ThNCl	P4/nmm	4.097	6.895				
Th	3/4	3/4	0.165	9⟨2.82⟩	4.8	4.0	−38.46
N	3/4	1/4	0	4⟨2.34⟩	4.0	−3.0	29.50
Cl	3/4	3/4	0.635	5⟨3.21⟩	0.8	−1.0	8.64
ThNI	P4/nmm	4.107	9.242				
Th	3/4	3/4	0.124	9⟨3.06⟩	4.8	4.0	−36.05
N	3/4	1/4	0	4⟨2.35⟩	3.9	−3.0	32.01
I	3/4	3/4	0.668	4⟨3.48⟩	0.9	−1.0	6.82
Th$_2$N$_2$O	P$\bar{3}$m1	3.8833	6.1870				
Th	1/3	2/3	1/4	7⟨2.52⟩	4.5	4.0	−39.29
N	1/3	2/3	0.631	4⟨2.36⟩	3.8	−3.0	27.93
O	0	0	0	6⟨2.72⟩	1.3	−2.0	23.34
Th$_3$N$_4$	R$\bar{3}$m	3.875	27.39	(hexagonal axes)			
Th(1)	0	0	0	6⟨2.53⟩	3.6	4.0	−35.76
Th(2)	0	0	0.2221	7⟨2.59⟩	4.6	4.0	−40.52
N(1)	0	0	0.1320	4⟨2.35⟩	4.0	−3.0	28.94
N(2)	0	0	0.3766	6⟨2.72⟩	2.4	−3.0	30.64

Table 10. (*Contd.*)

Atom	x	y	z	d	Σv_i	V	ϕ_i
TiNBr	*Pmmn*	3.927	3.349	8.332			
Ti	1/4	3/4	0.80	6⟨2.22⟩	4.2	4.0	−36.50
N	1/4	1/4	0.950	4⟨1.99⟩	3.6	−3.0	45.84
Br	1/4	1/4	0.330	2⟨2.67⟩	0.8	−1.0	5.96
TiNCl	*Pmmn*	3.937	3.258	7.803			
Ti	1/4	3/4	0.100	6⟨2.15⟩	4.3	4.0	−39.33
N	1/4	1/4	0.950	4⟨2.01⟩	3.3	−3.0	42.25
Cl	1/4	1/4	0.330	2⟨2.42⟩	1.0	−1.0	8.48
TiNI	*Pmmn*	3.941	3.515	8.955			
Ti	1/4	3/4	0.80	6⟨2.26⟩	4.3	4.0	−36.78
N	1/4	1/4	0.970	4⟨2.02⟩	3.2	−3.0	44.47
I	1/4	1/4	0.315	2⟨2.74⟩	1.1	−1.0	6.60
UNF	*P4/n*	3.951	5.724				
U	1/4	1/4	0.2024	8⟨2.45⟩	4.3	4.0	−40.45
N	1/4	3/4	0	4⟨2.29⟩	3.5	−3.0	29.38
F	1/4	3/4	1/2	4⟨2.61⟩	0.8	−1.0	12.65
UNBr	*P4/nmm*	3.944	7.950				
U	3/4	3/4	0.144	9⟨2.89⟩	4.4	4.0	−38.56
N	3/4	1/4	0	4⟨2.28⟩	3.6	−3.0	31.18
Br	3/4	3/4	0.650	5⟨3.37⟩	0.8	−1.0	7.94
UNCl	*P4/nmm*	3.979	6.811				
U	3/4	3/4	0.169	9⟨2.77⟩	4.2	4.0	−39.23
N	3/4	1/4	0	4⟨2.30⟩	3.4	−3.0	29.91
Cl	3/4	3/4	0.616	5⟨3.15⟩	0.8	−1.0	9.10
UNI	*P4/nmm*	3.990	9.206				
U	3/4	3/4	0.121	9⟨3.00⟩	4.4	4.0	−36.78
N	3/4	1/4	0	4⟨2.28⟩	3.5	−3.0	33.32
I	3/4	3/4	0.669	5⟨3.57⟩	0.8	−1.0	6.82
V_5P_3N	*P6_3/mcm*	6.880	4.771				
V(1)	1/3	2/3	0	6⟨2.43⟩	4.4		
V(2)	0.245	0	1/4	7⟨2.37⟩	4.4		
P	0.605	0	1/4	9⟨2.46⟩	6.1		
N	0	0	0	6⟨2.06⟩	3.4		
α-ZrNBr	*Pmmn*	4.116	3.581	8.701			
Zr	1/4	3/4	0.095	6⟨2.36⟩	4.6	4.0	−36.08
N	1/4	1/4	0.965	4⟨2.12⟩	3.9	−3.0	41.12
Br	1/4	1/4	0.350	2⟨2.85⟩	0.7	−1.0	5.64
ZrNI	*Pmmn*	4.114	3.724	9.431			
Zr	1/4	3/4	0.092	6⟨2.42⟩	4.5	4.0	−35.33
N	1/4	1/4	0.977	4⟨2.16⟩	3.5	−3.0	40.54
I	1/4	1/4	0.335	2⟨2.95⟩	1.0	−1.0	5.76
Zr_3AlN	*Cmcm*	3.3690	11.498	8.925			
Zr(1)	0	0.3719	0.420	1⟨1.52⟩	4.9		
Zr(2)	0	0.0446	1/4	4⟨2.84⟩	0.6		
Al	0	0.7474	1/4	8⟨2.90⟩		(Al–Zr only)	
N	1/2	0	0	6⟨2.40⟩	10.4		

Crystal Chemistry of Inorganic Nitrides

Table 11. Unit cell dimensions for materials without a refined structure. "angle" refers to α for rhombohedral crystals and β for monoclinic crystals

Compound	Spacegroup	a	b	c	angle
BaND	$I4/mmm$	4.062		6.072	
BaSiN$_2$	Cmc^*	5.585	7.541	11.340	
Ba$_3$N$_4$	Hexag	5.22		5.50	
Ba$_3$Ge$_3$N$_6$	Tetra	8.97		6.96	
Ba$_9$Re$_3$N$_{10}$	Ortho	10.94	8.09	30.40	
Ba$_9$Os$_3$N$_{10}$	Ortho	10.88	8.08	29.80	
BeThN$_2$	Hexag	10.501		3.955	
CaTaO$_2$N	Rhomb	5.583			60.6
γ-Ca$_3$N$_2$	Ortho	17.82	11.56	3.58	
Ca$_5$MoN$_5$	Hexag	11.40		7.45	
Ca$_5$WN$_5$	Hexag	11.40		7.45	
Ca$_{21}$Fe$_3$N$_{17}$	Mono	6.2081	6.103	18.7425	95.1793
Ca$_{21}$Ga$_3$N$_{17}$	Mono	6.29	6.13	18.94	95.3
Ce$_2$AlO$_3$N	$I4/mmm$	3.736		12.72	
Ce$_2$N$_2$O	Hexag	3.880		6.057	
Dy$_2$Si$_3$O$_3$N$_4$	$P42_1m$	7.60		4.93	
Dy$_4$Si$_2$O$_7$N$_2$	$P2_1/c$	7.60	10.49	10.81	111.7
Er$_2$Si$_3$O$_3$N$_4$	$P42_1m$	7.56		4.90	
Er$_4$Si$_2$O$_7$N$_2$	$P2_1/c$	7.53	10.40	10.80	111.1
Eu$_2$AlO$_3$N	$I4/mmm$	3.682		12.38	
Gd$_2$Si$_3$O$_3$N$_4$	$P42_1m$	7.63		4.98	
Gd$_4$Si$_2$O$_7$N$_2$	$P2_1/c$	7.66	10.52	10.93	111.0
Ho$_2$Si$_3$O$_3$N$_4$	$P42_1m$	7.59		4.92	
Ho$_4$Si$_2$O$_7$N$_2$	$P2_1/c$	7.57	10.46	10.81	111.0
La$_2$AlO$_3$N	$I4/mmm$	3.789		12.827	
La$_2$Si$_3$O$_3$N$_4$	$P42_1m$	7.89		5.11	
La$_4$Si$_2$O$_7$N$_2$	$P2_1/c$	8.03	10.99	11.05	110.1
LiBaN	Hexag	6.79		8.05	
LiBeN	Ortho	8.75	8.16	7.65	
LiCaN	$Pnma$	8.471	3.676	5.537	
β-LiSiON	Mono	9.190	4.779	5.428	90.51
Li$_2$HfN$_2$	Hexag	3.253		5.457	
Li$_2$UN$_2$	Hexag	5.902		5.324	
Li$_3$FeN$_2$	Ortho	9.65	8.66	8.38	
Li$_4$FeN$_2$	$Immm$	3.710	6.413	7.536	
Li$_4$SrN$_2$	$I4_1/amd$	3.822		27.042	
MgAlON	$Pna2_1$	5.361	6.94	5.08	
NaGeON	$Pca2_1$	5.584	7.403	5.037	
NaSiON	$Pca2_1$	5.346	7.186	4.823	
Nd$_2$Si$_3$O$_3$N$_4$	$P42_1m$	7.75		5.034	
Nd$_4$Si$_2$O$_7$N$_2$	$P2_1/c$	7.839	10.737	11.022	110.64
Pr$_2$AlO$_3$N	$I4/mmm$	3.715		12.60	
Si$_2$ON	Ortho	5.498	8.877	4.853	
Sm$_2$AlO$_3$N	$I4/mmm$	3.690		12.371	
Sm$_2$Si$_3$O$_3$N$_4$	$P42_1m$	7.69		4.99	
Sm$_3$Si$_6$N$_{11}$	Tetrag	9.9931		4.8361	
Sm$_4$Si$_2$O$_7$N$_2$	$P2_1/c$	7.75	10.60	10.99	111.1
Sm$_{10}$Si$_6$O$_{24}$N$_2$	Hexag	9.156		6.980	
Sr$_8$N$_5$	Mono	7.19	3.85	6.65	108
Sr$_9$Re$_3$N$_{10}$	Ortho	10.38	7.70	28.62	
Sr$_{21}$Ga$_3$N$_{17}$	Mono	6.62	6.48	20.06	94.7
Sr$_{27}$Mo$_5$N$_{28}$	Cubic	5.25			
Sr$_{27}$W$_5$N$_{28}$	Cubic	5.25			
Sr$_{27}$Re$_5$N$_{28}$	Cubic	5.25			

Table 11. (*Contd.*)

Th$_2$N$_2$ (NH)	Cm	7.167	3.860	6.242	92.21
UN	R$\bar{3}$m	4.657			85.8
Y$_2$Si$_3$O$_3$N$_4$	P42$_1$m	7.59		4.92	
Y$_4$Si$_2$O$_7$N$_2$	P2$_1$/c	7.59	10.44	10.89	111.1
Yb$_2$Si$_3$O$_3$N$_4$	P42$_1$m	7.55		4.88	
Yb$_4$Si$_2$O$_7$N$_2$	P2$_1$/c	7.47	10.32	10.77	111.3
ZnAlON	Pna2$_1$	5.332	7.006	5.109	
β-ZrNBr	Hex	2.100		9.751	
β-ZrNCl	Hex	2.081		9.234	
ZrN$_{0.906}$F$_{1.28}$	Abm2	5.186	5.374	5.368	(subcell)

Table 12. Bond valence parameters for nitrides and bond lengths for regular coordination geometries sorted by bond length. V is the valence and "4", "6" and "8" refer to coordination numbers

atom	V	R_{iN}	4	6	8
B	3	1.47	1.57	1.72	1.83
P	5	1.73	1.64	1.79	1.90
Be	2	1.50	1.75	1.90	2.01
Si	4	1.77	1.77	1.92	2.03
V	5	1.86	1.77	1.92	2.03
Ge	4	1.88	1.88	2.03	2.14
Mo	6	2.04	1.89	2.04	2.15
Al	3	1.79	1.89	2.04	2.15
W	6	2.06	1.91	2.06	2.17
Ta	5	2.01	1.92	2.07	2.18
Ti	4	1.93	1.93	2.08	2.19
Ga	3	1.84	1.94	2.09	2.20
Cr	3	1.85	1.95	2.10	2.21
Nb	5	2.06	1.97	2.12	2.23
Ni	2	1.75	2.00	2.15	2.26
Zn	2	1.77	2.02	2.17	2.28
Sc	3	1.98	2.08	2.23	2.34
Hf	4	2.09	2.09	2.24	2.35
Co	2	1.84	2.09	2.24	2.35
Mg	2	1.85	2.10	2.25	2.36
Zr	4	2.11	2.11	2.26	2.37
Fe	2	1.86	2.11	2.26	2.37
Cu	1	1.61	2.12	2.27	2.38
Li	1	1.61	2.12	2.27	2.38
Mn	2	1.87	2.12	2.27	2.38
In	3	2.03	2.13	2.28	2.39
Lu	3	2.11	2.21	2.36	2.47
Cd	2	1.96	2.21	2.36	2.47
Yb	3	2.12	2.22	2.37	2.48
U	4	2.24	2.24	2.39	2.50
Er	3	2.16	2.26	2.41	2.52
Y	3	2.17	2.27	2.42	2.53
Dy	3	2.18	2.28	2.43	2.54

Table 12. (*Contd.*)

Ho	3	2.18	2.28	2.43	2.54
Tb	3	2.20	2.30	2.45	2.56
Gd	3	2.22	2.32	2.47	2.58
Th	4	2.34	2.34	2.49	2.60
Eu	3	2.24	2.34	2.49	2.60
Sm	3	2.24	2.34	2.49	2.60
Ca	2	2.14	2.39	2.54	2.65
Nd	3	2.30	2.40	2.55	2.66
Pr	3	2.30	2.40	2.55	2.66
Na	1	1.93	2.44	2.59	2.70
Ce	3	2.34	2.44	2.59	2.70
La	3	2.34	2.44	2.59	2.70
Pb	2	2.22	2.47	2.62	2.73
Sr	2	2.23	2.48	2.63	2.74
Ba	2	2.47	2.72	2.87	2.98
K	1	2.26	2.77	2.92	3.03
Rb	1	2.37	2.88	3.03	3.14
Cs	1	2.53	3.04	3.19	3.30

References

1. Chase MW Jr, Davies CA, Downey JR Jr, Frurip DJ, McDonald RA, Syverud AN (1986) JANAF Thermochemical Tables, 3rd edn, American Institute of Physics, New York
2. Lang J (1976) NATO Adv Study Inst Ser, Series E 23: 89
3. Lang J, Laurent Y, Maunaye M, Marchand R (1979) Prog Crystal Growth Charact 2: 207
4. Schubert K (1982) Cryst Res Tech 17: 553
5. Hägg G (1930) Z Physik Chem, Abt B6: 221
6. Rundle RE (1948) Acta Cryst 1: 180
7. Gelatt CD Jr, Williams AR, Moruzzi VL (1983) Phys Rev B27: 2005
8. Pauling L (1960) The nature of the chemical bond, 3rd edn, Cornell University Press, Ithaca, NY
9. Brown ID (1981) In: O'Keeffe M, Navrotsky A (eds) Structure and bonding in crystals Academic, New York
10. O'Keeffe M (1989) Struct Bonding 71: 161
11. O'Keeffe M (1990) Acta Cryst A46: 138
12. Brown ID, Altermatt D (1985) Acta Cryst B41: 244
13. Shannon RD (1976) Acta Cryst A32: 751
14. Baur WH (1987) Cryst Rev 1: 59
15. Brese NE, O'Keeffe M (1991) Acta Cryst B47: 192
16. O'Keeffe M, Brese NE (1991) J Amer Chem Soc 113: 3226
17. O'Keeffe M (1990) J Solid State Chem 85: 108
18. Hoppe R (1970) Adv Fluorine Chem 6: 387
19. Tosi MP (1964) Solid State Phys 16: 1
20. Wells AF (1987) Structural Inorganic Chemistry, 5th edn, Oxford University Press
21. Hyde BG, Andersson S (1989) Inorganic Crystal Structures, New York, John Wiley
22. Reed JW, Harris PM (1961) J Chem Phys 35: 1730
23. Schomaker V, Spurr R (1942) J Amer Chem Soc 64: 1184
24. Winnewisser M, Cook RL (1964) J Chem Phys 41: 999
25. Lievin J, Breulet J, Verhaegen G (1979) Theor Chim Acta 52: 75

26. Gimarc BM, Woodcock DA (1981) J Mol Struct, THEOCHEM 85: 37
27. Prince E, Choi CS (1978) Acta Cryst B34: 2606
28. Chiglien G, Etienne J, Jaulmes S, Laruelle P (1974) Acta Cryst B30: 2229
29. Rabenau A, Schulz H (1976) J Less-Common Met 50: 155
30. Schulz H, Thiemann KH (1979) Acta Cryst A35: 309
31. Richards PM (1980) J Solid State Chem 33: 127
32. Schönherr E, Müller G, Winckler B (1978) J Cryst Growth 43: 469
33. Sachsze W, Juza R (1949) Z Anorg Allg Chem 259: 278
34. Gudat A, Kniep R, Rabenau A (1991) Angew Chem Int Ed 30: 199
35. Cordier G, Gudat A, Kniep R, Rabenau A (1989) Angew Chem Int Ed 28: 1702
36. Beister HJ, Haag S, Kniep R, Strössner K, Syassen K (1988) Angew Chem Int Ed 27: 1101
37. Pringle GE, Noakes DE (1968) Acta Cryst B24: 262
38. Juza R, Opp K (1951) Z Anorg Allg Chem 266: 313
39. Brice J, Motte J-P, Aubry J (1973) C R Acad Sci Paris C276: 1015
40. Brese NE, O'Keeffe M (unpublished results)
41. Juza R, Opp K (1951) Z Anorg Allg Chem 266: 325
42. Juza R, Langer K, Benda KV (1968) Angew Chem Int Ed 7: 360
43. Juza R, Uphoff W, Gieren W (1957) Z Anorg Allg Chem 292: 71
44. Schoch B, Hartmann E, Weppner W (1986) Solid State Ionics 18 & 19: 529
45. Juza R, Hund F (1948) Z Anorg Allg Chem 257: 1
46. Juza R, Hund F (1948) Z Anorg Allg Chem 257: 13
47. Gudat A, Kniep R, Rabenau A, Bronger W, Ruschewitz U (1990) J Less-Common Met 161: 31
48. Juza R, Weber HH, Meyer-Simon E (1953) Z Anorg Allg Chem 273: 48
49. Juza R, Anschutz E, Puff H (1959) Angew Chem 71: 161
50. Schnick W, Luecke J (1990) J Solid State Chem 87: 101
51. Juza R, Gieren W, Haug J (1959) Z Anorg Allg Chem 300: 61
52. Juza R, Haug J (1961) Z Anorg Allg Chem 309: 276
53. Gudat A, Haag S, Kniep R, Rabenau A (1990) Z Naturforsch 45b: 111
54. Palisaar A, Juza R (1971) Z Anorg Allg Chem 384: 1
55. Barker MG, Alexander IC (1974) J Chem Soc Dalton Trans 2166
56. O'Keeffe M, Hyde BG (1985) Struct Bonding 61: 77
57. Marchand R, L'Haridon P, Laurent Y (1982) J Solid State Chem 43: 126
58. Schnick W, Lücke J (1990) Z Anorg Allg Chem 588: 19
59. David J, Laurent Y, Charlot J, Lang J (1973) Bull Soc fr Miner Crist 96: 21
60. David J, Charlot J, Lang J (1974) Rev Chim Minér 11: 405
61. Laurent Y, Guyader J, Roult G (1981) Acta Cryst B37: 911
62. Laurent Y, Guyader J, Roult G (1982) Mater Sci Mongr React of Solids 2: 647
63. Cordier G, Gudat A, Kniep R, Rabenau A (1989) Angew Chem Int Ed 28: 201
64. Gerss MH, Jeitschko W (1986) Z Naturforsch B41: 946
65. Brese NE, Squattrito PJ, Ibers JA (1985) Acta Cryst C41: 1829
66. Yamane H, Kikkawa S, Horiuchi H, Koizumi M (1986) J Solid State Chem 65: 6
67. Yamane H, Kikkawa S, Koizumi M (1987) J Power Sources 20: 311
68. Yamane H, Kikkawa S, Koizumi M (1987) J Solid State Chem 71: 1
69. Iizuka E (1983) Chem Abs 99: 162956
70. Lang J, Charlot J (1970) Rev Chim Minér 7: 121
71. Dadd AT, Hubberstey P (1982) J Chem Soc Dalton Trans 2175
72. Le Mehaute A, Marchand R (1984) Mater Res Bull 19: 1251
73. Brice J-F, Aubry J (1970) C R Acad Sci Paris C271: 825
74. Brice J, Motte J-P, Streiff R (1969) C R Acad Sci Paris C269: 910
75. Wattenberg H (1930) Ber D Chem Gesellschaft 63: 1667
76. Janeff W (1955) Z Physik 142: 619
77. Choi CS, Prince E (1976) J Chem Phys 64: 4510
78. Zalkin A, Templeton DH (1956) J Phys Chem 60: 821
79. Nagib M, Kistrup H, Jacobs H (1975) Atomkernenerg 26: 87
80. Guyader J, L'Haridon P, Laurent Y, Jacquet R, Roult G (1984) J Solid State Chem 54: 251
81. Guyader J, Malhaire J, Laurent Y (1982) Rev Chim Minér 19: 701
82. Jacobs H, Pinkowski EV (1989) J Less-Common Met 146: 147
83. Evers J, Münsterkötter M, Oehlinger G, Polborn K, Sendlinger B (1990) J Less-Common Met 162: L17

84. Parsons RB, Yoffe AD (1966) Acta Cryst 20: 36
85. Müller U (1972) Z Anorg Allg Chem 392: 159
86. Ibers JA (1964) J Chem Phys 40: 402
87. Bohart GS (1914) J Phys Chem 19: 537
88. Juza R, Jacobs H, Klose W (1965) Z Anorg Allg Chem 338: 171
89. Jacobs H, Osten EV (1976) Z Naturforsch 31b: 385
90. Nagib M, Jacobs H, Osten EV (1977) Atomkernenerg 29: 303
91. Nagib M, Osten EV, Jacobs H (1977) Atomkernenerg 29: 41
92. Guyader J, Jacquet R, Malhaire JM, Roult G, Laurent Y (1983) Rev Chim Minér 20: 863
93. Schmitz-Dumont O, Kron N (1955) Z Anorg Allg Chem 280: 180
94. Laurent Y, Pastuszak R, L'Haridon P, Marchand R (1982) Acta Cryst B38: 914
95. Juza R, Mehne A (1959) Z Anorg Allg Chem 299: 33
96. Pastuszak R, L'Haridon P, Marchand R, Laurent Y (1982) Acta Cryst B38: 1427
97. Eckerlin P, Rabenau A (1960) Z Anorg Allg Chem 304: 218
98. Stackelberg MV, Paulus R (1933) Z Physik Chem, Abt B22: 305
99. Hall D, Gurr GE, Jeffrey GA (1969) Z Anorg Allg Chem 369: 108
100. Eckerlin P (1967) Z Anorg Allg Chem 353: 225
101. Shaw TM, Thomas G (1980) J Solid State Chem 33: 63
102. Juza R, Jacobs H, Gerke H (1966) Ber Beunsenges Phys Chem 70: 1103
103. Jacobs H (1976) Z Anorg Allg Chem 427: 1
104. Hägg G (1930) Z Krist 74: 95
105. David J, Laurent Y, Lang J (1971) Bull Soc fr Miner Crist 94: 340
106. Jacobs H (1971) Z Anorg Allg Chem 382: 97
107. Wintenberger M, Tcheou F, David J, Lang J (1980) Z Naturforsch 35B: 604
108. David J, Lang J (1967) C R Acad Sci Paris C265: 581
109. David J, Laurent Y, Lang J (1970) Bull Soc fr Miner Crist 93: 153
110. Marchand R, Laurent Y (1982) Mater Res Bull 17: 399
111. Shavers CL (1981) Theoretical and Experimental Studies in Crystal Chemistry (1981) thesis, Arizona State University
112. Andersson S (1970) J Solid State Chem 1: 306
113. David J, Lang J (1969) C R Acad Sci Paris C269: 771
114. Laurent PY, Lang J, LeBihan MT (1968) Acta Cryst B24: 494
115. Laurent Y, David J, Lang J (1964) C R Acad Sci Paris C259: 1132
116. Laurent Y, Lang J, Bihan MTL (1969) Acta Cryst B25: 199
117. Keve ET, Skapski AC (1968) Inorg Chem 7: 1757
118. Brice J-F, Motte J-P, Aubry J (1975) Rev Chim Minér 12: 105
119. Hadenfeldt C, Herdejürgen H (1987) Z Anorg Allg Chem 545: 177
120. Galy J, Jaccou M, Andersson S (1971) C R Acad Sci Paris C272: 1657
121. Ehrlich P, Linz W, Seifert HJ (1971) Naturwiss 58: 219
122. Emons H, Anders D, Roewer G, Vogt F (1964) Z Anorg Allg Chem 333: 99
123. Brice J, Motte J, Courtois A, Protas J, Aubry J (1976) J Solid State Chem 17: 135
124. Dutoit P, Schnorf A (1928) Compt Rend 187: 300
125. Hartmann H, Frohlich HJ, Ebert F (1934) Z Anorg Allg Chem 218: 181
126. Maunaye M, Guyader J, Laurent Y, Lang J (1971) Bull Soc fr Miner Crist 94: 347
127. David J, Lang J (1965) C R Acad Sci Paris C261: 1005
128. Verdier P, L'Haridon P, Maunaye M, Marchand R (1974) Acta Cryst B30: 226
129. Chern MY, DiSalvo FJ (1990) J Solid State Chem 88: 528
130. Chern MY, DiSalvo FJ (1990) J Solid State Chem 88: 459
131. Vennos DA, Badding ME, DiSalvo FJ (1990) Inorg Chem 29: 4059
132. Marchand R, Pors F, Laurent Y (1986) Rev Int Hautes Tempér Réfract 23: 11
133. Verdier P, Marchand R, Lang J (1976) Revue Chim Minér 13: 214
134. Linke K, Lingmann H (1969) Z Anorg Allg Chem 366: 89
135. Linke K, Lingmann H (1969) Z Anorg Allg Chem 366: 82
136. Laurent Y, Lang J (1966) C R Acad Sci Paris C262: 103
137. Karam R, Ward R (1970) Inorg Chem 9: 1849
138. Guntz A, Benoit F (1923) Ann Chim 20: 5
139. Goubeau J, Anselment W (1961) Z Anorg Allg Chem 310: 248
140. Gaudé J, L'Haridon P, Laurent Y, Lang J (1971) Rev Chim Minér 8: 287
141. Antropoff AV, Krüger KH (1933) Z Phys Chem A167: 49

142. Gaudé J, Lang J (1970) C R Acad Sci Paris C271: 510
143. Gaudé J, Lang J (1970) Rev Chim Minér 7: 1059
144. Gaudé J, L'Haridon P, Laurent Y, Lang J (1972) Bull Soc fr Miner Crist 95: 56
145. Gaudé J, Lang J (1972) Rev Chim Minér 9: 799
146. Motte J, Brice J, Aubry J (1972) C R Acad Sci Paris C274: 1814
147. Hulliger F (1976) Structural Chemistry of Layer-type Phases, Dordrecht, Holland, D Reidel
148. Brese NE, O'Keeffe M (1990) J Solid State Chem 87: 134
149. Brese NE, O'Keeffe M, Von Dreele RB (1990) J Solid State Chem 88: 571
150. Brese NE (1991) In: Davies PK, Roth RS (eds) Chemistry of electronic ceramic materials, NIST Special Publication 804: 275
151. Brice J, Motte J, Aubry J (1972) C R Acad Sci Paris C274: 2166
152. Llewellyn FJ, Whitmore FE (1947) J Chem Soc 881
153. Nagib M, Jacobs H, Kistrup H (1979) Atomkernenergie 33: 38
154. Pors F, Bacher P, Marchand R, Laurent Y, Roult G (1987-1988) Rev Int Haut Tempér Réfract 24: 239
155. Patterson FK, Ward R (1966) Inorg Chem 5: 1312
156. Gaudé J, Lang J (1974) Rev Chim Minér 11: 80
157. Ehrlich P, Hein HJ (1953) Z Elektrochem 57: 710
158. Gaudé J, Lang J (1969) C R Acad Sci Paris C268: 1785
159. Torkar K, Spath HT (1967) Monat Chem 98: 2020
160. Gaudé J, Lang J (1972) C R Acad Sci Paris C274: 521
161. Ariya SM, Prokofyeva EA, Matveeva II (1955) J Gen Chem USSR 25: 609
162. Brice J-F, Motte J-P, Aubry J (1973) C R Acad Sci Paris C276: 1093
163. Wegner B, Essmann R, Jacobs H, Fischer P (1990) J Less-Common Met 167: 81
164. Ehrlich P, Deissmann W (1958) Angew Chem 70: 656
165. Ehrlich P, Deissmann W, Koch E, Ullrich V (1964) Z Anorg Allg Chem 328: 243
166. Jacobs H, Hadenfeldt C (1975) Z Anorg Allg Chem 418: 132
167. Walitzi EM, Krischner H (1970) Z Krist 132: 19
168. Okamoto Y, Goswami JC (1966) Inorg Chem 5: 1281
169. Ariya SM, Prokofyeva EA (1955) J Gen Chem USSR 25: 813
170. Gudat A, Haag S, Kniep R, Rabenau A (1990) J Less-Common Met 159: L29
171. Gudat A, Milius W, Haag S, Kniep R, Rabenau A (1991) J Less-Common Met 168: 305
172. Pors F, Marchand R, Laurent Y, Bacher P, Roult G (1988) Mater Res Bull 23: 1447
173. Morgan PED (1984) J Mater Sci Lett 3: 131
174. Arbus A, Fournier M-T, Fournier J, Capestan M (1971) C R Acad Sci Paris C273: 751
175. Ehrlich P, Koch E, Ullrich V (1963) Angew Chem Int Ed 2: 97
176. Pease RS (1952) Acta Cryst 5: 356
177. Solozhenko VL, Chernyshev VV, Fetisov GV, Rybakov VB, Petrusha IA (1990) J Phys Chem Solids 51: 1011
178. Wentorf RH Jr (1957) J Chem Physics 26: 956
179. Singh BP (1986) Mater Res Bull 21: 85
180. Soma T, Sawaoka A, Saito S (1974) Mater Res Bull 9: 755
181. Onodera A, Miyazaki H, Fujimoto N (1981) J Chem Phys 74: 5814
182. Schulz H, Thiemann KH (1977) Solid State Comm 23: 815
183. Juza R, Hahn H (1938) Z Anorg Allg Chem 239: 282
184. Renner T (1959) Z Anorg Allg Chem 298: 22
185. Hahn H, Juza R (1940) Z Anorg Allg Chem 244: 111
186. Abriat N, Laval JP, Frit B, Roult G (1982) Acta Cryst B38: 1088
187. Jack KH (1976) J Mater Sci 11: 1135
188. Sorrell CC (1983) J Austr Ceram Soc 19: 48
189. Cao GZ, Metselaar R (1991) Chem Mater 3: 242
190. Borgen O, Seip HM (1961) Acta Chem Scand 15: 1789
191. Goodman P, O'Keeffe M (1980) Acta Cryst B36: 2891
192. Kato K, Inoue Z, Kijima K, Kawada I, Tanaka H (1975) J Amer Ceram Soc 58: 90
193. Popper P, Ruddlesden SN (1957) Nature 179: 1129
194. Harris RK, Leach MJ, Thompson DP (1990) Chem Mater 2: 320
195. Wild S, Grieveson P, Jack KH (1972) In: Popper P (ed) Special Ceramics 5, British Ceramic Research Association, Manchester
196. Wild S, Grieveson P, Jack KH, Latimer MJ (1972) In: Popper P (ed) Special Ceramics 5, British Ceramic Research Association, Manchester

197. Jack KH, Thompson DP (1980) Ann Chim 5: 645
198. Idrestedt I, Brosset C (1964) Acta Chem Scand 18: 1879
199. Srinivasa SR, Cartz L, Jorgensen JD, Worlton TG, Beyerlein RA, Billy M (1977) J Appl Cryst 10: 167
200. Billy M, Labbe J-C, Selvaraj A, Roult G (1980) Mater Res Bull 15: 1207
201. Forgeng WD, Decker BF (1958) Trans Met Soc AIME 343
202. Wild S, Grieveson P, Jack KH (1972) In: Popper P (ed) Special Ceramics 5, British Ceramic Research Association, Manchester
203. Juza R, Rabenau A (1956) Z Anorg Allg Chem 285: 212
204. Jorgensen JD, Srinivasa SR, Labbe JC, Roult G (1979) Acta Cryst B35: 141
205. Boller H (1969) Monat Chem 100: 1471
206. Bergstrom FW (1928) J Phys Chem 32: 433
207. Schwarz R, Jeanmaire A (1932) Ber Deut Chem Ges 65: 1662
208. Schwarz R, Jeanmaire A (1932) Ber Deut Chem Ges 65: 1443
209. Choi CS, Prince E, Garrett WL (1977) Acta Cryst B33: 3536
210. Huffman EO, Tarbutton G, Elmore GV, Smith AJ, Rountree MG (1957) J Amer Chem Soc 79: 1765
211. Hirota Y, Kobayashi T (1982) J Appl Phys 53: 5037
212. Bondars B, Millers T, Vitola A, Vilks J (1982) Latv PSR Zinat Akad Vestis Kim Ser 4: 498
213. Moureu H, Rocquet P (1934) Compt Rend 198: 1691
214. Schmulbach CD (1962) Prog Inorg Chem 4: 275
215. Franklin EC (1905) J Amer Chem Soc 27: 820
216. Schurman I, Fernelius WC (1930) J Amer Chem Soc 52: 2425
217. Zachariasen WH (1949) Acta Cryst 2: 388
218. Klemm W, Winkelmann G (1956) Z Anorg Allg Chem 288: 87
219. Eick HA, Baenziger NC, Eyring L (1956) J Amer Chem Soc 78: 5987
220. Akimoto Y (1967) J Inorg Nucl Chem 29: 2650
221. Child HR, Wilkinson MK, Cable JW, Koehler WC, Wollan EO (1963) Phys Rev 131: 922
222. Brooks MSS (1982) J Mag Mag Mater 29: 257
223. Brown RC, Clark NJ (1974) J Inorg Nucl Chem 36: 1777
224. Brown RC, Clark NJ (1974) J Inorg Nucl Chem 36: 2287
225. Mourgout C, Chevalier B, Étourneau J, Portier J, Hagenmuller P, Georges R (1977) Rev Int Haut Temp Réfract 14: 89
226. Chevalier B, Etourneau J, Hagenmuller P (1977) Mater Res Bull 12: 473
227. Schuster JC (1985) J Less-Common Met 105: 327
228. Bacher P, Antoine P, Marchand R, L'Haridon P, Laurent Y, Roult G (1988) J Solid State Chem 77: 67
229. Antoine P, Marchand R, Laurent Y, Michel C, Raveau B (1988) Mater Res Bull 23: 953
230. Labbe J, Jardin JP (1982) Ann Chim Fr 7: 505
231. García J, Bartolomé J, González D, Navarro R, Fruchart D (1983) J Chem Thermodyn 15: 1041
232. García J, Bianconi A, Marcelli A, Davoli I, Bartolome J (1986) Nuovo Cimento Soc Ital Fis D7: 493
233. Guyader J, Marchand R, Gaudé J, Lang J (1975) C R Acad Sci Paris C281: 307
234. Gaudé J, Guyader J, Lang J (1975) C R Acad Sci Paris C280: 883
235. Marchand R, Jayaweera A, Verdier P, Lang J (1976) C R Acad Sci Paris C283: 675
236. Antoine P, Marchand R, Laurent Y (1987) Rev Int Hautes Temper Réfract Fr 24: 43
237. Liu G, Eick HA (1990) J Solid State Chem 89: 366
238. Marchand R (1976) C R Acad Sci Paris C282: 329
239. Marchand R, Pastuszak R, Laurent Y (1982) Rev Chim Minér 19: 684
240. Kieffer R, Ettmayer P, Pajakoff S (1972) Monat Chem 103: 1285
241. Benz R, Zachariasen WH (1969) Acta Cryst B25: 294
242. Benz R, Zachariasen WH (1970) Acta Cryst B26: 823
243. Benz R, Zachariasen WH (1966) Acta Cryst 21: 838
244. Juza R, Gerke H (1968) Z Anorg Allg Chem 363: 245
245. Despotovi'c Z, Ban Z (1969) Croat Chem Acta 41: 25
246. Juza R, Sievers R (1965) Naturwiss 52: 538
247. Juza R, Meyer W (1969) Z Anorg Allg Chem 366: 43
248. Yoshihara K, Yamagami S, Kanno M, Mukaibo T (1971) J Inorg Nucl Chem 33: 3323
249. Juza R, Sievers R (1968) Z Anorg Allg Chem 363: 258

250. Jung W, Juza R (1973) Z Anorg Allg Chem 399: 148
251. Zachariasen WH (1951) Acta Cryst 4: 231
252. Mann AW, Bevan DJM (1970) Acta Cryst B26: 2129
253. Blunck H, Juza R (1974) Z Anorg Allg Chem 410: 9
254. Schmitz-Dumont O, Raabe F (1954) Z Anorg Allg Chem 277: 297
255. Bowman AL, Arnold GP (1971) Acta Cryst B27: 243
256. Mueller MH, Knott HW (1958) Acta Cryst 11: 751
257. Rundle RE, Baenziger NC, Wilson AS, McDonald RA (1948) J Amer Chem Soc 70: 99
258. Tro'c R (1975) J Solid State Chem 13: 14
259. Vogt T, Schweda E, Laval JP, Frit B (1989) J Solid State Chem 83: 324
260. Hadenfeldt C, Jacobs H, Juza R (1970) Z Anorg Allg Chem 379: 144
261. Gaudé J, L'Haridon P, Guyader J, Lang J (1985) J Solid State Chem 59: 143
262. Marchand R, Lemarchand V (1981) J Less-Common Met 80: 157
263. Waldhart J, Ettmayer P (1979) Monat Chem 110: 21
264. Hadenfeldt C, Gieger B, Jacobs H (1974) Z Anorg Allg Chem 410: 104
265. Gaudé J, Lang J, Louër D (1983) Rev Chim Minér 20: 523
266. Hardy GF, Hulm JK (1954) Phys Rev 93: 1004
267. Matthias BT (1953) Phys Rev 92: 874
268. Papaconstantopoulos DA, Pickett WE, Klein BM, Boyer LL (1984) Nature 308: 494
269. Papaconstantopoulos DA, Pickett WE (1985) Phys Rev B31: 7093
270. Schwarz K (1987) CRC Critical Reviews Solid State Mat 13: 211
271. Ivanovskii AL, Novikov DL, Gubanov VA (1987) Phys Stat Sol (b) 141: 9
272. Marksteiner P, Weinberger P, Neckel A, Zeller R, Dederichs PH (1986) Phys Rev B33: 6709
273. Dawson PT, Stazyk SAJ (1982) J Vac Sci Technol 20: 966
274. Archer NJ (1984) Brit Ceram Proc 34: 187
275. Juza R, Hahn H (1940) Z Anorg Allg Chem 244: 133
276. Juza R (1966) Advances Inorg Chem Radiochem 9: 81
277. Franzen HF (1978) Prog Solid State Chem 12: 1
278. Schwarzkopf P, Kieffer R (1953) Refractory hard metals: Borides, carbides, nitrides and silicides, MacMillan
279. Toth LE (1971) Transition metal carbides and nitrides, Academic, New York
280. Holmberg B (1962) Acta Chem Scand 16: 1255
281. Christensen AN, Alamo A, Landesman JP (1985) Acta Cryst C41: 1009
282. Khidirov I, Karimov I, Ém VT, Loryan VÉ (1986) Inorg Mater 22: 1467
283. Ruff O, Treidel O (1912) Ber 45: 1364
284. Schuster JC, Bauer J (1984) J Solid-State Chem 53: 260
285. Juza R, Heners J (1964) Z Anorg Allg Chem 332: 159
286. Juza R, Gabel A, Rabenau H, Klose W (1964) Z Anorg Allg Chem 329: 136
287. Juza R, Rabenau A, Nitschke I (1964) Z Anorg Allg Chem 332: 1
288. Schwarz K, Williams AR, Cuomo JJ, Harper JHE, Hentzell HTG (1985) Phys Rev B32: 8312
289. Juza R, Klose W (1964) Z Anorg Allg Chem 327: 207
290. Blunck H, Juza R (1974) Z Anorg Allg Chem 406: 145
291. Juza R, Friedrichsen H (1964) Z Anorg Allg Chem 332: 173
292. Ohashi M, Yamanaka S, Sumihara M, Hattori M (1988) J Solid State Chem 75: 99
293. Jung W, Juza R (1973) Z Anorg Allg Chem 399: 129
294. Schuster JC (1986) Z Krist 175: 211
295. Rudy E (1970) Metallurg Trans 1: 1249
296. Demyashev GM, Koshchug EE, Repnikov NN, Chuzhko RK (1985) J Less-Common Met 107: 59
297. Christensen AN, Lebech B (1979) Acta Cryst B35: 2677
298. Boller H, Nowotny H (1968) Monat Chem 99: 721
299. Boller H, Nowotny H (1968) Monat Chem 99: 672
300. Brauer G, Jander J (1952) Z Anorg Allg Chem 270: 160
301. Schönberg N (1954) Acta Chem Scand 8: 208
302. Brauer G (1960) J Less-Common Met 2: 131
303. Brauer G (1961) Z Anorg Allg Chem 309: 151
304. Guard RW, Savage JW, Swarthout DG (1967) Trans Metall Soc AIME 239: 643
305. Brauer G, Kern W (1984) Z Anorg Allg Chem 512: 7
306. Christensen AN (1976) Acta Chem Scand A30: 219

307. Christensen AN (1977) Acta Chem Scand A31: 77
308. Terao N (1965) Jap J Appl Phys 4: 353
309. Weishaupt M, Strähle J (1977) Z Anorg Allg Chem 429: 261
310. Terao N (1971) Jap J Appl Phys 10: 248
311. Petrunin VF, Sorokin NI (1983) Inorg Mater 18: 1733
312. Terao N (1977) C R Acad Sci Paris C285: 17
313. Strähle J (1973) Z Anorg Allg Chem 402: 47
314. Brese NE, O'Keeffe M, Rauch P, DiSalvo FJ (1991) Acta Cryst in press
315. Armytage D, Fender BEF (1974) Acta Cryst B30: 809
316. Kiessling R, Liu YH (1951) J Metals 3: 639
317. Jehn H, Ettmayer P (1978) J Less-Common Met 58: 85
318. Hägg G (1930) Z Physik Chem Abt B7: 339
319. Schönberg N (1954) Acta Chem Scand 8: 204
320. Langmuir I (1913) J Chem Soc 931
321. Langmuir I (1914) Z Anorg Allg Chem 85: 261
322. Smithells CJ, Rooksby HP (1927) J Chem Soc 1882
323. Khitrova VI (1962) Sov Phys Cryst 6: 439
324. Hägg G (1929) Z Physik Chem Abt B4: 346
325. Juza R, Puff H, Wagenknecht F (1957) Z Elektrochem 61: 804
326. Juza R, Puff H (1957) Z Elektrochem 61: 810
327. Juza R, Deneke K, Puff H (1959) Z Elektrochem 63: 551
328. Takei WJ, Shirane G, Frazer BC (1960) Phys Rev 119: 122
329. Mekata M, Haruna J, Takaki H (1968) J Phys Soc Jap 25: 234
330. Nasr-Eddine M, Bertaut EF, Maunaye M (1977) Acta Cryst B33: 2696
331. Jacobs H, Stüve C (1984) J Less-Common Met 96: 323
332. Wintenberger M, Marchand R, Maunaye M (1977) Solid State Comm 21: 733
333. Wintenberger M, Guyader J, Maunaye M (1972) Solid State Comm 11: 1485
334. Hahn H, Konrad A (1951) Z Anorg Allg Chem 264: 174
335. Hägg G (1934) J Iron Steel Inst 130: 439
336. Hägg G (1928) Nature 121: 826
337. Jack KH (1950) Acta Cryst 3: 392
338. Hägg G (1930) Z Physik Chem B8: 455
339. Jack KH (1952) Acta Cryst 5: 404
340. Andriamandroso D, Demazeau G, Pouchard M, Hagenmuller P (1984) J Solid State Chem 54: 54
341. Dvoriankina GG, Pinsker ZG (1958) Sov Phys Cryst 3: 439
342. Goodenough JB, Wold A, Arnott RJ (1960) J Appl Physics 31: 342
343. Arnott RJ, Wold A (1960) J Phys Chem Solids 15: 152
344. Bergstrom W (1924) J Amer Chem Soc 46: 2631
345. Juza R, Sachsze W (1945) Z Anorg Allg Chem 253: 95
346. Terao N (1960) Chem Abs 54: 17188
347. Schmitz-Dumont O, Broja H, Piepenbrink HF (1947) Z Anorg Allg Chem 253: 118
348. Juza R, Sachsze W (1943) Z Anorg Allg Chem 251: 201
349. Terao N (1959) Naturwiss 46: 204
350. Watt GW, Davies DD (1948) J Amer Chem Soc 70: 3753
351. Benabdoun A, Rémy F, Bernard J (1968) C R Acad Sci Paris C266: 1579
352. Wilsdorf H (1948) Acta Cryst 1: 115
353. Söderquist R (1968) Acta Cryst B24: 450
354. West CD (1936) Z Krist 95: 421
355. Marr III HE, Stanford RH Jr (1962) Acta Cryst 15: 1313
356. Ennis JL, Shanley ES (1991) J Chem Ed 68: A6
357. Juza R (1941) Z Anorg Allg Chem 248: 118
358. Hahn H, Gilbert E (1949) Z Anorg Allg Chem 258: 77
359. Juza R, Neuber A, Hahn H (1938) Z Anorg Allg Chem 239: 273
360. Juza R, Hahn H (1940) Z Anorg Allg Chem 244: 125
361. Juza R, Fasold K, Kuhn W (1937) Z Anorg Allg Chem 234: 86
362. Müller U (1973) Z Anorg Allg Chem 399: 183
363. Wintenberger M, Maunaye M, Laurent Y (1973) Mater Res Bull 8: 1049
364. O'Keeffe M, Hyde BG (1984) Nature 309: 411

365. O'Keeffe M, Brese NE (1991) Acta Cryst in press
366. Abegg R (1904) Z Anorg Chem 39: 330
367. Bonnemay A, Daudel R (1950) Compt Rend 230: 2300
368. Shidharan K, Walter KC, Conrad JR (1991) Mater Res Bull 26: 367
369. Foct J, Herdy A (eds) (1989) High nitrogen steels, The Institute of Metals, London
370. Massidda A, Pickett WE, Posternak M (1991) Phys Rev B44: 1258
371. Cordier G, Höhn P, Kniep R, Rabenau A (1991) Z anorg allg Chem 591: 58
372. Höhn P, Kniep R, Rabenau A (1991) Z Kristallogr 196: 153
373. Zachwieja U, Jacobs H (1991) Eur J Solid State Inorg Chem 28: 1055
374. Schnick W, Berger U (1991) angew Chem 103: 857
375. Tenten A, Jacobs H (1991) J Less-common Mets 170: 145
376. Hiraguchi H, Hashizume H, Fukunaga O, Takenaka A, Sakata M (1991) J Appl Crystallogr 24: 286
377. Sjöberg J, Helgesson G, Idrestedt I (1991) Acta Crystallogr C47: 2438

Author Index Volumes 1–79

Ahrland, S.: Factors Contributing to (b)-behavior in Acceptors, Vol. 1, pp. 207–220.
Ahrland, S.: Thermodynamics of Complex Formation between Hard and Soft Acceptors and Donors. Vol. 5, pp. 118–149.
Ahrland, S.: Thermodynamics of the Stepwise Formation of Metal-Ion Complexes in Aqueous Solution. Vol. 15, pp. 167–188.
Allen, G. C., Warren, K. D.: The Electronic Spectra of the Hexafluoro Complexes of the First Transition Series. Vol. 9, pp. 49–138.
Allen, G. C., Warren, K. D.: The Electronic Spectra of the Hexafluoro Complexes of the Second and Third Transition Series. Vol. 19, pp. 105–165.
Alonso, J. A., Balbás, L. C.: Simple Density Functional Theory of the Electronegativity and Other Related Properties of Atoms and Ions. Vol. 66, pp. 41–78.
Andersson, L. A., Dawson, J. H.: EXAFS Spectroscopy of Heme-Containing Oxygenases and Peroxidases. Vol. 74, pp. 1–40.
Ardon, M., Bino, A.: A New Aspect of Hydrolysis of Metal Ions: The Hydrogen-Oxide Bridging Ligand ($H_3O_2^-$). Vol. 65, pp. 1–28.
Armstrong, F. A.: Probing Metalloproteins by Voltammetry. Vol. 72, pp. 137–221.
Augustynski, J.: Aspects of Photo-Electrochemical and Surface Behavior of Titanium(IV) Oxide. Vol. 69, pp. 1–61.
Averill, B. A.: Fe–S and Mo–Fe–S Clusters as Models for the Active Site of Nitrogenase. Vol. 53, pp. 57–101.
Babel, D.: Structural Chemistry of Octahedral Fluorocomplexes of the Transition Elements. Vol. 3, pp. 1–87.
Bacci, M.: The Role of Vibronic Coupling in the Interpretation of Spectroscopic and Structural Properties of Biomolecules. Vol. 55, pp. 67–99.
Baker, E. C., Halstead, G. W., Raymond, K. N.: The Structure and Bonding of 4f and 5f Series Organometallic Compounds. Vol. 25, pp. 21–66.
Balsenc, L. R.: Sulfur Interaction with Surfaces and Interfaces Studied by Auger Electron Spectrometry. Vol. 39, pp. 83–114.
Banci, L., Bencini, A., Benelli, C., Gatteschi, D., Zanchini, C.: Spectral-Structural Correlations in High-Spin Cobalt(II) Complexes. Vol. 52, pp. 37–86.
Banci, L., Bertini, I., Luchinat, C.: The ^1H NMR Parameters of Magnetically Coupled Dimers — The Fe_2S_2 Proteins as an Example. Vol. 72, pp. 113–136.
Bartolotti, L. J.: Absolute Electronegativities as Determined from Kohn-Sham Theory. Vol. 66, pp. 27–40.
Baughan, E. C.: Structural Radii, Electron-cloud Radii, Ionic Radii and Solvation. Vol. 15, pp. 53–71.
Bayer, E., Schretzmann, P.: Reversible Oxygenierung von Metallkomplexen. Vol. 2, pp. 181–250.
Bearden, A. J., Dunham, W. R.: Iron Electronic Configuration in Proteins: Studies by Mössbauer Spectroscopy. Vol. 8, pp. 1–52.
Bergmann, D., Hinze, J.: Electronegativity and Charge Distribution. Vol. 66, pp. 145–190.
Berners-Price, S. J., Sadler, P. J.: Phosphines and Metal Phosphine Complexes: Relationship of Chemistry to Anticancer and Other Biological Activity. Vol. 70, pp. 27–102.
Bertini, I., Luchinat, C., Scozzafava, A.: Carbonic Anhydrase: An Insight into the Zinc Binding Site and into the Active Cavity Through Metal Substitution. Vol. 48, pp. 45–91.
Bertrand, P.: Application of Electron Transfer Theories to Biological Systems. Vol. 75, pp. 1–48.
Blasse, G.: The Influence of Charge-Transfer and Rydberg States on the Luminescence Properties of Lanthanides and Actinides. Vol. 26, pp. 43–79.
Blasse, G.: The Luminescence of Closed-Shell Transition Metal-Complexes. New Developments. Vol. 42, pp. 1–41.
Blasse, G.: Optical Electron Transfer Between Metal Ions and its Consequences. Vol. 76, pp. 153–188.
Blauer, G.: Optical Activity of Conjugated Proteins. Vol. 18, pp. 69–129.
Bleijenberg, K. C.: Luminescence Properties of Uranate Centres in Solids. Vol. 42, pp. 97–128.
Bŏca, R., Breza, M., Pelikán, P.: Vibronic Interactions in the Stereochemistry of Metal Complexes. Vol. 71, pp. 57–97.

Boeyens, J. C. A.: Molecular Mechanics and the Structure Hypothesis. Vol. 63, pp. 65–101.
Bonnelle, C.: Band and Localized States in Metallic Thorium, Uranium and Plutonium, and in Some Compounds, Studied by X-ray Spectroscopy. Vol. 31, pp. 23–48.
Bradshaw, A. M., Cederbaum, L. S., Domcke, W.: Ultraviolet Photoelectron Spectroscopy of Gases Adsorbed on Metal Surfaces. Vol. 24, pp. 133–170.
Braterman, P. S.: Spectra and Bonding in Metal Carbonyls. Part A: Bonding. Vol. 10, pp. 57–86.
Braterman, P. S.: Spectra and Bonding in Metal Carbonyls. Part B: Spectra and Their Interpretation. Vol. 26, pp. 1–42.
Bray, R. C., Swann, J. C.: Molybdenum-Containing Enzymes. Vol. 11, pp. 107–144.
Brese, N. E., O'Keeffe, M.: Crystal Chemistry of Inorganic Nitrides. Vol. 79, pp. 307–378.
Brooks, M. S. S.: The Theory of 5f Bonding in Actinide Solids. Vol. 59/60, pp. 263–293.
van Bronswyk, W.: The Application of Nuclear Quadrupole Resonance Spectroscopy to the Study of Transition Metal Compounds. Vol. 7, pp. 87–113.
Buchanan, B. B.: The Chemistry and Function of Ferredoxin. Vol. 1, pp. 109–148.
Buchler, J. W., Kokisch, W., Smith, P. D.: Cis, Trans, and Metal Effects in Transition Metal Porphyrins. Vol. 34, pp. 79–134.
Bulman, R. A.: Chemistry of Plutonium and the Transuranics in the Biospere. Vol. 34, pp. 39–77.
Bulman, R. A.: The Chemistry of Chelating Agents in Medical Sciences. Vol. 67, pp. 91–141.
Burdett, J. K.: The Shapes of Main-Group Molecules; A Simple Semi-Quantitative Molecular Orbital Approach. Vol. 31, pp. 67–105.
Burdett, J. K.: Some Structural Problems Examined Using the Method of Moments. Vol. 65, pp. 29–90.
Campagna, M., Wertheim, G. K., Bucher, E.: Spectroscopy of Homogeneous Mixed Valence Rare Earth Compounds. Vol. 30, pp. 99–140.
Ceulemans, A., Vanquickenborne, L. G.: The Epikernel Principle. Vol. 71, pp. 125–159.
Chasteen, N. D.: The Biochemistry of Vanadium, Vol. 53, pp. 103–136.
Cheh, A. M., Neilands, J. P.: The γ-Aminolevulinate Dehydratases: Molecular and Environmental Properties. Vol. 29, pp. 123–169.
Ciampolini, M.: Spectra of 3d Five-Coordinate Complexes. Vol. 6, pp. 52–93.
Chimiak, A., Neilands, J. B.: Lysine Analogues of Siderophores. Vol. 58, pp. 89–96.
Clack, D. W., Warren, K. D.: Metal-Ligand Bonding in 3d Sandwich Complexes. Vol. 39, pp. 1–141.
Clark, R. J. H., Stewart, B.: The Resonance Raman Effect. Review of the Theory and of Applications in Inorganic Chemistry. Vol. 36, pp. 1–80.
Clarke, M. J., Fackler, P. H.: The Chemistry of Technetium: Toward Improved Diagnostic Agents. Vol. 50, pp. 57–58.
Cohen, I. A.: Metal–Metal Interactions in Metalloporphyrins, Metalloproteins and Metalloenzymes. Vol. 40, pp. 1–37.
Connett, P. H., Wetterhahn, K. E.: Metabolism of the Carcinogen Chromate by Cellular Constituents. Vol. 54, pp. 93–124.
Cook, D. B.: The Approximate Calculation of Molecular Electronic Structures as a Theory of Valence. Vol. 35, pp. 37–86.
Cooper, S. R., Rawle, S. C.: Crown Thioether Chemistry. Vol. 72, pp. 1–72.
Cotton, F. A., Walton, R. A.: Metal–Metal Multiple Bonds in Dinuclear Clusters. Vol. 62, pp. 1–49.
Cox, P. A.: Fractional Parentage Methods for Ionisation of Open Shells of d and f Electrons. Vol. 24, pp. 59–81.
Crichton, R. R.: Ferritin. Vol. 17, pp. 67–134.
Daul, C., Schläpfer, C. W., von Zelewsky, A.: The Electronic Structure of Cobalt(II) Complexes with Schiff Bases and Related Ligands. Vol. 36, pp. 129–171.
Dehnicke, K., Shihada, A.-F.: Structural and Bonding Aspects in Phosphorus Chemistry-Inorganic Derivates of Oxohalogeno Phosphoric Acids. Vol. 28, pp. 51–82.
Denning, R. G.: Electronic Structure and Bonding in Actinyl Ions. Vol. 79, pp. 215–276.
Dobiáš, B.: Surfactant Adsorption on Minerals Related to Flotation. Vol. 56, pp. 91–147.
Doi, K., Antanaitis, B. C., Aisen, P.: The Binuclear Iron Centers of Uteroferrin and the Purple Acid Phosphatases. Vol. 70, pp. 1–26.
Doughty, M. J., Diehn, B.: Flavins as Photoreceptor Pigments for Behavioral Responses. Vol. 41, pp. 45–70.
Drago, R. S.: Quantitative Evaluation and Prediction of Donor-Acceptor Interactions. Vol. 15, pp. 73–139.

Drillon, M., Darriet, J.: Progress in Polymetallic Exchange-Coupled Systems, some Examples in Inorganic Chemistry. Vol. 79, pp. 55–100.
Dubhghaill, O. M. Ni, Sadler, P. J.: The Structure and Reactivity of Arsenic Compounds. Biological Activity and Drug Design. Vol. 78, pp. 129–190.
Duffy, J. A.: Optical Electronegativity and Nephelauxetic Effect in Oxide Systems. Vol. 32, pp. 147–166.
Dunn, M. F.: Mechanisms of Zinc Ion Catalysis in Small Molecules and Enzymes. Vol. 23, pp. 61–122.
Emsley, E.: The Composition, Structure and Hydrogen Bonding of the β-Diketones. Vol. 57, pp. 147–191.
Englman, R.: Vibrations in Interaction with Impurities. Vol. 43, pp. 113–158.
Epstein, I. R., Kustin, K.: Design of Inorganic Chemical Oscillators. Vol. 56, pp. 1–33.
Ermer, O.: Calculations of Molecular Properties Using Force Fields. Applications in Organic Chemistry. Vol. 27, pp. 161–211.
Ernst, R. D.: Structure and Bonding in Metal-Pentadienyl and Related Compounds. Vol. 57, pp. 1–53.
Erskine, R. W., Field, B. O.: Reversible Oxygenation. Vol. 28, pp. 1–50.
Evain, M., Brec, R.: A New Approach to Structural Description of Complex Polyhedra Containing Polychalcogenide Anions. Vol. 79, pp. 277–306.
Fajans, K.: Degrees of Polarity and Mutual Polarization of Ions in the Molecules of Alkali Fluorides, SrO, and BaO. Vol. 3, pp. 88–105.
Fee, J. A.: Copper Proteins – Systems Containing the "Blue" Copper Center. Vol. 23, pp. 1–60.
Feeney, R. E., Komatsu, S. K.: The Transferrins. Vol. 1, pp. 149–206.
Felsche, J.: The Crystal Chemistry of the Rare-Earth Silicates. Vol. 13, pp. 99–197.
Ferreira, R.: Paradoxical Violations of Koopmans' Theorem, with Special Reference to the 3d Transition Elements and the Lanthanides. Vol. 31, pp. 1–21.
Fidelis, I. K., Mioduski, T.: Double-Double Effect in the Inner Transition Elements. Vol. 47, pp. 27–51.
Fournier, J. M.: Magnetic Properties of Actinide Solids. Vol. 59/60, pp. 127–196.
Fournier, J. M., Manes, L.: Actinide Solids. 5f Dependence of Physical Properties. Vol. 59/60, pp. 1–56.
Fraga, S., Valdemoro, C.: Quantum Chemical Studies on the Submolecular Structure of the Nucleic Acids. Vol. 4, pp. 1–62.
Fraústo da Silva, J. J. R., Williams, R. J. P.: The Uptake of Elements by Biological Systems. Vol. 29, pp. 67–121.
Fricke, B.: Superheavy Elements. Vol. 21, pp. 89–144.
Fricke, J., Emmerling, A.: Aerogels–Preparation, Properties, Applications. Vol. 77, pp. 37–88.
Frenking, G., Cremer, D.: The Chemistry of the Noble Gas Elements Helium, Neon, and Argon – Experimental Facts and Theoretical Predictions, Vol. 73, pp. 17–96.
Fuhrhop, J.-H.: The Oxidation States and Reversible Redox Reactions of Metalloporphyrins. Vol. 18, pp. 1–67.
Furlani, C., Cauletti, C.: He(I) Photoelectron Spectra of d-metal Compounds. Vol. 35, pp. 119–169.
Gázquez, J. L., Vela, A., Galván, M.: Fukui Function, Electronegativity and Hardness in the Kohn-Sham Theory. Vol. 66, pp. 79–98.
Gerloch, M., Harding, J. H., Woolley, R. G.: The Context and Application of Ligand Field Theory. Vol. 46, pp. 1–46.
Gillard, R. D., Mitchell, P. R.: The Absolute Configuration of Transition Metal Complexes. Vol. 7, pp. 46–86.
Gleitzer, C., Goodenough, J. B.: Mixed-Valence Iron Oxides. Vol. 61, pp. 1–76.
Gliemann, G., Yersin, H.: Spectroscopic Properties of the Quasi One-Dimensional Tetracyanoplatinate(II) Compounds. Vol. 62, pp. 87–153.
Golovina, A. P., Zorov, N. B., Runov, V. K.: Chemical Luminescence Analysis of Inorganic Substances. Vol. 47, pp. 53–119.
Green, J. C.: Gas Phase Photoelectron Spectra of d- and f-Block Organometallic Compounds. Vol. 43, pp. 37–112.
Grenier, J. C., Pouchard, M., Hagenmuller, P.: Vacancy Ordering in Oxygen-Deficient Perovskite-Related Ferrites. Vol. 47, pp. 1–25.
Griffith, J. S.: On the General Theory of Magnetic Susceptibilities of Polynuclear Transitionmetal Compounds. Vol. 10, pp. 87–126.

Gubelmann, M. H., Williams, A. F.: The Structure and Reactivity of Dioxygen Complexes of the Transition Metals. Vol. 55, pp. 1–65.

Guilard, R., Lecomte, C., Kadish, K. M.: Synthesis, Electrochemistry, and Structural Properties of Porphyrins with Metal–Carbon Single Bonds and Metal–Metal Bonds. Vol. 64, pp. 205–268.

Gütlich, P.: Spin Crossover in Iron(II)-Complexes. Vol. 44, pp. 83–195.

Gutmann, V., Mayer, U.: Thermochemistry of the Chemical Bond. Vol. 10, pp. 127–151.

Gutmann, V., Mayer, U.: Redox Properties: Changes Effected by Coordination. Vol. 15, pp. 141–166.

Gutmann, V., Mayer, H.: Application of the Functional Approach to Bond Variations Under Pressure. Vol. 31, pp. 49–66.

Hall, D. I., Ling, J. H., Nyholm, R. S.: Metal Complexes of Chelating Olefin-Group V Ligands. Vol. 15, pp. 3–51.

Harnung, S. E., Schäffer, C. E.: Phase-fixed 3-Γ Symbols and Coupling Coefficients for the Point Groups. Vol. 12, pp. 201–255.

Harnung, S. E., Schäffer, C. E.: Real Irreducible Tensorial Sets and their Application to the Ligand-Field Theory. Vol. 12, pp. 257–295.

Hathaway, B. J.: The Evidence for "Out-of-the Plane" Bonding in Axial Complexes of the Copper(II) Ion. Vol. 14, pp. 49–67.

Hathaway, B. J.: A New Look at the Stereochemistry and Electronic Properties of Complexes of the Copper(II) Ion. Vol. 57, pp. 55–118.

Hellner, E. E.: The Frameworks (Bauverbände) of the Cubic Structure Types. Vol. 37, pp. 61–140.

von Herigonte, P.: Electron Correlation in the Seventies. Vol. 12, pp. 1–47.

Hemmerich, P., Michel, H., Schug, C., Massey, V.: Scope and Limitation of Single Electron Transfer in Biology. Vol. 48, pp. 93–124.

Henry, M., J. P. Jolivet, Livage, J.: Aqueous Chemistry of Metal Cations: Hydrolysis, Condensation and Complexation. Vol. 77, pp. 153–206.

Hider, R. C.: Siderophores Mediated Absorption of Iron. Vol. 58, pp. 25–88.

Hill, H. A. O., Röder, A., Williams, R. J. P.: The Chemical Nature and Reactivity of Cytochrome P-450. Vol. 8, pp. 123–151.

Hilpert, K.: Chemistry of Inorganic Vapors. Vol. 73, pp. 97–198.

Hogenkamp, H. P. C., Sando, G. N.: The Enzymatic Reduction of Ribonucleotides. Vol. 20, pp. 23–58.

Hoffman, B. M., Natan, M. J. Nocek, J. M., Wallin, S. A.: Long-Range Electron Transfer Within Metal-Substituted Protein Complexes. Vol. 75, pp. 85–108.

Hoffmann, D. K., Ruedenberg, K., Verkade, J. G.: Molecular Orbital Bonding Concepts in Polyatomic Molecules – A Novel Pictorial Approach. Vol. 33, pp. 57–96.

Hubert, S., Hussonnois, M., Guillaumont, R.: Measurement of Complexing Constants by Radiochemical Methods. Vol. 34, pp. 1–18.

Hudson, R. F.: Displacement Reactions and the Concept of Soft and Hard Acids and Bases. Vol. 1, pp. 221–223.

Hulliger, F.: Crystal Chemistry of Chalcogenides and Pnictides of the Transition Elements. Vol. 4, pp. 83–229.

Ibers, J. A., Pace, L. J., Martinsen, J., Hoffman, B. M.: Stacked Metal Complexes: Structures and Properties. Vol. 50, pp. 1–55.

Iqbal, Z.: Intra- und Inter-Molecular Bonding and Structure of Inorganic Pseudohalides with Triatomic Groupings. Vol. 10, pp. 25–55.

Izatt, R. M., Eatough, D. J., Christensen, J. J.: Thermodynamics of Cation-Macrocyclic Compound Interaction. Vol. 16, pp. 161–189.

Jain, V. K., Bohra, R., Mehrotra, R. C.: Structure and Bonding in Organic Derivatives of Antimony(V). Vol. 52, pp. 147–196.

Jerome-Lerutte, S.: Vibrational Spectra and Structural Properties of Complex Tetracyanides of Platinum, Palladium and Nickel. Vol. 10, pp. 153–166.

Jørgensen, C. K.: Electric Polarizability, Innocent Ligands and Spectroscopic Oxidation States. Vol. 1, pp. 234–248.

Jørgensen, C. K.: Heavy Elements Synthesized in Supernovae and Detected in Peculiar A-type Stars. Vol. 73, pp. 199–226.

Jørgensen, C. K.: Recent Progress in Ligand Field Theory. Vol. 1, pp. 3–31.

Jørgensen, C. K.: Relationship Between Softness, Covalent Bonding, Ionicity and Electric Polarizability. Vol. 3, pp. 106–115.

Jørgensen, C. K.: Valence-Shell Expansion Studied by Ultra-violet Spectroscopy. Vol. 6, pp. 94–115.
Jørgensen, C. K.: The Inner Mechanism of Rare Earths Elucidated by Photo-Electron Spectra. Vol. 13, pp. 199–253.
Jørgensen, C. K.: Partly Filled Shells Constituting Anti-bonding Orbitals with Higher Ionization Energy than Their Bonding Counterparts. Vol. 22, pp. 49–81.
Jørgensen, C. K.: Photo-electron Spectra of Non-metallic Solids and Consequences for Quantum Chemistry. Vol. 24, pp. 1–58.
Jørgensen, C. K.: Narrow Band Thermoluminescence (Candoluminescence) of Rare Earths in Auer Mantles. Vol. 25, pp. 1–20.
Jørgensen, C. K.: Deep-lying Valence Orbitals and Problems of Degeneracy and Intensities in Photo-electron Spectra. Vol. 30, pp. 141–192.
Jørgensen, C. K.: Predictable Quarkonium Chemistry. Vol. 34, pp. 19–38.
Jørgensen, C. K.: The Conditions for Total Symmetry Stabilizing Molecules, Atoms, Nuclei and Hadrons. Vol. 43, pp. 1–36.
Jørgensen, C. K., Frenking, G.: Historical, Spectroscopic and Chemical Comparison of Noble Gases. Vol. 73, pp. 1–16.
Jørgensen, C. K., Kauffmann, G. B.: Crookes and Marignac – A Centennial of an Intuitive and Pragmatic Appraisal of "Chemical Elements" and the Present Astrophysical Status of Nucleosynthesis and "Dark Matter". Vol. 73, pp. 227–254.
Jørgensen, C. K., Reisfeld, R.: Uranyl Photophysics. Vol. 50, pp. 121–171.
O'Keeffe, M.: The Prediction and Interpretation of Bond Lengths in Crystals. Vol. 71, pp. 161–190.
O'Keeffe, M., Hyde, B. G.: An Alternative Approach to Non-Molecular Crystal Structures with Emphasis on the Arrangements of Cations. Vol. 61, pp. 77–144.
Kahn, O.: Magnetism of the Heteropolymetallic Systems. Vol. 68, pp. 89–167.
Keppler, B. K., Friesen, C., Moritz, H. G., Vongerichten, H., Vogel, E.: Tumor-Inhibiting Bis (β-Diketonato) Metal Complexes. Budotitane, cis-Diethoxybis (1-phenylbutane-1,3-dionato) titanium (IV). Vol. 78, pp. 97–128.
Kimura, T.: Biochemical Aspects of Iron Sulfur Linkage in None-Heme Iron Protein, with Special Reference to "Adrenodoxin". Vol. 5, pp. 1–40.
Kitagawa, T., Ozaki, Y.: Infrared and Raman Spectra of Metalloporphyrins. Vol. 64, pp. 71–114.
Kiwi, J., Kalyanasundaram, K., Grätzel, M.: Visible Light Induced Cleavage of Water into Hydrogen and Oxygen in Colloidal and Microheterogeneous Systems. Vol. 49, pp. 37–125.
Kjekshus, A., Rakke, T.: Considerations on the Valence Concept. Vol. 19, pp. 45–83.
Kjekshus, A., Rakke, T.: Geometrical Considerations on the Marcasite Type Structure. Vol. 19, pp. 85–104.
König, E.: The Nephelauxetic Effect. Calculation and Accuracy of the Interelectronic Repulsion Parameters I. Cubic High-Spin d^2, d^3, d^7 and d^8 Systems. Vol. 9, pp. 175–212.
König, E.: Nature and Dynamics of the Spin-State Interconversion in Metal Complexes. Vol. 76, pp. 51–152.
Köpf-Maier, P., Köpf, H.: Transition and Main-Group Metal Cyclopentadienyl Complexes: Preclinical Studies on a Series of Antitumor Agents of Different Structural Type. Vol. 70, pp. 103–185.
Koppikar, D. K., Sivapullaiah, P. V., Ramakrishnan, L., Soundararajan, S.: Complexes of the Lanthanides with Neutral Oxygen Donor Ligands. Vol. 34, pp. 135–213.
Krause, R.: Synthesis of Ruthenium(II) Complexes of Aromatic Chelating Heterocycles: Towards the Design of Luminescent Compounds. Vol. 67, pp. 1–52.
Krumholz, P.: Iron(II) Diimine and Related Complexes. Vol. 9, pp. 139–174.
Kuki, A.: Electronic Tunneling Paths in Proteins. Vol. 75, pp. 49–84.
Kustin, K., McLeod, G. C., Gilbert, T. R., Briggs, LeB. R., 4th.: Vanadium and Other Metal Ions in the Physiological Ecology of Marine Organisms. Vol. 53, pp. 137–158.
Labarre, J. F.: Conformational Analysis in Inorganic Chemistry: Semi-Empirical Quantum Calculation vs. Experiment. Vol. 35, pp. 1–35.
Lammers, M., Follmann, H.: The Ribonucleotide Reductases: A Unique Group of Metalloenzymes Essential for Cell Proliferation. Vol. 54, pp. 27–91.
Lehn, J.-M.: Design of Organic Complexing Agents. Strategies Towards Properties. Vol. 16, pp. 1–69.
Linarès, C., Louat, A., Blanchard, M.: Rare-Earth Oxygen Bonding in the LnMO$_4$ Xenotime Structure. Vol. 33, pp. 179–207.

Lindskog, S.: Cobalt(II) in Metalloenzymes. A Reporter of Structure-Function Relations. Vol. 8, pp. 153–196.
Liu, A., Neilands, J. B.: Mutational Analysis of Rhodotorulic Acid Synthesis in *Rhodotorula pilimanae*. Vol. 58, pp. 97–106.
Livorness, J., Smith, T.: The Role of Manganese in Photosynthesis. Vol. 48, pp. 1–44.
Llinás, M.: Metal-Polypeptide Interactions: The Conformational State of Iron Proteins. Vol. 17, pp. 135–220.
Lucken, E. A. C.: Valence-Shell Expansion Studied by Radio-Frequency Spectroscopy. Vol. 6, pp. 1–29.
Ludi, A., Güdel, H. U.: Structural Chemistry of Polynuclear Transition Metal Cyanides. Vol. 14, pp. 1–21.
Lutz, H. D.: Bonding and Structure of Water Molecules in Solid Hydrates. Correlation of Spectroscopic and Structural Data. Vol. 69, pp. 125.
Maggiora, G. M., Ingraham, L. L.: Chlorophyll Triplet States. Vol. 2, pp. 126–159.
Magyar, B.: Salzebullioskopie III. Vol. 14, pp. 111–140.
Makovicky, E., Hyde, B. G.: Non-Commensurate (Misfit) Layer Structures. Vol. 46, pp. 101–170.
Manes, L., Benedict, U.: Structural and Thermodynamic Properties of Actinide Solids and Their Relation to Bonding. Vol. 59/60, pp. 75–125.
Mann, S.: Mineralization in Biological Systems. Vol. 54, pp. 125–174.
Mason, S. F.: The Ligand Polarization Model for the Spectra of Metal Complexes: The Dynamic Coupling Transition Probabilities. Vol. 39, pp. 43–81.
Mathey, F., Fischer, J., Nelson, J. H.: Complexing Modes of the Phosphole Moiety. Vol. 55, pp. 153–201.
Mauk, A. G.: Electron Transfer in Genetically Engineered Proteins. The Cytochrome *c* Paradigm. Vol. 75, pp. 131–158.
Mayer, U., Gutmann, V.: Phenomenological Approach to Cation-Solvent Interactions. Vol. 12, pp. 113–140.
McLendon, G.: Control of Biological Electron Transport via Molecular Recognition and Binding: The "Velcro" Model. Vol. 75, pp. 159–174.
Mehrotra, R. C.: Present Status and Future Potential of the Sol–Gel Process. Vol. 77, pp. 1–36.
Mildvan, A. S., Grisham, C. M.: The Role of Divalent Cations in the Mechanism of Enzyme Catalyzed Phosphoryl and Nucleotidyl. Vol. 20, pp. 1–21.
Mingos, D. M. P., Hawes, J. C.: Complementary Spherical Electron Density Model. Vol. 63, pp. 1–63.
Mingos, D. M. P., Johnston, R. L.: Theoretical Models of Cluster Bonding. Vol. 68, pp. 29–87.
Mingos, D. M. P., Zhenyang, L.: Non-Bonding Orbitals in Co-ordination Hydrocarbon and Cluster Compounds. Vol. 71, pp. 1–56.
Mingos, D. M. P., Zhenyang, L.: Hybridization Schemes for Co-ordination and Organometallic Compounds. Vol. 72, pp. 73–112.
Mingos, D. M. P., McGrady, J. E., Rohl, A. L.: Moments of Inertia in Cluster and Coordination Compounds. Vol. 79, pp. 1–54.
Moreau-Colin, M. L.: Electronic Spectra and Structural Properties of Complex Tetracyanides of Platinum, Palladium and Nickel. Vol. 10, pp. 167–190.
Morgan, B., Dophin, D.: Synthesis and Structure of Biometric Porphyrins. Vol. 64, pp. 115–204.
Morris, D. F. C.: Ionic Radii and Enthalpies of Hydration of Ions. Vol. 4, pp. 63–82.
Morris, D. F. C.: An Appendix to Structure and Bonding. Vol. 4 (1968). Vol. 6, pp. 157–159.
Mortensen, O. S.: A Noncommuting-Generator Approach to Molecular Symmetry. Vol. 68, pp. 1–28.
Mortier, J. W.: Electronegativity Equalization and its Applications. Vol. 66, pp. 125–143.
Müller, A., Baran, E. J., Carter, R. O.: Vibrational Spectra of Oxo-, Thio-, and Selenometallates of Transition Elements in the Solid State. Vol. 26, pp. 81–139.
Müller, A., Diemann, E., Jørgensen, C. K.: Electronic Spectra of Tetrahedral Oxo, Thio and Seleno Complexes Formed by Elements of the Beginning of the Transition Groups. Vol. 14, pp. 23–47.
Müller, U.: Strukturchemie der Azide. Vol. 14, pp. 141–172.
Müller, W., Spirlet, J.-C.: The Preparation of High Purity Actinide Metals and Compounds. Vol. 59/60, pp. 57–73.
Mullay, J. J.: Estimation of Atomic and Group Electronegativities. Vol. 66, pp. 1–25.
Murrell, J. N.: The Potential Energy Surfaces of Polyatomic Molecules. Vol. 32, pp. 93–146.

Naegele, J. R., Ghijsen, J.: Localization and Hybridization of 5f States in the Metallic and Ionic Bond as Investigated by Photoelectron Spectroscopy. Vol. 59/60, pp. 197–262.
Nag, K., Bose, S. N.: Chemistry of Tetra- and Pentavalent Chromium. Vol. 63, pp. 153–197.
Neilands, J. B.: Naturally Occurring Non-porphyrin Iron Compounds. Vol. 1, pp. 59–108.
Neilands, J. B.: Evolution of Biological Iron Binding Centers. Vol. 11, pp. 145–170.
Neilands, J. B.: Methodology of Siderophores. Vol. 58, pp. 1–24.
Nieboer, E.: The Lanthanide Ions as Structural Probes in Biological and Model Systems. Vol. 22, pp. 1–47.
Novack, A.: Hydrogen Bonding in Solids. Correlation of Spectroscopic and Crystallographic Data. Vol. 18, pp. 177–216.
Nultsch, W., Häder, D.-P.: Light Perception and Sensory Transduction in Photosynthetic Prokaryotes. Vol. 41, pp. 111–139.
Odom, J. D.: Selenium Biochemistry. Chemical and Physical Studies. Vol. 54, pp. 1–26.
Oelkrug, D.: Absorption Spectra and Ligand Field Parameters of Tetragonal 3d-Transition Metal Fluorides. Vol. 9, pp. 1–26.
Oosterhuis, W. T.: The Electronic State of Iron in Some Natural Iron Compounds: Determination by Mössbauer and ESR Spectroscopy. Vol. 20, pp. 59–99.
Orchin, M., Bollinger, D. M.: Hydrogen-Deuterium Exchange in Aromatic Compounds. Vol. 23, pp. 167–193.
Peacock, R. D.: The Intensities of Lanthanide $f \leftrightarrow f$ Transitions. Vol. 22, pp. 83–122.
Penneman, R. A., Ryan, R. R., Rosenzweig, A.: Structural Systematics in Actinide Fluoride Complexes. Vol. 13, pp. 1–52.
Powell, R. C., Blasse, G.: Energy Transfer in Concentrated Systems. Vol. 42, pp. 43–96.
Que, Jr., L.: Non-Heme Iron Dioxygenases. Structure and Mechanism. Vol. 40, pp. 39–72.
Ramakrishna, V. V., Patil, S. K.: Synergic Extraction of Actinides. Vol. 56, pp. 35–90.
Raymond, K. N., Smith, W. L.: Actinide-Specific Sequestering Agents and Decontamination Applications. Vol. 43, pp. 159–186.
Reedijk, J., Fichtinger-Schepman, A. M. J., Oosterom, A. T. van, Putte, P. van de: Platinum Amine Coordination Compounds as Anti-Tumour Drugs. Molecular Aspects of the Mechanism of Action. Vol. 67, pp. 53–89.
Reinen, D.: Ligand-Field Spectroscopy and Chemical Bonding in Cr^{3+}-Containing Oxidic Solids. Vol. 6, pp. 30–51.
Reinen, D.: Kationenverteilung zweiwertiger $3d^n$-Ionen in oxidischen Spinell-, Granat- und anderen Strukturen. Vol. 7, pp. 114–154.
Reinen, D., Friebel, C.: Local and Cooperative Jahn-Teller Interactions in Model Structures. Spectroscopic and Structural Evidence. Vol. 37, pp. 1–60.
Reisfeld, R.: Spectra and Energy Transfer of Rare Earths in Inorganic Glasses. Vol. 13, pp. 53–98.
Reisfeld, R.: Radiative and Non-Radiative Transitions of Rare Earth Ions in Glasses. Vol. 22, pp. 123–175.
Reisfeld, R.: Excited States and Energy Transfer from Donor Cations to Rare Earths in the Condensed Phase. Vol. 30, pp. 65–97.
Reisfeld, R., Jørgensen, C. K.: Luminescent Solar Concentrators for Energy Conversion. Vol. 49, pp. 1–36.
Reisfeld, R., Jørgensen, C. K.: Excited States of Chromium(III) in Translucent Glass-Ceramics as Prospective Laser Materials. Vol. 69, pp. 63–96.
Reisfeld, R., Jørgensen, Ch. K.: Optical Properties of Colorants or Luminescent Species in Sol–Gel Glasses. Vol. 77, pp. 207–256.
Russo, V. E. A., Galland, P.: Sensory Physiology of *Phycomyces Blakesleeanus*. Vol. 41, pp. 71–110.
Rüdiger, W.: Phytochrome, a Light Receptor of Plant Photomorphogenesis. Vol. 40, pp. 101–140.
Ryan, R. R., Kubas, G. J., Moody, D. C., Eller, P. G.: Structure and Bonding of Transition Metal-Sulfur Dioxide Complexes. Vol. 46, pp. 47–100.
Sadler, P. J.: The Biological Chemistry of Gold: A Metallo-Drug and Heavy-Atom Label with Variable Valency. Vol. 29, pp. 171–214.
Sakka, S., Yoko, T.: Sol–Gel-Derived Coating Films and Applications. Vol. 77, pp. 89–118.
Schäffer, C. E.: A Perturbation Representation of Weak Covalent Bonding. Vol. 5, pp. 68–95.
Schäffer, C. E.: Two Symmetry Parameterizations of the Angular-Overlap Model of the Ligand-Field. Relation to the Crystal-Field Model. Vol. 14, pp. 69–110.
Scheidt, W. R., Lee, Y. J.: Recent Advances in the Stereochemistry of Metallotetrapyrroles. Vol. 64, pp. 1–70.

Schmid, G.: Developments in Transition Metal Cluster Chemistry. The Way to Large Clusters. Vol. 62, pp. 51–85.
Schmidt, P. C.: Electronic Structure of Intermetallic B 32 Type Zintl Phases. Vol. 65, pp. 91–133.
Schmidt, H.: Thin Films, the Chemical Processing up to Gelation. Vol. 77, pp. 115–152.
Schmidtke, H.-H., Degen, J.: A Dynamic Ligand Field Theory for Vibronic Structures Rationalizing Electronic Spectra of Transition Metal Complex Compounds. Vol. 71, pp. 99–124.
Schneider, W.: Kinetics and Mechanism of Metalloporphyrin Formation. Vol. 23, pp. 123–166.
Schubert, K.: The Two-Correlations Model, a Valence Model for Metallic Phases. Vol. 33, pp. 139–177.
Schultz, H., Lehmann, H., Rein, M., Hanack, M.: Phthalocyaninatometal and Related Complexes with Special Electrical and Optical Properties. Vol. 74, pp. 41–146.
Schutte, C. J. H.: The Ab-Initio Calculation of Molecular Vibrational Frequencies and Force Constants. Vol. 9, pp. 213–263.
Schweiger, A.: Electron Nuclear Double Resonance of Transition Metal Complexes with Organic Ligands. Vol. 51, pp. 1–122.
Sen, K. D., Böhm, M. C., Schmidt, P. C.: Electronegativity of Atoms and Molecular Fragments. Vol. 66, pp. 99–123.
Shamir, J.: Polyhalogen Cations. Vol. 37, pp. 141–210.
Shannon, R. D., Vincent, H.: Relationship Between Covalency, Interatomic Distances, and Magnetic Properties in Halides and Chalcogenides. Vol. 19, pp. 1–43.
Shriver, D. F.: The Ambident Nature of Cyanide. Vol. 1, pp. 32–58.
Siegel, F. L.: Calcium-Binding Proteins. Vol. 17, pp. 221–268.
Simon, A.: Structure and Bonding with Alkali Metal Suboxides. Vol. 36, pp. 81–127.
Simon, W., Morf, W. E., Meier, P. Ch.: Specificity of Alkali and Alkaline Earth Cations of Synthetic and Natural Organic Complexing Agents in Membranes. Vol. 16, pp. 113–160.
Simonetta, M., Gavezzotti, A.: Extended Hückel Investigation of Reaction Mechanisms. Vol. 27, pp. 1–43.
Sinha, S. P.: Structure and Bonding in Highly Coordinated Lanthanide Complexes. Vol. 25, pp. 67–147.
Sinha, S. P.: A Systematic Correlation of the Properties of the f-Transition Metal Ions. Vol. 30, pp. 1–64.
Schmidt, W.: Physiological Bluelight Reception. Vol. 41, pp. 1–44.
Smith D. W.: Ligand Field Splittings in Copper(II) Compounds. Vol. 12, pp. 49–112.
Smith D. W., Williams, R. J. P.: The Spectra of Ferric Haems and Haemoproteins, Vol. 7, pp. 1–45.
Smith, D. W.: Applications of the Angular Overlap Model. Vol. 35, pp. 87–118.
Solomon, E. I., Penfield, K. W., Wilcox, D. E.: Active Sites in Copper Proteins. An Electric Structure Overview. Vol. 53, pp. 1–56.
Somorjai, G. A., Van Hove, M. A.: Adsorbed Monolayers on Solid Surfaces. Vol. 38, pp. 1–140.
Speakman, J. C.: Acid Salts of Carboxylic Acids, Crystals with some "Very Short" Hydrogen Bonds. Vol. 12, pp. 141–199.
Spiro, G., Saltman, P.: Polynuclear Complexes of Iron and Their Biological Implications. Vol. 6, pp. 116–156.
Strohmeier, W.: Problem und Modell der homogenen Katalyse. Vol. 5, pp. 96–117.
Sugiura, Y., Nomoto, K.: Phytosiderophores – Structures and Properties of Mugineic Acids and Their Metal Complexes. Vol. 58, pp. 107–135.
Sykes, A. G.: Plastocyanin and the Blue Copper Proteins. Vol. 75, pp. 175–224.
Tam, S.-C., Williams, R. J. P.: Electrostatics and Biological Systems. Vol. 63, pp. 103–151.
Teller, R., Bau, R. G.: Crystallographic Studies of Transition Metal Hydride Complexes. Vol. 44, pp. 1–82.
Therien, M. J., Chang, J., Raphael, A. L., Bowler, B. E., Gray, H. B.: Long-Range Electron Transfer in Metalloproteins. Vol. 75, pp. 109–130.
Thompson, D. W.: Structure and Bonding in Inorganic Derivatives of β-Diketones. Vol. 9, pp. 27–47.
Thomson, A. J., Williams, R. J. P., Reslova, S.: The Chemistry of Complexes Related to cis-Pt(NH$_3$)$_2$Cl$_2$. An Anti-Tumor Drug. Vol. 11, pp. 1–46.
Tofield, B. C.: The Study of Covalency by Magnetic Neutron Scattering. Vol. 21, pp. 1–87.
Trautwein, A.: Mössbauer-Spectroscopy on Heme Proteins. Vol. 20, pp. 101–167.
Tressaud, A., Dance, J.-M.: Relationships Between Structure and Low-Dimensional Magnetism in Fluorides. Vol. 52, pp. 87–146.

Trautwein, A. X., Bill, E., Bominaar, E. L., Winkler, H.: Iron-Containing Proteins and Related Analogs–Complementary Mössbauer, EPR and Magnetic Susceptibility Studies. Vol. 78, pp. 1–96.
Tributsch, H.: Photoelectrochemical Energy Conversion Involving Transition Metal d-States and Intercalation of Layer Compounds. Vol. 49, pp. 127–175.
Truter, M. R.: Structures of Organic Complexes with Alkali Metal Ions. Vol. 16, pp. 71–111.
Umezawa, H., Takita, T.: The Bleomycins: Antitumor Copper-Binding Antibiotics. Vol. 40, pp. 73–99.
Vahrenkamp, H.: Recent Results in the Chemistry of Transition Metal Clusters with Organic Ligands. Vol. 32, pp. 1–56.
Valach, F., Koreň, B., Sivý, P., Melník, M.: Crystal Structure Non-Rigidity of Central Atoms for Mn(II), Fe(II), Fe(III), Co(II), Co(III), Ni(II), Cu(II) and Zn(II) Complexes. Vol. 55, pp. 101–151.
Wallace, W. E., Sankar, S. G., Rao, V. U. S.: Field Effects in Rare-Earth Intermetallic Compounds. Vol. 33, pp. 1–55.
Warren, K. D.: Ligand Field Theory of Metal Sandwich Complexes. Vol. 27, pp. 45–159.
Warren, K. D.: Ligand Field Theory of f-Orbital Sandwich Complexes. Vol. 33, pp. 97–137.
Warren, K. D.: Calculations of the Jahn-Teller Coupling Constants for d^x Systems in Octahedral Symmetry via the Angular Overlap Model. Vol. 57, pp. 119–145.
Watson, R. E., Perlman, M. L.: X-Ray Photoelectron Spectroscopy. Application to Metals and Alloys. Vol. 24, pp. 83–132.
Weakley, T. J. R.: Some Aspects of the Heteropolymolybdates and Heteropolytungstates. Vol. 18, pp. 131–176.
Wendin, G.: Breakdown of the One-Electron Pictures in Photoelectron Spectra. Vol. 45, pp. 1–130.
Weissbluth, M.: The Physics of Hemoglobin. Vol. 2, pp. 1–125.
Weser, U.: Chemistry and Structure of some Borate Polyol Compounds. Vol. 2, pp. 160–180.
Weser, U.: Reaction of some Transition Metals with Nucleic Acids and Their Constituents. Vol. 5, pp. 41–67.
Weser, U.: Structural Aspects and Biochemical Function of Erythrocuprein. Vol. 17, pp. 1–65.
Weser, U.: Redox Reactions of Sulphur-Containing Amino-Acid Residues in Proteins and Metalloproteins, an XPS-Study. Vol. 61, pp. 145–160.
West, D.X., Padhye, S.B., Sonawane, P.B.: Structural and Physical Correlations in the Biological Properties of Transitions Metal Heterocyclic Thiosemicarbazone and S-alkyldithiocarbazate Complexes. Vol. 76, pp. 1–50.
Willemse, J., Cras, J. A., Steggerda, J. J., Keijzers, C. P.: Dithiocarbamates of Transition Group Elements in "Unusual" Oxidation State. Vol. 28, pp. 83–126.
Williams, R. J. P.: The Chemistry of Lanthanide Ions in Solution and in Biological Systems. Vol. 50, pp. 79–119.
Williams, R. J. P., Hale, J. D.: The Classification of Acceptors and Donors in Inorganic Reactions. Vol. 1, pp. 249–281.
Williams, R. J. P., Hale, J. D.: Professor Sir Ronald Nyholm. Vol. 15, pp. 1 and 2.
Wilson, J. A.: A Generalized Configuration-Dependent Band Model for Lanthanide Compounds and Conditions for Interconfiguration Fluctuations. Vol. 32, pp. 57–91.
Winkler, R.: Kinetics and Mechanism of Alkali Ion Complex Formation in Solution. Vol. 10, pp. 1–24.
Wood, J. M., Brown, D. G.: The Chemistry of Vitamin B_{12}-Enzymes. Vol. 11, pp. 47–105.
Woolley, R. G.: Natural Optical Activity and the Molecular Hypothesis. Vol. 52, pp. 1–35.
Wüthrich, K.: Structural Studies of Hemes and Hemoproteins by Nuclear Magnetic Resonance Spectroscopy. Vol. 8, pp. 53–121.
Xavier, A. V., Moura, J. J. G., Moura, I.: Novel Structures in Iron-Sulfur Proteins. Vol. 43, pp. 187–213.
Zanello, P.: Stereochemical Aspects Associated with the Redox Behaviour of Heterometal Carbonyl Clusters. Vol. 79, pp. 101–214.
Zumft, W. G.: The Molecular Basis of Biological Dinitrogen Fixation. Vol. 29, pp. 1–65.